Handbook of CCL Microbes in Drinking Water

American Water Works Association

Science and Technology

AWWA unites the drinking water community by developing and distributing authoritative scientific and technological knowledge. Through its members, AWWA develops industry standards for products and processes that advance public health and safety. AWWA also provides quality improvement programs for water and wastewater utilities.

Handbook of *CCL Microbes* *in* Drinking Water

Martha A. Embrey, MPH
Rebecca T. Parkin, PhD, MPH
John M. Balbus, MD, MPH
The George Washington University School of Public Health
and Health Services, Washington, D.C.

Managing Editor: Mary Kay Kozyra
Proofreader: Mart Kelle
Design and Composition: Carol Stearns
Cover Design: Karen Staab
Print Buyer: Susan Anderson
Cover photo courtesy of Dr. B.V.V. Prasad, Baylor College of Medicine

American Water Works Association
6666 West Quincy Avenue
Denver, CO 80235

Library of Congress Cataloging-in-Publication Data.
Embrey, Martha A., 1961 -
Handbook of CCL microbes in drinking water / Martha A. Embrey, Rebecca T. Parkin, John M. Balbus.
p. cm.
Includes bibliographical references and index.
ISBN 1-58321-160-8
1. Drinking water--Microbiology. I. Parkin, Rebecca. II. Balbus, John M., 1961- III. Title.

QR105.5.E435 2002
628.1'6--dc21 2002026202

ISBN: 1-58321-160-8

Contents

Chapter 4

Chapter 5

Chapter 6

CHAPTER 9

CHAPTER 10

Chapter 11

Chapter 12

Chapter 13

Chapter 14

Chapter 15

Appendix A

Preface

As a requirement of the Safe Drinking Water Act Amendments of 1996, the US Environmental Protection Agency (USEPA) created a list of contaminants to be used when setting priorities for possible rulemaking in their drinking water program. The resulting contaminant candidate list (CCL), published in 1998, included 10 microbiological contaminants, 9 of which were being considered for further regulatory action.

Through a cooperative agreement with USEPA's Office of Water, we compiled and analyzed published data on the nine microbiological contaminants. The data were organized within a microbial risk assessment framework based on human health effects, occurrence in the environment, and the efficacy of water treatment methods. These literature reviews comprise the greater part of this publication. Each microbial review chapter begins with an executive summary of our conclusions, based on the available data, and ends with a broad list of references. We found most of the literature through extensive searches in the National Library of Medicine's Medline database, supplemented with books and other reports. We also include insights from three workshops that we convened over the last 2 years to define and incorporate susceptibility in microbial risk assessment and the practice of risk communication in water-related issues.

To help put this risk assessment framework in context, we have included additional chapters on topics of interest in microbial risk assessment, such as the development of microbial risk assessment, the incorporation of susceptible subpopulations, and risk communication. In light of the events of Sept. 11, 2001, and the subsequent anthrax incidents, we appended the book with a brief introduction of bioterrorism agents that could be used to contaminate drinking water, along with additional readings.

These chapters are not meant to be a comprehensive description of microbial risk assessment, and indeed, plenty of resources are available for readers who are interested in learning more about a particular topic. However, we hope the additional material that accompanies the CCL microbe reviews will be a helpful and appropriate introduction to the complex area of microbial risk assessment.

We acknowledge that much of this work was accomplished under Cooperative Agreement #CX8236396-01-0 between the USEPA Office of Water and The George Washington University's Center for Risk Science and Public Health.

Martha A. Embrey, MPH
Rebecca T. Parkin, PhD, MPH
John M. Balbus, MD, MPH

CHAPTER 1

Contaminant Candidate List: Summary of Regulations

by John M. Balbus

Introduction

Under the 1996 amendments of the Safe Drinking Water Act (SDWA), the US Environmental Protection Agency (USEPA) was ordered to create a contaminant candidate list (CCL) of water contaminants that they intended to consider for regulation. The 1996 SDWA amendments further specified that USEPA must consult with experts, including their Science Advisory Board, in this process and that the public must have an opportunity to comment on the list. This extra step in the creation of drinking water regulations accomplished several goals: It added transparency to the process by which contaminants are selected for regulation, and it allowed additional time for public debate on the merits of regulating different contaminants via different methods or different standards.

Background

The 1996 SDWA amendments represent an attempt to improve the process by which drinking water regulations are promulgated. The initial SDWA of 1974 established a two-step process for regulating contaminants in water. The initial step consisted of determining the recommended maximum contaminant level (MCL), which was meant to be the level at which no health effects would result, with an additional margin of safety, which would protect susceptible subpopulations. The recommended MCL represented a public health goal and was not by itself enforceable. In the second step of the process, the maximum contaminant level was established. The MCL was based on the recommended MCL and, in addition, considered the technological and economical feasibility of achieving a defined concentration of a contaminant in drinking water. The process by which contaminants would be selected was left to USEPA to decide; initially, the agency adopted the 1962 US Public Health Service drinking water standards as interim National Primary Drinking Water Regulations (NPDWRs). Subsequently, contaminants were considered for regulation based on internal agency processes that were not standardized.

The slow pace at which USEPA was adding contaminants to be regulated led Congress to enact amendments to the SDWA in 1986. These amendments required USEPA to adopt standards for 83 contaminants by June 1989 and then to adopt 25 additional contaminants every 3 years. Contaminants to be regulated were selected based on the Drinking Water Priority List.* At that time, there was no provision to require data showing that the regulated contaminants were found to a significant degree in the nation's drinking water.

Process of Selecting Contaminants for the Current CCL

Identifying Contaminants

Although creation of a CCL was not formally requested until the SDWA amendments of 1996 were passed, USEPA had begun developing criteria for prioritizing chemical and microbial contaminants prior to 1996 as part of its National Drinking Water Program Redirection Strategy (USEPA 1996). This approach, labeled the contaminant identification method, became the initial strategy used by USEPA after passage of the 1996 SDWA amendments. The goal was to use risk-based criteria to rank and prioritize water contaminants; different methods were developed for chemicals and microbes. The selection of chemical contaminants was envisioned as a four-step process, beginning with the initial identification of contaminants. In this step, multiple resources would be used to identify contaminants worthy of further consideration. Potential sources for identifying chemicals included stakeholders, such as researchers, industry representatives, and environmental advocacy groups, the general US public, other USEPA programs, and other

* SDWA required USEPA to publish a list of contaminants every 3 years that either occurred or were anticipated to occur in drinking water. USEPA published lists in 1988 and in 1991.

federal agencies. Criteria for including a chemical contaminant for initial identification included

- known occurrence or documented toxicity,
- suspected endocrine disruption,
- structural similarity to known toxicants,
- availability of new data filling critical gaps on chemicals previously not regulated,
- high-volume industry production or environmental release,
- use of the chemical as a direct or indirect additive to the water treatment process,
- application to land, and
- high potential for leaching or runoff into water supplies.

In addition, chemicals of high interest globally were included. The second step, preliminary screening, would consider existing release, use, production, toxicity, and occurrence data to exclude chemicals unlikely to pose a significant threat to water supplies. The third step would involve more formal risk ranking and risk assessment. The last step, program decisions, would involve consideration of the range of options for managing the risks posed by specific chemicals to help determine whether those chemicals should be selected for standard setting (Bergeson 1997). For microbial contaminants, the process of selecting chemicals was envisioned to involve either expert judgment or more formal risk assessment using weighting criteria.

Gathering Additional Data

After passage of the SDWA amendments in 1996, the National Drinking Water Advisory Council recommended the formation of an Occurrence and Contaminant Selection Working Group to help create the CCL. This working group comprised stakeholders and experts involved with both chemical and microbial contaminants. For chemical contaminants, the working group initially considered chemicals derived from eight other lists: (1) the 1991 Drinking Water Priority List, the list of health advisories previously released by USEPA; (2) a list of chemicals culled from the Integrated Risk Information System (IRIS) based on level of risk and concentration in water as noted in STORET (USEPA's environmental data system); (3) level of production and discharge into water, as noted in the toxics release inventory (TRI); (4) a survey of water analysis laboratories for commonly found nonregulated chemicals; (5) the Comprehensive Environmental Response, Compensation, and Liability Act (CERCLA or Superfund) priority list;

(6) stakeholder responses; (7) high release chemicals for the TRI; and (8) pesticides identified by the Office of Pesticides. The group decided not to consider chemicals based solely on potential endocrine disruption or mention on Office of Water hotlines.

Screening Criteria for Chemicals

After creating this initial list, the working group decided on two screening criteria for consideration of health risks from the chemicals. The first was that the chemical had to be on one of the initial USEPA lists and could not already have a NPDWR standard. The second was that the chemical had to be known or anticipated to occur in drinking water, as documented in the STORET, IRIS, Hazardous Substance, and other databases.

Once these two screening criteria were met, chemicals were evaluated on the basis of eight health criteria. Fulfilling any one of the following criteria resulted in inclusion on the CCL:

- is listed on California Proposition 65 (a list of chemicals determined by the state of California to cause cancer, birth defects, or other reproductive harm),
- has a USEPA health advisory,
- is listed by the International Agency for Research on Cancer or USEPA as a known or likely carcinogen,
- has been linked with adverse health effects in more than one epidemiological study,
- has an oral reference dose in USEPA's IRIS database,
- is regulated in drinking water by another industrial country,
- is a member of a chemical class or family of known toxicity, or
- has a structure–activity relationship that indicates toxicity.

Although a negative answer to any of the above criteria resulted in exclusion from the draft CCL, USEPA noted that criteria that included a concentration at which toxicity was felt to occur were more valuable than criteria that simply noted the chemical as having structural similarity or belonging to a chemical class of known toxicity (NRC 1999). From an initial list of 262 chemicals, 58 were selected for the draft CCL.

Screening Criteria for Microbials

To help select microbial pathogens for the CCL, the Occurrence and Contaminant Selection Working Group recommended convening a group of microbiological experts. The meeting took place May 20–21, 1997, and included representatives of academia, USEPA and other federal agencies, and the water industry. In preparation for the meeting, USEPA scientists prepared a list of 23 microbial pathogens based on outbreak data, occurrence in water, and potential public health significance.

Participants in the Workshop on Microbiology and Public Health considered the pathogens listed by USEPA as well as several other pathogens and selected 13 for the draft CCL. The workshop participants established a set of criteria for selection that included

- public health significance,
- known waterborne transmission,
- documentation of occurrence in source water,
- effectiveness of current water treatment, and
- adequacy of analytical methods to detect the organism.

Several pathogens, such as hepatitis E, were excluded from the list because they were not commonly transmitted in the United States; others, such as rotavirus and several enteric bacteria, were excluded because current water treatment practices were deemed sufficient to control these pathogens (USEPA 1997).

The proposed CCL was published in the *Federal Register* on Oct. 6, 1997. USEPA sought comments both on the general process of compiling the CCL and on the inclusion or exclusion of specific contaminants. Seventy-one comments were recorded and addressed during the ensuing comment period. Most of the comments that addressed the overall method were supportive, although the need for more rigorous criteria was noted for future lists (NRC 1999). The CCL was revised, and a final list of 50 chemicals and 10 microbial pathogens published in the *Federal Register* on Mar. 2, 1998 (USEPA 1998).

Selection of Contaminants From CCL for Regulatory Action

Categorizing the Contaminants

After an initial evaluation of the available data on the CCL contaminants, the contaminants were placed into one of three categories: regulatory determination, occurrence, or research. Contaminants judged to have sufficient data on which to base a standard were classified as a regulatory determination priority. Contaminants selected for regulatory action are anticipated to come from this category. Additional data gathered between the development of the CCL and the deadline for regulatory action could move contaminants from an occurrence or research priority category to the regulatory determination priority category.

To help the agency develop a process for selecting contaminants for regulatory action, USEPA commissioned the Committee on Drinking Water Contaminants of the National Research Council to provide a report on prioritization methods. This report, "Setting Priorities for Drinking Water Contaminants," was released in 1999 (NRC 1999). Recognizing that it was unlikely that the proposed framework would be implemented in time for the August 2001 selection of CCL contaminants for regulatory action, the committee developed a framework and timeline for making such decisions in the future. Under this framework, a decision would be made within 12 to 15 months, either to drop the contaminant from the CCL, select the contaminant for regulation, request additional research, or a combination of the latter two actions. A decision document would be released detailing the rationale for the decision, and a health advisory would be issued as a preliminary step for any contaminant going on to a regulatory determination. The decision would be based on a preliminary risk assessment, considering only data on exposure and health effects. Issues related to treatment and analytic techniques would not enter into the decision on whether to drop the contaminant from the list or select it for further action. Nonetheless, the committee recommended gathering data on treatment and analytic techniques as part of an initial evaluation, which would avoid later delays in deciding upon regulatory action.

Data Gathering for Regulatory Decision Making

Two programs are expected to provide a sounder database on which to make decisions about inclusion and exclusion of contaminants for future CCLs. The 1996 amendments of SDWA required that both the National Contaminant Occurrence Database and the Unregulated Contaminant Monitoring Regulation be established by August 1999. The National Contaminant Occurrence Database was developed with stakeholder input obtained through a series of meetings. It lists occurrence data for both regulated and unregulated contaminants and is expected to be used not only in selecting contaminants for future CCLs but also for setting standards for regulated contaminants. Under the Unregulated Contaminant Monitoring Regulation, no more than 30 unregulated contaminants are to be selected for monitoring, with results stored in the National Contaminant Occurrence Database. Initial contaminants selected for the Unregulated Contaminant

Monitoring Regulation were obtained from the CCL's category of occurrence priority contaminants.

Creating a Regulatory Framework

In making recommendations for a framework to be used by USEPA for future CCLs, the National Research Council Committee on Drinking Water Contaminants recognized two key issues. The first is that the high degree of uncertainty surrounding much of the data on waterborne contaminants limits the value of mechanized prioritization schemes. Expert judgment and consideration of both science and policy information were regarded as essential for the agency to make appropriate decisions in the face of uncertainty. If data are not sufficient for making more unequivocal regulatory decisions, frameworks and algorithms will be used to target research initiatives that will provide answers to the high-priority questions necessary for regulatory decision-making purposes. In addition, the committee noted that flexibility in decision making was necessary for effective public health protection. Specifically, the framework established for making regulatory decisions must allow for expedited standards to be set under the following circumstances: (1) when a contaminant is demonstrated to have a significant adverse health effect; (2) when the contaminant is demonstrated to occur in public drinking water supplies at levels high enough to cause a significant adverse health effect; and (3) when the possibility for significant risk reduction exists (NRC 1999).

In mandating the creation of the CCL, the SDWA amendments of 1996 both reflected ongoing efforts within USEPA to improve the scientific basis of regulatory contaminant screening and facilitated the development of better methods and information. Regulatory processes, as practiced in the twenty-first century, may not allow for the rapid setting of standards for numerous contaminants as envisioned in the 1986 SDWA amendments. Nonetheless, improvements in stakeholder involvement and the collection and use of data should ensure that the regulatory decisions made are the soundest decisions possible.

Bibliography

Bergeson, L.L. 1997. SDWA contaminant identification method developed. *Pollution Engineering,* 29(2):33–36.

National Research Council (NRC). 1999. *Setting Priorities for Drinking Water Contaminants.* Washington, DC: National Academy Press.

US Environmental Protection Agency (USEPA). 1996. *National Drinking Water Program Redirection Strategy.* EPA 810-R-96-003.

US Environmental Protection Agency (USEPA). 1997. Announcement of the Draft Drinking Water Contaminant Candidate List. *Fed. Reg.,* 62:52193–52219.

US Environmental Protection Agency (USEPA). 1998. Announcement of the Drinking Water Contaminant Candidate List. *Fed. Reg.,* 63:0273–10287.

CHAPTER 2

Development of Risk Assessment

by Rebecca T. Parkin

Historical Background

People have been concerned about the risks of disease and have sought ways to reduce these risks since prehistoric times. Early communities tended to see diseases, particularly epidemics and remedies to them, in a supernatural context. According to one public health historian, the earliest effort to address the natural causes of disease in a rational framework appears to have occurred in ancient Greece in the fifth century B.C.E. (Rosen 1993).

For more than 4,000 years, human populations have taken steps to ensure the safety and adequacy of their water supplies (Rosen 1993; Winslow 1952). Archaeological evidence reveals that home and community-level solutions were developed in ancient Egypt, Jerusalem, Crete, Troy, the Etruscan region of Italy, Asia Minor, India, China, and Rome. In these societies, people used storage, boiling, and filtering strategies to improve the quality of their water. The Greeks built clay pipelines to sustain supplies and installed fountains and cisterns to provide water in public and private settings. More than 2,300 years ago, the Romans recognized that public water supplies were essential to civic life. They designed the most advanced systems of that time, using aqueducts and lead pipes to protect their water supplies. The Romans also appointed a water commissioner and created a water board responsible for assuring the quantity and quality of water needed to serve the city of Rome. Water distribution and the maintenance of aqueducts, reservoirs, and pipes were accomplished by the commissioner's administration and oversight of many managers and hundreds of skilled slaves. The commissioner granted permits for access to private water supplies to prominent citizens upon the payment of fees or a royalty to the treasury. By the second century A.D., many other cities in the Roman Empire had adopted this model (Rosen 1993). Other peoples, such as the Incas, learned to adapt to differing seasonal water conditions by changing their behaviors or water sources.

Not until cities grew in the late Middle Ages, from the late thirteenth century, did the need for more advanced infrastructures for safe and sufficient water supplies become urgent again (Rosen 1993; Winslow 1952). Systems for distributing water became more complex, and authorities developed regulations with penalties to keep people from contaminating the water supply. The economic growth of the Renaissance (particularly the sixteenth and seventeenth centuries) brought even larger cities, new technologies to pump water, and private companies to distribute water. In the late seventeenth and eighteenth centuries, municipal commissions were established in Europe to address water supply issues. By the late eighteenth century, physicians and social reformers, influenced by the Age of Enlightenment, began to state that it was the government's responsibility to protect people from sources of contamination (Walker 1968).

In the early nineteenth century, the Industrial Revolution resulted in large influxes of people to urban centers that were poorly prepared to provide adequate water supplies. Only after water supply and pollution were recognized as critical issues during this period did new materials (chiefly iron) and technologies (such as steam-powered pumps; high-pressure systems; and slow sand, stone, and gravel filtration) become more widely available. Cities used these methods to improve both water quantity and quality (Rosen 1993; Webster 1993; Winslow 1952). By mid-century, several investigators in England and the United States had documented high levels of disease and mortality and associated them with different water supplies. However, beliefs rather than scientific studies about the causal relationship between water and health primarily motivated remedial actions during the Sanitary Reform Movement of the late 1800s (Rosen 1993; Walker 1968; Webster 1993).

Not only did industrialization change the nature of water contamination at this time, but the discovery of bacteria and more sophisticated scientific frameworks provided new knowledge and approaches (such as chlorination) to address water quality problems (Rosen 1993; Webster 1993; Winslow 1952). However, it was the discovery of the health effects of radiation and concerns about adulterated foods that led to the first use of formal risk assessment strategies that are employed today.

Development of Risk Assessment Concepts and Methods

Risk assessment involves a series of data gathering, evaluation, and synthesis steps that produce new information to support risk management decision making. This iterative method is based on defining a problem, describing the problem scope, and asking specific health questions that need to be addressed by the risk assessment. Its conduct relies on the availability of data sufficient to characterize the risks and related uncertainties, a set of increasingly technical methods and tools, complex assumptions, and interactive public processes. Today, the key questions that influence the risk assessment process are: Who is to be protected and at what level? and How much certainty is necessary to support the decision process? (Rhomberg 1995). However, the questions that frame risk assessments, the interpretations of legislative intent, and the public's expectations have expanded and shifted into new areas over time.

Food Safety

When the colonists came to America, they brought their traditional practices of protecting food and medicines. It was not until two reports on adulterated foods and drugs were released in the mid-1800s that municipal and state safety laws were developed. Only when Upton Sinclair's book *The Jungle* (Sinclair 1906) raised concerns about food safety at the national level was the Federal Food and Drug Act passed and a federal agency established to ensure food and drug safety (the US Food and Drug Administration [FDA]). The original act was revised several times, eventually becoming the Federal Food, Drug, and Cosmetics Act of 1938 (FFDCA) (Hutt 1997).

Section 409 of the FFDCA required a health-based standard of regulation and directed the FDA to demonstrate what uses of a chemical are safe; i.e., what uses provide "a reasonable certainty of no harm," such as injury or poisoning (Hutt 1997; Rhomberg 1995). The goals of this directive were to avoid spoilage and chemical contamination, consider cumulative health effects, and prevent the risks related to actual exposures, particularly among frequent users of a food or drug. From the first few decades of experience under this law and its many revisions, scientists and decision makers developed their earliest approaches to formal risk assessment, their concept of "safety" as "the avoidance of harm," and new methods for animal testing.

In the 1930s, statisticians began to design methods to quantify health risks associated with carcinogens. By 1937, experts used a "safety factor" of 100, based on available animal data (Hutt 1997), in their risk estimations. A few years later, there was a growing awareness that carcinogens and noncarcinogens operate differently in the human body. The FDA created the "no observable effect level" (NOEL) metric with a safety factor of 100 and later revised and renamed it the acceptable daily intake level for noncarcinogens. The FDA also adopted a zero-tolerance policy for carcinogens. This decision and the 1958 Delaney Clause, which banned carcinogens from foods and drugs, were based on the belief that any level of exposure to a carcinogen would result in harm (NRC 1983). As the public's perceptions and fears about health risks from involuntary exposures expanded, there was more pressure to address risks at ever-lower levels of exposure. In response, new statistical methods to extrapolate findings from higher to lower doses were developed (Hutt 1997; NRC 1983; Rhomberg 1995).

By the early 1970s, methods to detect chemicals had improved so much that more chemicals could be found at even lower levels, making bans more infeasible. When these new methods detected an animal drug (diethylstilbestrol [DES]) in meat that was becoming suspected as a human carcinogen, the FDA banned the use of DES. The agency was then required to adopt quantitative risk assessment as the method for determining levels of acceptable risk for carcinogens. In 1979, the FDA adopted a *de minimus* interpretation of the Delaney Clause and began using 10^{-6} (one adverse health event in one million people exposed) as the standard for acceptable risk for carcinogens (Rhomberg 1995). The outright ban on carcinogens was essentially removed by the Food Quality Protection Act (FQPA 1996).

The increasing use of quantitative methods in risk assessment required measurement of human factors as well as contaminants. One important, but poorly characterized, factor was food and water consumption. There was little evidence on which to base consumption averages or ranges for all or parts of the population. To improve food and water exposure estimates, FDA, in the 1970s and 1980s, collected national data on the population's consumption patterns and analyzed the data to identify the average and upper-tenth percentile levels for several demographic groups. More recently, FDA has begun using a defined protocol for toxicity parameters (the lowest observable adverse effect level [LOAEL]). The LOAEL includes 10 as the factor of uncertainty for acute effects; 100 for developmental and reversible effects; and 1,000 for more serious, irreversible effects (Hutt 1997; Rhomberg 1995).

Radiation

As health hazards of radiation became apparent in the early twentieth century, radiation safety standards were set for workers who handled radioactive materials. By the 1930s, scientists agreed that there were levels of radiation exposure that did not induce acute, adverse health effects and that a no-effect threshold could be identified. Based on this belief and methods available to measure radiation ionization in air, the US National Commission on Radiation Protection and Measurements (NCRP) set an occupational tolerance dose for radiation that included a safety factor of 100 to reduce acute health effects (Kocher 1991; NCRP 1987).

After World War II, there were new technologies to measure absorbed dose, an increasing number of radiation sources, epidemiological results linking leukemia with workers' exposures, and a growing recognition that long-term health problems and genetic risks needed to be addressed. During the Cold War of the 1950s, the general public became more concerned about the potential health risks of radiation exposures. In response, the NCRP set exposure levels for the general public that were lower than occupational levels on the grounds that the public did not receive any direct benefits from their exposures, unlike workers who earned income related to their exposures. During this period, more scientists became convinced that the threshold concept was not appropriate for radiation or for chemicals associated with cancer (Morgan and Turner 1967; Tran, Locke, and Burke 2000).

As public concerns grew throughout the 1970s, the National Academy of Sciences convened committees of experts to assess the health effects of radiation using the data from the Atomic Bomb Casualty Commission. Quantitative estimates of radiation-related health risks have become more technically complex since the 1960s, and comparisons of radiation risks to other socially accepted risks have been made since the late 1970s. Workers' radiation exposure levels were viewed as acceptable if they were correlated to no more than one adverse health event in 10,000 (10^{-4}) exposed (ICRP 1977; Kocher 1991). These are annual average levels for all types of radiation exposure.

More recently, the NCRP set an exposure limit for radiation for the general public at a level of risk (10^{-5}) believed to be reasonable compared to risks from other sources (NCRP 1987). This standard-setting organization reduced its recommended level even further after the International Commission on Radiation Protection (ICRP) set a lower exposure limit for the general public that considered radiation risks from all sources combined (ICRP 1991; Presidential/Congressional Commission on Risk Assessment and Risk Management [P/C] 1997). By 1997, the Nuclear Regulatory Commission and the Conference on Radiation Control Program Directors adopted an exposure level (100 mrem) for the general public, including pregnant women and developing fetuses, that is one fiftieth of the occupational exposure level (P/C 1997). In 1999, the National Research Council (NRC) released a major new analysis of data on the population health effects of radiation and an extensive sensitivity analysis, but the NRC committee could not identify a low-dose threshold (NRC 1999).

Chemicals

Chemicals were first seen in a public policy framework as potentially harmful substances when they were recognized as adulterants of food in the mid-1800s and addressed as a food safety issue in the early 1900s (Hutt 1997). Early regulations focused on reducing risks of poisoning and other acute health effects, relying on animal studies to determine NOELs (NRC 1983). Based on the prevailing threshold concept of chemicals and acute health effects, scientists determined "safe" levels for occupational and environmental settings, known respectively as threshold limit values (TLVs) and no observable adverse effect levels (NOAELs). In the 1940s and 1950s, the public became increasingly aware of the large amount of chemicals being produced and used (Wargo 1996). After *Silent Spring* was published (Carson 1962), the public became concerned and wanted to know how much chemical exposure was "safe" enough to prevent cancer, as well as noncancer outcomes, such as birth defects and long-term neurological effects.

In 1970, the US Environmental Protection Agency (USEPA) was established and began to use increasingly sophisticated monitoring and laboratory methods, thereby identifying even more chemicals in the environment. This led to greater public concerns and the need for clearer prioritization of chemical risk management activities. When it became known that most people were exposed to many chemicals much of the time, the public policies and methods that were designed when chemically related health effects were thought to be rare no longer fit. Public pressures and policy changes evolved rapidly and served as important stimuli for new epidemiological, toxicological, and statistical methods. New quantitative methods had to be designed to deal with more extensive data and sources of statistical uncertainty to understand the potential health impacts of pervasive low doses.

In the absence of knowledge about causal, biological mechanisms and the inability to link effects definitively with single chemicals, many assumptions had to be made to conduct low-dose risk assessments. This situation opened up new and often bitter science-policy debates, because scientists and policymakers did not agree on the assumptions that needed to be made (NRC 1983). In response to the public's wide range of health concerns and the need for consistent procedures and assumptions, USEPA released cancer risk assessment guidelines (USEPA 1976). Three years later, the president's Regulatory Council adopted the Interagency Regulatory Liaison Group's risk assessment principles (IRLG 1979; Latin 1997; NRC 1983) to foster consistency across agencies.

The Safe Drinking Water Act (SDWA) was passed in 1974 in response to growing public concerns about the presence of hazardous substances found in drinking water. The Act focused protection on the portion of the population that had high-end exposures to contaminants that were known or anticipated to occur in public drinking water supplies and that had or may have had adverse health impacts. The SDWA required USEPA to consider not only what chemicals should be assessed but also how chemical concentrations in environmental media should be linked with estimated doses and effects. Although the maximum contaminant level goal (MCLG) had to include an "adequate margin of safety" to protect the public's health, the recommended maximum total exposure level was based on all sources of direct, but not indirect, exposures over an average adult's lifetime. More recently, the USEPA Office of Water has focused on exposures from all sources, including indirect sources of exposure (Rhomberg 1995).

Statistical methods developed for risk assessment included the one-hit linear model for carcinogens, the nonthreshold carcinogen model (IRLG 1979), the fitted dose–response model (Gaylor and Kodell 1980), and the benchmark dose method for noncarcinogens (Crump 1984). However, these approaches were not uniformly adopted or applied across or even within agencies.

In another attempt to foster consistency, the NRC was commissioned by Congress and the FDA to examine federal institutional arrangements and processes for risk assessments. The NRC's committee released a landmark book of recommendations about assessments of both carcinogens and noncarcinogens, setting forth principles and procedures for implementing a four-step risk assessment paradigm that is fundamentally still in use today (NRC 1983). The four steps, as shown in Figure 2-1, are

1. hazard identification,
2. dose response,
3. exposure assessment, and
4. risk characterization.

For chemicals, the first two steps typically rely on chemical structure or activity information and/or animal toxicology; the exposure assessment step requires human data. In the late 1980s and early 1990s, USEPA published guidelines for exposure assessment and a handbook on exposure factors. These guidelines considered both direct and indirect sources of exposure (USEPA 1989, 1992). The final step, risk characterization, requires compilation, interpretation, and application of the results of the first three steps to the specific risk problem being assessed. The results of this step include estimation of risks, with the sources and levels of uncertainty and the potential impacts of assumptions stated, and recommendations for risk management options.

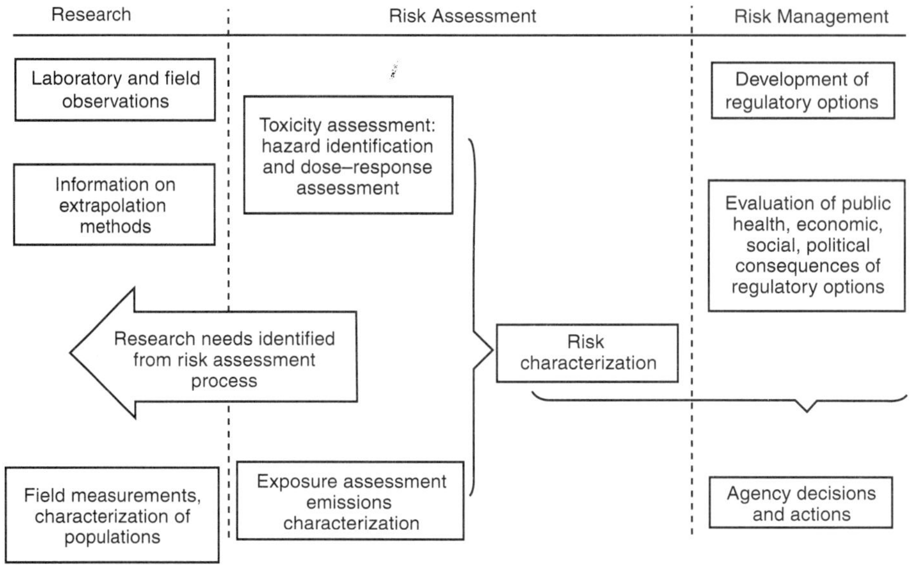

Figure 2-1 National Academy of Science/National Research Council risk assessment/management paradigm

Reprinted with permission from Science and Judgment in Risk Assessment. *Copyright 1994 by the National Academy of Sciences. Courtesy of the National Academy Press, Washington, D.C.*

As these methods became more widely used, new questions and issues were recognized and efforts made to address them. For example, debates ensued about the linearity or nonlinearity of dose–response relationships, e.g., the value of one-hit versus multi-hit carcinogen models. Other controversies included the factors to consider when determining how safe is "safe," the magnitude and impact of uncertainty in risk assessments, and the appropriate assumptions for a "typical" or "maximally exposed" person in the population of concern. The 1994 NRC publication commissioned by USEPA addressed many of these issues in depth and emphasized that risk assessment is an iterative, not a rigidly sequential process (NRC 1994). In 1997, both the Presidential/Congressional Commission on Risk Assessment and Risk Management and the Canadian Standards Association (CSA) confirmed the NRC's perspective in the structure and rationales for their own risk management paradigms (CSA 1997; P/C 1997).

The strategy for chemical exposure limits has been to keep the risk from cancer over a 45-year worklife to below 1 case per 1,000 (10^{-3}) workers exposed and the risk over a 70-year lifetime for the general public to below 1 per 10,000 (10^{-4}) population exposed. Unlike radiation exposures, chemical exposure limits are not annual averages and are for individual chemicals, not all chemical exposures (P/C 1997).

In the 1990s, USEPA attempted to harmonize its risk assessment methods for chemicals and radioactive substances in drinking water under the SDWA, to utilize more recent consumption data in order to develop more reliable exposure estimates, and to clarify its assumptions and the uncertainties embedded in risk assessments. Some issues that continue to challenge the agency include addressing the exposures and health effects in susceptible subpopulations, characterizing the variations in exposure within and among populations, clarifying the sources of variability and uncertainty in the data and results, harmonizing cancer and noncancer risk assessment approaches, and estimating cumulative and aggregated exposures. Clearly, many complex issues remain to be resolved.

Microbial Pathogens in Drinking Water

For decades, the FDA has estimated the health effects of chemicals and microbial pathogens in foods. Until recently, little effort had been made to transfer the agency's experience with foodborne pathogens to drinking water. In the mid-1990s, USEPA's Office of Water and the International Life Sciences Institute (ILSI) convened a workshop of experts to design a conceptual framework for assessing risks associated with waterborne microbial pathogens. Building on chemical and ecological risk assessment methods and lessons learned, the workshop participants designed a four-component process that was organized into three steps: (1) problem formulation, (2) problem analysis (with two substeps for exposure and human health effects), and (3) risk characterization (ILSI 1996) (Figure 2-2).

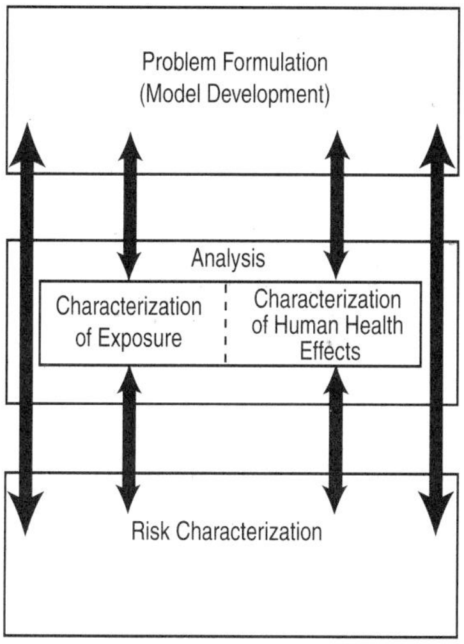

Figure 2-2 Generalized framework for assessing the risks of human disease following exposure to water- and foodborne pathogens

Source: ILSI 2000. Reprinted with permission from the publisher.

In the first step (problem formulation), the fundamental questions to be addressed are stated, the scope and scale of the assessment is defined, and the necessary variables and data are identified. Risk assessors, managers, and stakeholders consider these issues collaboratively to ensure that the end users' needs will be met. In the second step (analysis), the independent and related variabilities of the organism, host, and environment are characterized; the impacts of the variations are discussed; and the nature of the dose–response relationship is decided. For example, subpopulations are identified, specific pathogen–host relationships are considered, exposure pathways are determined, the impacts of dose variability are evaluated, and uncertainties in the field and laboratory data are examined. In the final step (risk characterization), the exposure and health findings are integrated, biological plausibility of the estimated effects is reviewed, sensitivity analyses are conducted, uncertainties and assumptions are documented, management options (including the use of surrogates and indicators in monitoring programs) are analyzed, and the use of measures of effectiveness (such as disability adjusted life years [DALYs]) are considered (ILSI 2000).

This model has been tested in two case studies—one for *Cryptosporidium* (Teunis and Havelaar 1999) and one for rotavirus (Soller, Eisenberg, and Oliveri 1999). While conducting these studies, the authors identified modeling issues, information limitations, and data needs. Their results were used at a second workshop to review and revise the initial pathogen risk assessment paradigm. The deliberations of this group led to reorganization of some elements of the model and enhancements and clarifications of several other elements (ILSI 2000). More recently, Haas and coworkers (1999) systematically described quantitative methods appropriate for microbial risk assessments and for the development of guidelines for human exposure to microbes.

The microbial risk assessment paradigm addresses several issues that are unique to microorganisms (e.g., their capacity to grow and remain viable in drinking water [as an environmental medium], agent–host specificity, and secondary or person-to-person spread of disease). Most data available to implement this framework come from literature on human individuals and populations, because specific strains of microbes typically do not have the same effects in animals as in people. A common modeling assumption is that once an individual is infected, the probability of illness is independent of the ingested dose of the microorganism.

Despite the progress that has been made, many issues remain to be addressed. For example, more knowledge is needed to reduce microbial risk assessment uncertainties. In addition, data are needed to clarify the distribution of pathogens in water supplies, the probability of infection following ingestion of organisms at a specific dose, the nature of the dose–response relationship, the determinants of illness following infection, as well as the impact of factors such as immune status, prior exposures, and medical conditions (ILSI 2000).

Comparison of Current Risk Assessment Strategies

Similarities

Radiological, chemical, and microbial risk assessments in the United States share the same societal origins from the early 1900s. Although they were originally shaped by different ideas about risks and the adverse outcomes related to them, it is interesting to note that over time, all three areas of risk assessment have moved toward no-threshold, multistep conceptualizations of risks in terms of conditional probability. All areas of risk assessment started with limited, high-dose data and have required assumptions and exposure monitoring surrogates to be used until knowledge gaps could be narrowed. All forms have become dynamic, iterative, science-based paradigms that are focused on protecting the environment and the public's health from a wide range of adverse outcomes, particularly at low levels of exposure. Most assessments are based on common principles, processes, and assumptions that have gained acceptance over time.

Increasingly sophisticated analytic tools and techniques have shaped risk assessment strategies and methods and have led to increasingly more detailed questions that demand new data. However, the models continue to be limited by the lack of appropriate data for one or more variables, ensuring that assumptions will continue to play an important role in risk assessments. Important knowledge gaps remain in the areas of background exposures, biological mechanisms, dose–response relationships, and the impacts of variability and uncertainty.

The risk manager's goal is typically focused on utilizing the risk assessment results to reduce or eliminate exposures over the long term, especially among persons who may be at increased risk due to exposure or inherent personal characteristics. The best risk assessments are those that meet the needs of both the risk managers and the stakeholders who will be directly affected by the assessment's results.

Differences

Risk assessments vary in scope and conduct in response to the legislation that mandates them, the data that are available, the underlying assumptions used to fill knowledge gaps, and specific questions that must be addressed.

For example, chemical and radiological risk assessments began by using occupational health data, but microbial and radiological risk assessments have relied on human population rather than animal or occupational data. Human health effects data are predominantly used in microbial risk assessment because of the pathogen–host specificity of most pathogens. While microbial and radiological risk assessments consider all sources of exposure, microbial and chemical risk assessments tend to address the relative contribution of sources.

In recent years, microbial and chemical risk assessments have focused on preventing exposures, addressed one substance at a time, and targeted the risks of susceptible subpopulations. Because of the assumption that exposure to one viable organism causes infection, high- to low-dose extrapolation is not an issue in microbial risk assessment. However, it is an issue in both chemical and radiological risk analyses. Although this assumption is somewhat similar to the assumption that there is no safe level of exposure to a carcinogen, microbial risk assessment on the individual level does not require the dose–response gradient that is used in chemical and radiological risk assessment. However, shedding of organisms by the initial host and secondary spread to susceptible individuals make microbial pathogen risk assessment more valuable for public health purposes when conducted on a population or subpopulation scale. Particularly in cases where secondary spread increases public health risks, dose–response concerns need to be addressed.

Unique Aspects of Microbial Pathogen Risk Assessment

The most recent version of the microbial pathogen risk assessment process indicates that the paradigm is very flexible in that it is applicable to different media, including food, drinking water, recreational water, and occupational contexts (ILSI 2000).

Modeling concerns unique to microbial risk assessment result primarily from the living nature of microbial pathogens and their interactions with human populations. Special issues arise from the following conditions:

- the ecological characteristics of the pathogen and its life cycle (e.g., seasonality, temporal and spatial distribution);
- the ability of microorganisms to change their character and functions in the presence of like organisms (e.g., increasing their infectivity to humans);

- the specificity of pathogen–human relationships, to the level of genotypes for a specific organism;
- the types and levels of susceptibility that vary within individuals over their lifespans and within populations;
- the wide range of factors that may change a host's susceptibility (e.g., medications, concurrent disease or medical state, genetics, behavioral traits);
- the diversity of adverse health responses to microbial pathogens (from raised antibody titers to potentially chronic diseases);
- the ability of microorganisms to be transformed and shed by the human body for variable time periods;
- the ability of infected humans to shed and thus spread pathogens to susceptible individuals;
- the distributions of and potential for contact between infected and susceptible persons;
- the repeated exposure events and multiple routes of exposure that are related to the life-cycle characteristics of microorganisms;
- the ability of pathogens to change over time, multiply, and remain viable for a long time throughout a water supply and distribution system;
- humans' variable ability to detect, quantify, and reduce microorganisms in bodies of water and distribution systems; and
- the very limited value of animal data for risk assessment purposes.

Each of these issues requires data and modeling methods that would not be useful in chemical or radiological assessments. Many of these issues require collection of new data and development of new statistical approaches. Typically, those who assess risks from microbial pathogens are learning from the methods used to model infectious disease processes and sometimes chemical risks and are designing approaches more suitable for the dynamic natures of microbial and human populations.

Summary

The concepts, frameworks, and methods of risk assessment are a twentieth-century response to growing public concerns about health effects associated with environmental hazards. These methods were designed to characterize the sources and extent of risks and to identify ways to prevent exposures so that related health effects can be reduced or eliminated. Over time, risk assessment has evolved from largely qualitative approaches to increasingly quantitative methods. Societal concepts of "safety," "acceptable risk," and "health effects" have changed as well. Despite the gaps in fundamental knowledge and uncertainties in the assessments' results, decision makers and the public have expected risk assessments to address more health concerns and to answer more complex questions. For example, it is no longer sufficient to determine what level of exposure is "safe"; assessors must also determine how safe it is for whom and under what circumstances. One expert argues that risk assessment is now stretched beyond its original purpose into detailed issues for which it was never intended and is not well suited to address (Rhomberg 1995).

Following improvements in technology, more data on very low environmental concentrations and complex exposure pathways have become available. Statistical methods have advanced to take advantage of these data for risk assessment purposes. Although concepts of safety and dose–response relationships have changed, the scope of protection has expanded from occupational groups to the general public and then to susceptible subpopulations among the public. As risk assessment questions, methods, and interpretations have become more complicated and public concern has grown, the public has demanded an active part in transparent, replicable processes.

Microbial pathogen risk assessment is developing in the context of the prior history of risk assessment. Although there are unique issues involved in addressing microbial pathogen risks, there are also important questions, principles, and aspects shared across chemical, radiological, and microbial risk assessment processes.

Bibliography

Canadian Standards Association (CSA). 1997. *Risk Management: Guideline for Decision-Makers.* CAN/CSA-Q850-97. Toronto: Canadian Standards Association.

Carson, R. 1962. *Silent Spring.* Greenwich, CT: Fawcett.

Crump, K.S. 1984. A new method for determining allowable daily intakes. *Fundam. Appl. Toxicol.*, 4:854–871.

Food Quality Protection Act (FQPA). 1996. Pub. L. 104-170, 110. Food Quality Protection Act of 1996 (Aug. 3, 1996). Section 1489. Public Notification.

Gaylor, D.W., and R.L. Kodell. 1980. Linear interpolation algorithm for low-dose risk assessment of toxic substances. *J. Envir. Path. Toxicol.*, 4:305–312.

Haas, C.N., J.B. Rose, and C.P. Gerba. 1999. *Quantitative Microbial Risk Assessment.* New York: John Wiley & Sons.

Hutt, P.B. 1997. Law and Risk Assessment in the United States. In *Fundamentals of Risk Analysis and Risk Management.* Edited by V. Molak. Boca Raton, FL: CRC Lewis Publishers.

International Commission on Radiation Protection (ICRP). 1977. *Recommendations of the ICRP.* ICRP Publication No. 26, Ann. ICRP 1:1–53. Oxford: Pergamon Press.

International Commission on Radiation Protection (ICRP). 1991. *1990 Recommendations of the ICRP.* ICRP Publication No. 60, Ann. ICRP 21:1–201. Oxford: Pergamon Press.

International Life Sciences Institute (ILSI). 1996. A conceptual framework for assessing the risks of human disease following exposure to waterborne pathogens. *Risk Anal.*, 16:841–848.

International Life Sciences Institute (ILSI). 2000. *Revised Framework for Microbial Risk Assessment.* Washington, DC: ILSI Press.

International Regulatory Liaison Group (IRLG). 1979. Scientific bases of identification of potential carcinogens and estimation of risks, 44 *Fed. Reg.*, 39:858–875.

Kocher, D.C. 1991. Perspective on the historical development of radiation standards. *Health Phys.*, 61:519–27.

Latin, H. 1997. Science, Regulation, and Toxic Risk Assessment. In *Fundamentals of Risk Analysis and Risk Management.* Edited by V. Molak. Boca Raton, FL: CRC Lewis Publishers.

Morgan, K.Z., and J.E. Turner. 1967. *Principles of Radiation Protection.* New York: John Wiley & Sons.

National Council on Radiation Protection and Measurements (NCRP). 1987. *Recommendations on Limits for Exposure to Ionizing Radiation.* NCRP Report No. 9, Bethesda, MD: NCRP Press.

National Research Council (NRC). 1983. *Risk Assessment in the Federal Government: Managing the Process.* Washington, DC: National Academy Press.

National Research Council (NRC). 1994. *Science and Judgment in Risk Assessment.* Washington, DC: National Academy Press.

National Research Council (NRC). 1999. *Health Effects of Radon. BEIR VI.* Washington, DC: National Academy Press.

Presidential/Congressional Commission on Risk Assessment and Risk Management (P/C). 1997. *Risk Assessment and Risk Management in Regulatory Decision-Making: Final Report.* Vol. 2. Washington, DC: The Presidential/Congressional Commission on Risk Assessment and Risk Management.

Rhomberg, L.R. 1995. *A Survey of Methods for Chemical Health Risk Assessment Among Federal Regulatory Agencies.* Washington, DC: National Commission on Risk Assessment and Risk Management.

Rosen, G. 1993. *The History of Public Health.* Baltimore, MD: Johns Hopkins University Press.

Safe Drinking Water Act (SDWA). 42 *US Code* Sec. 300f–300j-26.

Sinclair, U.B. 1906. *The Jungle.* New York: Doubleday, Page & Company.

Soller, J.A., J.N. Eisenberg, and A.W. Oliveri. 1999. *Evaluation of Pathogen Risk Assessment Framework.* Washington, DC: ILSI Research Foundation.

Teunis, P.F.M., and A.H. Havelaar. 1999. *Cryptosporidium in Drinking Water: Evaluation of the ILSI/IRSI Quantitative Risk Assessment Framework.* RIVM Report No. 284 550 006. Bilthoven, The Netherlands: National Institute of Public Health and the Environment.

Tran, N.L., P.A. Locke, and T.A. Burke. 2000. Chemical and radiation environmental risk management: differences, commonalities, and challenges. *Risk Anal.,* 20(2):163–172.

US Environmental Protection Agency (USEPA). 1976. Interim procedures and guidelines for health risk and economic impact assessment of suspected carcinogens. *Fed. Reg.,* 41:21402–21405.

US Environmental Protection Agency (USEPA). 1989. *Exposure Factors Handbook,* Final, Vol. I–III. EPA/600/8-89/043. Springfield, VA: National Technical Information Service.

US Environmental Protection Agency (USEPA). 1992. Guidelines for exposure assessment. EPA/600Z-92/001. *Fed. Reg.*, 57:22888–22938.

Walker, M.E.M. 1968. *Pioneers for Public Health*. Freeport, NY: Books for Libraries Press.

Wargo, J. 1996. *Our Children's Toxic Legacy*. New Haven, CT: Yale University Press.

Webster, C., ed. 1993. Caring For Health: History and Diversity. Norwich, UK: Open University.

Winslow, C.E.A. 1952. Man and Epidemics. Princeton, NJ: Princeton University Press.

CHAPTER 3

Evaluating Existing Studies and Data Gaps

by Martha A. Embrey

Introduction

The 1996 amendments to the Safe Drinking Water Act (SDWA) require the US Environmental Protection Agency (USEPA) to use the "best available, peer-reviewed science and supporting studies conducted in accordance with sound and objective scientific practices" in regulatory decision making. The SDWA also requires the agency to use data collected with accepted methods or the best available methods. Science is clearly pivotal in environmental rulemaking, and understandably, there are disagreements about what comprises "good" science.

Because the risk assessment paradigm has traditionally centered on chemical risks, the evaluation of data to quantify that risk has also followed a chemical exposure protocol. In the traditional risk assessment framework, some elements transfer to the assessment of microbial risk, and some do not. In this section, we will summarize the types of studies that assessors use to evaluate chemical risk, then juxtapose them with the types of data and studies available to evaluate microbial risk. The section will define the usefulness of each type of study for evaluating public health risks from chemical and microbial agents in drinking water, detail the quality criteria used to assess the effectiveness of specific studies, and outline the general interpretational considerations to place individual studies into scientific perspective.

Research Methods and Designs

Toxicological Studies

The science of toxicology is concerned with how chemicals interact with and harm organisms. Researchers conduct studies using laboratory animals and in vitro test systems to augment data available from human experience or to provide a surrogate for unavailable human data. Available human and animal toxicological evidence regarding an agent is combined to produce a synthesized judgment of the scientific "weight of evidence." Human (i.e., epidemiological) data generally prevail over animal studies (Rhomberg 1995; NRC 1994) in the weight-of-evidence approach to hazard identification; however, experiments can help evaluate likely mechanisms of action through which toxicologists observe adverse effects (Chan and Hayes 1989; Doull and Bruce 1986; IARC 1992). The same premise applies to evaluation of data other than chemical data (e.g., microbial), though it is most applicable to toxicology.

Key differences exist between toxicology and epidemiology data, and interpreting results requires knowledge of their differences and limitations. Epidemiological studies are not exact enough to prove the carcinogenicity or toxicity of a given substance—except in the cases of a very high risk or a very unusual type of cancer. In these cases, toxicology studies, both in vitro and in animal models, provide the best way to assess the properties of a particular agent. However, in toxicology studies, animals are given much greater doses than what humans would normally experience, so any responses must be extrapolated to a risk to the general population. In simple terms, animal studies can predict what may happen in humans under a given set of conditions. These studies, however, are not always determinants of human health risk. Animals under test may experience harmful effects that are exclusive to their particular species or a test species may be more or less sensitive to a specific agent than humans (Goodman 1995).

In chemical risk assessment, the role of toxicology studies is large. For many reasons, the greatest of which are ethical considerations, animal studies are relied on extensively in chemical evaluation. Usually, the only type of data on human exposure to chemicals are occupational in origin or the result of an accidental spill or release. In the absence of human data, toxicological evaluation usually begins with the simplest, fastest, and most economical tests and becomes more complex only as warranted by the initial results (NRC 1994). In the case of microbial infection, it is difficult to develop animal models for most pathogens because of their specificity to humans, but some do exist. On the other hand, the high incidence of many waterborne diseases gives us human exposure and health effects data in the form of surveillance and outbreak reports that would not typically be available for chemicals.

Epidemiologic Studies

Epidemiology has been defined as the study of the distribution and determinants of disease and health in specific, human populations (Last 1995). Unlike clinicians who study disease in individuals, epidemiologists focus on what is common about members of populations. Though epidemiologists study individuals, their interpretations are based on combining information collected on many persons.

Environmental epidemiology then is the study of health-related states or events that are caused by environmental factors. The Committee on Environmental Epidemiology, which was convened by the National Research Council (NRC 1991), defines the term as follows:

> Environmental epidemiology is the study of the effect on human health of physical, biologic, and chemical factors in the external environment, broadly conceived. By examining specific populations or communities exposed to different ambient environments, it seeks to clarify the relationship between physical, biologic, or chemical factors and human health.

Epidemiological studies allow for direct observation of effects in humans but are often less precise than animal experiments. For example, quantitatively characterizing exposure in epidemiological studies is difficult, and unlike long-term animal bioassays, human subjects are not followed over the course of their lifetimes (Rhomberg 1995).

Because epidemiology is usually observational, drawing conclusions about causation is problematic. People engage in many different behaviors that affect their exposures: choice of food, drinking water, recreational activities, workplace, and contact with others. All of these affect a person's susceptibility to becoming infected. Epidemiologists must sort through these many factors in order to draw inferences about their hypotheses. Because epidemiology is observational and inductive, no single study can provide a definite answer about cause and effect, even if bias is minimal (Pontius 2000).

The study of health in humans takes many forms, from informal case reports to strictly controlled clinical trials. Of course, the more an investigator is able to control the external factors in a study, the more valid the results. However, research that is carefully planned and followed up is also very resource intensive—in terms of both time and money. Therefore, risk assessors must take everything available and evaluate its worth to the final assessment. Clinical trials are the gold standard, but even a case report can offer useful information to develop a hypothesis. A description of the types of epidemiological research designs, from least formal to most formal, follows.

Cross-Sectional Surveys or Prevalence Studies

A cross-sectional survey looks at the prevalence of exposure or intervention and outcome in a population at one point in time. In effect, it is like a snapshot of the population at a given time. A cross-sectional study looks at only the current status of a population, so it cannot establish a temporal association between the exposure or intervention and the outcome under study. An epidemiologist can use this type of study to detect the magnitude of a disease or effect in a population, though the results cannot prove cause and effect.

Case-Control Studies

Case-control studies are useful epidemiologic tools for the study of rare conditions and diseases with latent expression. Because the case-control study compares past exposure in affected and unaffected individuals, fewer numbers of subjects are necessary to complete a rigorous study than are required by a cohort study. However, background prevalence of exposure must be high enough for significant numbers of study participants to have been exposed (Feinstein and Horowitz 1982; Monson 1990). Additionally, case-control studies have shorter study periods than prospective cohort studies, which in turn makes them less expensive and less time consuming. However, case-control studies are limited methodologically because epidemiologists obtain information about the exposure or intervention (often years) after its occurrence; they take cases and controls from two different populations, which introduces possible selection and recall bias, and, therefore, changes in the pattern of excess risk over time are difficult to estimate (Peto 1992).

Waterborne disease investigations essentially follow a case-control approach. Epidemiologists compare the drinking water exposure of people who are ill with people who are not ill.

Cohort Studies

Cohort or follow-up studies compare incidence of disease in groups of people based on their exposure, or lack of it, to the agent or intervention under study. Epidemiologists follow the exposed and unexposed cohorts through time and compare the differences in rate of disease or mortality. Cohort studies can be retrospective, with the point of initial follow-up beginning in the past and ending at some later time, or prospective, with the initiation of surveillance in the present and follow-up through disease monitoring into the future (Monson 1990).

A cohort study is not a good study design for research into rare diseases or conditions, because information has to be collected on a large population in order to capture enough people with a disease of low incidence (Kleinbaum, Morgenstern, and Kupper 1981). Conversely, the cohort study design is a valuable instrument for studying diseases that occur commonly in the population; however, they are expensive and time consuming. Nevertheless, well-conducted cohort studies are among the most preferred sources of epidemiologic data because information about the causal agent is gathered before the health outcome is known, and this is not subject to observer or participant reporting bias.

Clinical Studies

Clinical studies are related to patients in a clinical setting, such as a physician submitting a case report, or in a clinical research setting, such as a clinical trial, where drugs or devices are tested for their safety and efficacy.

Case Reports and Series

Case reports describe interesting variations of diseases or conditions, experiences with new treatments, or unexpected findings (Roht et al. 1982). They can be the result of a clinician's suspicion that a particular agent or intervention may have been responsible for a disease outcome or resolution in one or more patients. The clinician usually describes his or her hypothesis and observations in a letter or brief report in a scientific journal. A case series from multiple physicians is considered useful in developing qualitative databases or registries. For example, reports of adverse effects are important aspects of postmarketing (after the drug has been put on the market) surveillance or monitoring (Carson and Strom 1986). These types of data are limited, however, because they describe the experience of an individual or a group, are not population based, and usually do not offer themselves for important comparisons to control groups.

Clinical Trials

Clinical trials are a type of epidemiological study unique for their rigorous design and execution of experimental interventions in human subjects. Clinical trials are mainly used to test pharmaceuticals, where subjects are randomly assigned to an intervention group and compared with a control group.

Randomized designs provide the closest methodological analog to animal experiments. Commonly, researchers design clinical studies specifically to investigate issues of safety and efficacy of a drug, device, or intervention. The chief advantage of conducting a clinical study is the ability to control many aspects of the study environment, which allows the investigator to better isolate the effect(s) of the study factor. Control of the study environment distinguishes the true cause of the outcome from possible confounders that mask, limit, or overestimate the effect of the true cause of an outcome. Although clinical studies are highly desirable from a scientific perspective, ethical concerns come into play when addressing identified risks, and clinical studies can be deemed inappropriate.

Surveillance

Formal surveillance data that are collected through collaboration between USEPA, the Council of State and Territorial Epidemiologists, and the Centers for Disease Control and Prevention (CDC) track occurrences and causes of waterborne disease outbreaks in this country. The goals of this surveillance program include characterizing the epidemiology of outbreaks, identifying the agents causing outbreaks, and identifying how an outbreak occurred (Barwick et al. 2000). CDC defines an outbreak as two people experiencing a similar illness after exposure to drinking water or recreational water. Evidence must indicate water as the probable source of the outbreak, which is rarely simple to determine. The CDC uses both epidemiologic data and water-quality data, and they weight epidemiologic data more heavily.

Waterborne disease surveillance data are useful in evaluating our treatment and supply of safe drinking and recreational water; however, the reported data no doubt underestimate the incidence of outbreaks. The extent of that underestimation varies widely from locale to locale and is unknown overall. Recognition of outbreaks relies on a number of factors, including public awareness, ill people seeking not only health care but also going to the same providers, providers' reporting practices, the extent of laboratory testing, and local infrastructure available for investigating and reporting outbreaks.

In an attempt to improve public health surveillance data, CDC has instituted a program to standardize the way data are collected, managed, transmitted, analyzed, accessed, and disseminated (National Electronic Disease Surveillance System). Data are critical in evaluating the adequacy of current regulations for water treatment and monitoring, as well as giving important information on the pathology and epidemiology of waterborne diseases. Timely and accurate surveillance data are crucial to identifying an outbreak's etiology and therefore methods for correction. Surveillance studies are perhaps the most important source of information we have on waterborne disease.

Epidemiologic Study Design Summary

Each type of epidemiologic study generates information that can be used to evaluate the cause of a disease or adverse effect, to develop interventions to control or prevent that effect, and to evaluate the efficacy of interventions. Roht et al. (1982) summarize the purpose of each study type:

- "Case reports and case series call attention to the problem.
- Cross-sectional studies indicate the extent of the problem.
- Retrospective (case-control) studies test hypotheses of association or identify factors that may be related to the disease or health problem in ill and non-ill persons.
- Prospective (cohort) studies of exposed and nonexposed persons test hypotheses of 'causal' association that have been derived from case reports, cross-sectional studies or retrospective studies."

Evaluating Data

The validity of risk assessment depends on how well it predicts health effects in human populations; therefore, epidemiologic data are required to validate predictions. The value of the overall risk assessment depends on the validity of its individual components (NRC 1991).

Nine basic characteristics are widely used when evaluating epidemiologic data: strength, specificity, consistency of the association; temporality, relationship between dose and response, effects of removal of suggested cause, biologic plausibility of association, study precision, and study validity (Mausner and Kramer 1985; NRC 1991; Pontius 2000). The characteristics are described here:

- *Strength of association.* This refers to the ratio of outcome rates for those with and without the exposing factor. The larger the relative risk or odds ratio, the more likely the study factor and outcome will be causally related, and the less likely the association will be due to confounding bias. A causal association cannot be ruled out simply because a weak association is observed.
- *Specificity of association.* A putative cause or exposure leads to a specific effect. Examples of specificity include cases of cryptosporidiosis requiring ingestion of *Cryptosporidium* oocysts. One manifestation follows from one cause. This is less clear in cases of cancer, for example, which are usually multifactorial. The presence of specificity argues for causality, but its absence does not rule it out.

- *Consistency of association.* Repeated observation of an association under different study conditions, different populations, and different study methods supports an inference of causality. The more an association appears under diverse circumstances, the more likely it is to be causal; however, the absence of consistency does not rule out causality.
- *Temporal association or period of exposure.* Exposure must precede the disease. This can be inferred from most epidemiological studies. However, when exposure and disease are measured simultaneously, it is possible that the exposure has been modified by the presence of disease. It is easier to establish cases of waterborne disease outbreaks with temporal relationships than with relationships that have a long latency period such as exposures leading to cancer.
- *Dose–response relationship.* A causal interpretation is more plausible when a risk gradient is found (e.g., increasing levels of exposure to the factor in question results in a corresponding rise in the associated disease).
- *Effects of removal of suggested cause or reversibility.* An observed association leads to some preventive action, and removal of the possible cause leads to a reduction of disease or risk of disease.
- *Biological plausibility.* An inference of causality is strengthened if the association is supported by evidence from clinical research or toxicology about biological behavior or mechanisms. This is dependent on the state of scientific knowledge on the particular factor and disease at the time.
- *Study precision.* Individual studies that provide evidence of an association must be well designed with an adequate number of study participants.
- *Study validity.* Individual studies must be conducted well with valid results (i.e., the association is not likely a result of systematic bias).

Statistical Significance

Conventionally, epidemiologists and toxicologists follow the concept of statistical significance in evaluating research results. If the likelihood of study results occurring by chance is 5% or less, they are usually considered statistically significant. This is determined through any one of a number of statistical tests and is designated by the use of a *p*-value (e.g., $p < .05$). Sometimes, confidence intervals are used. This is a broader application that shows a result having a 95% chance of falling between a high value and a low value (e.g., CI = 0.5 – 2.7, $p < .105$) (NRC 1991). However, in their report on setting priorities for drinking water contaminants, the NRC committee on drinking water contaminants warned against using statistical significance to assess a study, because it depends on the size of the study and the size of the association. This may make a large, yet poorly executed study appear more important than a smaller, yet scientifically more sound, study (NRC 1999). Statistical significance also focuses on values of the measure of association that are unlikely to exist given the study results (NRC 1999).

The NRC's Drinking Water Contaminant Committee was tasked to develop a way to prioritize contaminants on the contaminant candidate list (CCL). The committee's report, *Setting Priorities for Drinking Water Contaminants,* is the response to that request (NRC 1999). The committee identified a number of factors to look at when evaluating microbial assessment data—those related to exposure and those related to health effects. They defined exposure factors as

- Transmission
- Environmental source
- Survival potential
- Regrowth potential
- Occurrence in raw water
- Resistance to treatment
- Environmental transport
- Availability of methods*

Examples of health effects factors that must be evaluated are

- Waterborne outbreak data
- Endemic disease rates
- Immune status of host

* Sampling and measurement methods must be reliable and well documented. Methods are available for detecting the presence of almost any microorganism, but difficulties lie in collecting samples, determining frequency, and judging the impact on public health from positive samples.

- Description of microbial pathogens
- Description of disease
- Methods for diagnosing disease

The committee states that positive epidemiologic studies are the highest value in evaluating health effects data—even in the presence of negative toxicologic studies. Positive toxicologic studies will take priority, but negative or inconclusive epidemiologic studies should also be considered in an evaluation.

A spectrum of beliefs and data must typically be deciphered and evaluated to extract those scientific facts, assumptions, and beliefs that will stand the test of time. Determining what constitutes sound science requires time to gather information and consider opposing views. Context, source, presumptions, and biases of scientific studies must be evaluated.

Summary of CCL Microbe Data Gaps

When USEPA published its final drinking water CCL in 1998, they placed the contaminants into three priority categories: rulemaking consideration, research, and occurrence data. More data on the contaminants placed on the occurrence and research priorities lists will be needed before USEPA can determine if these contaminants should be regulated.

The categories below were based on USEPA's information in 1997 and 1998, when the list was compiled. USEPA expects that contaminants will move in and out of the categories as more data become available.

1. Regulatory determination priorities: This category does not contain any of the microbes on the list, except *Acanthamoeba*, which is slated for guidance only.
2. Research priorities:
   a. **Health research**: *Aeromonas hydrophila*, cyanobacteria, caliciviruses, *Helicobacter pylori*, microsporidia, *Mycobacterium avium* complex (MAC).
   b. **Treatment research:** Adenoviruses, *A. hydrophila*, cyanobacteria, caliciviruses, coxsackieviruses, echoviruses, *H. pylori*, microsporidia, MAC.
   c. **Analytical methods research:** Adenoviruses, cyanobacteria, caliciviruses, *H. pylori*, microsporidia.
3. Occurrence priorities: Adenoviruses, *A. hydrophila*, cyanobacteria, caliciviruses, coxsackieviruses, echoviruses, *H. pylori*, microsporidia.

In an attempt to address the data gaps for the microbes that are listed above, USEPA established a National Drinking Water Contaminant Occurrence Database and an Unregulated Contaminant Monitoring Regulation program. These programs are discussed more fully in chapter 1. Currently, USEPA is designing a risk-based research plan to inform officials on the health risks posed by contaminants on the drinking water CCL. In phase one of their two-phase plan, a workgroup will review the available data to determine whether the contaminant poses a public health hazard. If the working group considers it a potential hazard, they will determine if the contaminant is treatable under current drinking water treatment practices.

USEPA has published a research plan for microbial pathogens (USEPA 1997) detailing what they see as the primary research questions to evaluate the public health impact of waterborne pathogens. They divide the plan into four sections and define the research questions and research needs as follows:

- Health effects research for microbial pathogens: USEPA characterizes the research needs under this category as dose–response relationships, pathogen- and host-specific factors involved in infection and disease, and effects on susceptible subpopulations. Also important is the characterization of endemic and epidemic illness rates through epidemiological studies.

- Exposure research for microbial pathogens: Data needs include the development of analytical methods to detect protozoa and viruses in source and finished waters; the occurrence and sources of waterborne pathogens in environmental water and in distribution systems; the fate and transport of pathogens in groundwater; and an assessment of the methods to protect groundwater.

- Risk assessment research for microbial pathogens: The focus of this data gathering is on modeling microbial risks using dose–response models and other statistical methods.

- Risk management research for microbial pathogens: This section addresses the protection of source waters and the effectiveness of drinking water treatment options. Special emphases will be placed on viral contamination of groundwater and the factors affecting microbial growth in distribution systems.

For our data gap analysis, we took the current USEPA listing of the CCL microbes and their categorization by research priority and occurrence priority and looked at them in a more detailed way. (See Table 3-1.) The result is a microbial risk assessment framework that addresses specific issues of microbial contamination of drinking water. Following are examples of data gaps we have identified for the 1998 list of CCL microbes.

Health Research

Microsporidia was one of six microbes that were listed as lacking data in health research on the CCL.

Occurrence of Illness: Microsporidia

- Most of the knowledge available on microsporidia-related illness is in the form of case reports. Serological testing is still being developed to determine exposure, but one study reported a sero-prevalence for *Encephalitozoon* spp. as 8% in Dutch blood donors and 5% in French pregnant women.

Degree of Morbidity and Mortality: Microsporidia

- In the immunocompromised (especially AIDS patients), microsporidiosis most often causes chronic diarrhea with wasting syndrome. However, microsporidial infection can occur any place in the body. Mortality is difficult to measure because those affected are generally end-stage AIDS patients, who are affected by a myriad other opportunistic infections. The disease appears to be uncommon in the immunocompetent, but infections in this group have been associated with self-limited diarrhea and kerato-conjunctivitis that is not severe.

Infectious Dose: Adenovirus, Cyanobacteria, *H. pylori*, Microsporidia, MAC

- Unknown for adenovirus.
- Unknown for cyanobacteria in humans. There is information on effect levels in mice for microcystin-LR and anatoxin-a.
- The smallest documented infectious dose for *H. pylori* in an adult is 3×10^5. The infectious dose for a child is unknown, and the majority of people are infected in childhood. Data suggest that some people may be genetically more susceptible than others.
- Unknown for microsporidia.
- For MAC, the dose is known for mice but not for humans.

Table 3-1: Identified data gaps for CCL microbes

	Aeromonas	Adenovirus	Cyanobacteria	Calicivirus	Coxsackievirus	Echovirus	*H. pylori*	Microsporidia	MAC
Health	■		■	■			■	■	■
Occurrence of disease								x	
Morbidity/ mortality								x	
Infectious dose		x	x				x	x	x
Illness/infection	x							x	x
Medical treatment			x					x	
Chronic sequelae			x		x				
Analytic methods		■	■	■			■	■	
Detection: Environ-mental/clinical			x	x	x	x	x		x
Occurrence	■	■	■	■	■	■	■	■	
Concentration in source water				x			x	x	
Waterborne exposure	x	x			x	x	x		
Treatment	■	■	■	■	■	■	■	■	■
Water		x	x	x			x	x	x
Distribution system	x	x	x	x	x	x		x	

■ Identified by USEPA as data gaps.
x Identified by authors as data gaps.

Probability of Illness Based on Infection: *Aeromonas* spp., Microsporidia, MAC

- Unknown for *Aeromonas*-related gastroenteritis.
- Asymptomatic microsporidial infection occurs in immunocompetent and immunosuppressed hosts, but the rate is unknown.
- MAC is generally an opportunistic infection; however, the incidence of pulmonary MAC in people without risk factors and lymphadenitis in children has been increasing.

Efficacy of Medical Treatment: Cyanobacteria, Microsporidia

- Unknown for cyanobacteria; however, antihistamines and steroids may be helpful for allergic reactions. If given in a timely manner, activated charcoal or an emetic could have a positive effect on the toxic response.
- No standard treatment exists for any microsporidial infection.

Chronic Sequelae: Coxsackievirus, Cyanobacteria

- Coxsackievirus infection in children has been a suggested precursor to insulin-dependent diabetes. Also, coxsackievirus B5 is a major cause of viral myocarditis, which is primarily acute. However, myocarditis can become chronic, and infection is also associated with chronic dilated cardiomyopathy. More research is needed on the possibility of infection from coxsackievirus through drinking or recreational water.
- Outcomes from cyanobacterial toxins are mostly acute. However, the microcystins have been shown to be tumor promoters, and epidemiological evidence suggests that low levels of microcystins in drinking water are associated with an increase in hepatocellular cancer.

Analytical Methods Research

It is difficult to evaluate how often a pathogen occurs in drinking water sources if appropriate and efficacious methods to detect them do not exist. Likewise, the ability to identify pathogens in clinical samples as part of a diagnosis determines the prevalence of pathogens' ability to cause illness. The cause of most waterborne gastrointestinal outbreaks remains unknown, because clinical detection methods—for viruses especially—are too complex and expensive to use routinely.

Detection Methods in Water/Clinical Specimens: Calicivirus, Coxsackievirus, Cyanobacteria, Echovirus, *H. pylori*, MAC

For environmental samples of calicivirus, reverse transcriptase polymerase chain reaction (RT-PCR), conventional electron microscopy, or immune electron microscopy are used. Calicivirus cannot be cultured in vitro or passaged in animal models. Consequently, infectivity cannot be evaluated through available sampling methods.

- Isolation of coxsackieviruses and echoviruses in cell culture, followed by serotyping using neutralization, is still the "gold standard" of detection. Most coxsackievirus group A species are difficult to culture and are traditionally detected using a newborn mouse model. Molecular methods like PCR are being used increasingly in clinical and environmental research, but although molecular detection methodologies are more sensitive, they are not standardly used in clinical or environmental settings. PCR is appealing for clinical applications because of its rapid results; however, for environmental samples, culturing is currently the only way to show infectivity.

- The most obvious way to detect cyanobacteria in water is visually—seeing evidence of algal blooms in surface water. Standard microscopy can be used to speciate environmental samples. Quick and easy ways to detect toxicity are still under development. Mouse bioassays have traditionally been used for this purpose, but this method is undesirable for many reasons, including expense. Newer methods that are seen more frequently include the use of liquid chromatography and enzyme-linked immunosorbent assays. PCR is also under development.

- Because *H. pylori* is such a fastidious organism, culturing in environmental and clinical samples can be challenging. PCR has been used more successfully to detect *H. pylori* DNA in environmental samples.

- MAC strains need dedicated growth media and extended incubation. Because they grow slowly, isolation is made difficult by overgrowth of other organisms. Samples are usually treated to remove other organisms and physically concentrated to improve isolation.

Occurrence Priorities

Eight microbes are listed as lacking data on occurrence in water sources on the CCL. Besides determining a pathogen's concentration in drinking water source water, we also include the importance of determining the role of waterborne exposure in the transmission of a pathogen. Because it is present in source water, does not necessarily mean that water is an important source for disease transmission.

Concentrations at Intake: Calicivirus, *H. pylori,* Microsporidia

- Unknown for calicivirus but dependent on the level of human fecal contamination in the source water. Outbreaks have been associated with wet weather events.
- Unknown for *H. pylori.*
- Concentration of microsporidia in source water is dependent on the level of human and possibly animal fecal contamination. In published reports, surface water, groundwater, and sewage effluent samples were positive for pathogenic microsporidia.

Role of Waterborne Exposure: Adenovirus, *Aeromonas* spp., Coxsackievirus, Echovirus, *H. pylori*

- Adenovirus has not been positively associated with any drinking water outbreaks. A number of outbreaks have occurred as a result of exposure to recreational water.
- *Aeromonas* spp. are found in all types of water: lakes, rivers, marine environments, wastewater, and chlorinated drinking water. Because *Aeromonas* is also found frequently in all types of food and is ubiquitous in the environment, the role of water ingestion in the development of gastroenteritis is unknown. Exposure to fresh water is the primary risk factor associated with wound infection.
- Coxsackieviruses and echoviruses are found in all types of water: lakes, rivers, marine environments, wastewater, and chlorinated drinking water. Theoretically, coxsackieviruses and echoviruses can be transmitted by water; however, little evidence supports this. Most outbreaks appear to result from person-to-person spread.

- Evidence is limited on *H. pylori*, but the waterborne route of exposure is probably important in some populations.

Treatment Research

USEPA has placed every microbe on the CCL in the category of needing more information on drinking water treatment. A microbe's susceptibility to chlorination is important, but issues regarding alternative treatments and especially how a microbe lives in the distribution system have also gained attention.

Efficacy of Water Treatment: Viruses, Cyanobacteria, *H. pylori*, Microsporidia, MAC

- No direct data exist on the efficacy of chlorine disinfection on adenovirus or calicivirus; generally, viruses can be inactivated with 0.1 mg/L of free chlorine.
- There is some question as to the efficacy of standard drinking water treatment for removing all but large concentrations of cyanobacterial toxins, although current methods are effective enough to prevent any acute effects. Evidence on the efficacy of chlorine on microcystins is equivocal; chlorine is ineffective on anatoxin-a. Activated carbon treatment appears to be the best method for removing them from treated water.
- Equivocal data exist on the effectiveness of chlorine on *H. pylori*.
- Unknown for microsporidia, but their small size may decrease the efficacy of sand filtration, and disinfection is generally not effective in cyst-forming protozoa. Coagulation and flocculation may decrease their concentration.
- Chlorine is a relatively ineffective disinfectant for MAC—especially at residual levels. Physical treatment, such as sand filtration and coagulation–sedimentation, is effective at removing most, but not all, organisms.

Survival/Amplification in Distribution: *Aeromonas* spp., Viruses, Cyanobacteria, Microsporidia

- *Aeromonas* can colonize wells and distribution systems for months and years. Substantial regrowth can occur after disinfection, and the aeromonads colonize biofilm, which makes them resistant to disinfectant residuals.
- Not applicable in viruses, though survival within biofilms is an untested possibility.
- Unknown for cyanobacteria.
- Unknown for microsporidia.

Bibliography

Barwick, R.S. D.A. Levy, G.F. Craun, M.J. Beach, R.L. Calderon. 2000. Surveillance for waterborne-disease outbreaks—United States, 1997–1998. *MMWR*, 49(4):1–21.

Carson, J.L., and B.L. Strom. 1986. Techniques of postmarketing surveillance: an overview. *Med. Toxicol.*, 1(4):1–237.

Chan, P.K., and A.W. Hayes. 1989. Principles and Methods for Acute Toxicity and Eye Irritancy. In *Principles and Methods of Toxicology*, 2nd ed. Edited by A.W. Hayes. New York: Raven Press.

Doull, J., and M.C. Bruce. 1986. Origin and scope of toxicology. In *Casarett and Doull's Toxicology, the Basic Science of Poisons.* 3rd ed. Edited by C.D. Klaassen, M.O. Amdur, and J. Doull. New York: Macmillan Publishing Company.

Feinstein, A.R., and R.I. Horowitz. 1982. Double standards, scientific methods, and epidemiologic research. *N. Engl. J. Med.*, 307(26):1611–1617.

Goodman, G. 1995. Use of Available Epidemiological Data to Validate Rodent-Based Carcinogenicity Models. In *The Role of Epidemiology in Regulatory Risk Assessment, Proc. of the Conference on the Proper Role of Epidemiology in Risk Analysis.* Edited by J.D. Graham. The Netherlands: Elsevier Science B.V. pp. 133–144.

International Agency for Research on Cancer (IARC). 1992. *Long-term and Short-term Screening Assays for Carcinogens: A Critical Appraisal.* World Health Organization. IARC Monographs on the Evaluation of the Carcinogenic Risk of Chemicals to Humans. Lyon, France: IARC.

Kleinbaum, D.G., H. Morgenstern, and L.L. Kupper. 1981. Selection bias in epidemiological studies. *Am. J. Epidemiol.*, 113(4):452–463.

Last, J.M., ed. 1995. *A Dictionary of Epidemiology, 3rd ed.* New York: Oxford University Press.

Mausner, J.S., and S. Kramer. 1985. *Mausner & Bahn Epidemiology: An Introductory Text.* 2nd ed. Philadelphia: W.B. Saunders.

Monson, R.R. 1990. *Occupational Epidemiology.* Boca Raton, FL: CRC Press, Inc.

National Research Council (NRC). 1991. *Environmental Epidemiology: Public Health and Hazardous Wastes.* Vol .1. Edited by National Research Council. Washington, DC: National Academy Press.

National Research Council (NRC). 1994. *Science and Judgment in Risk Assessment.* Washington, DC: National Academy Press.

National Research Council (NRC). 1999. *Setting Priorities for Drinking Water Contaminants.* Washington, DC: National Academy Press.

Peto, J. 1992. Meta-analysis of Epidemiological Studies of Carcinogenesis. In *Mechanisms of Carcinogenesis in Risk Identification.* Edited by H. Vainio, P.N. Magee, and D.B. McGregor. Lyon, France: IARC.

Pontius, F.W. 2000. Defining sound science. *Jour. AWWA,* 92(10):16–20, 92.

Rhomberg, L. 1995. Using Both Animal and Human Data in Risk Assessment: Notes on Current Practice. In *The Role of Epidemiology in Regulatory Risk Assessment, Proc. of the Conference on the Proper Role of Epidemiology in Risk Analysis.* Edited by J.D. Graham. The Netherlands: Elsevier Science B.V. pp. 125–132.

Roht, L.H., B.J. Selwyn, A.H. Holguin, and B.L. Christensen. 1982. *Principles of Epidemiology: A Self-Teaching Guide.* San Diego, CA: Academic Press, Inc.

US Environmental Protection Agency (USEPA), Office of Research and Development. 1997. *Research Plan for Microbial Pathogens and Disinfection By-Products in Drinking Water.* EPA 600-R-97-122. Washington DC: USEPA.

CHAPTER 4

Susceptible Subpopulations

by Martha A. Embrey and Rebecca T. Parkin

In 1996, the US Congress called for the health protection of susceptible subpopulations in the areas of food and drinking water quality (FQPA 1996; SDWA 1996). These mandates led to an increased scientific interest in incorporating susceptibility in risk assessment methodologies. From 1995 to 1999, the Centers for Disease Control and Prevention (CDC), the International Life Sciences Institute (ILSI), and the Codex Committee on Food Hygiene held interdisciplinary workshops and conferences to explore ways to address the differential risks of potentially vulnerable subpopulations.

Definitions of Susceptibility

There are many different definitions of "susceptibility." As a result, risk assessors and stakeholders may not always realize that they may be focused on addressing different concerns. If variations in conceptualizations of susceptible subpopulations are not understood, risk assessors can expend a lot of time and effort, but still not meet stakeholders' needs. We have published previously on this issue (Parkin and Balbus 2000) and summarize our key findings and recommendations here.

Definitions found in major English dictionaries describe susceptibility as an intrinsic characteristic of some entity, living or not, that is made manifest by an external factor interacting with that entity and changing it in some way. Although all definitions include these fundamental relational elements, some dictionaries imply that the relationship can only be injurious or adverse.

Although many scientific fields define susceptibility in individual terms, others use a population scale. There are some notable differences between scientific definitions that relate to the underlying perspectives and methods of each field. For example, ecological approaches note that susceptible hosts respond in different ways and to different extents. An environmental engineering definition lists a range of potential adverse health effects and mentions several aspects of the host–agent relationship that affect the host's susceptibility. Medical and biological definitions tend to focus on the physiological nature of the individual human, plant, or animal. An environmental medicine text includes the dose–response concept; that is, susceptible individuals respond at doses much lower than those that induce response in the general population. Toxicological definitions are often similar to medical ones, but they add concepts of conditionality on exposure and variability of response. Some toxicologists conceive of susceptibility in a population, rather than an individual, framework. Epidemiologists tend to describe susceptibility as an individual characteristic. While some include the notion of risk in their approaches, others see the host characteristic of susceptibility only as a genetic trait and separate it from exposure. Still others include acquired, and not just intrinsic, host traits as modifiers of susceptibility.

When professionals from different disciplines and sectors have convened to assess health risks, they have at times developed consensus definitions of susceptibility or susceptible people. As in recent workshops on microbial risk assessment (Balbus, Parkin, and Embrey 2000; ILSI 2000), interdisciplinary definitions incorporate a range, but not necessarily a full range, of concepts and views that dominate the fields of the participants. For example, the ILSI (2000) framework for microbial risk assessment defines susceptibility as, "The extent to which a host is vulnerable to infection by a pathogen, taking into account a host's intrinsic and/or acquired traits that modify the risk of infection or illness."

Environmental laws and public policies typically address susceptibility in one of two ways. They either use a brief phrase to describe people of concern ("highly exposed *and* highly susceptible" or "highly exposed *or* highly susceptible") or they use a list of subpopulations of concern. The most frequently mentioned groups are children, the elderly, pregnant women, immunocompromised persons, and people with chronic illnesses. Risk assessment guidelines issued by public sector organizations tend to describe susceptibility in terms of variability and distributions, as comparisons of subgroups' responses to the general population's average response, or in both statistical and comparative terms.

Definitions of Susceptible Subpopulations

Typically, a list of at-risk subpopulations, based on readily observed characteristics, is used as a default option to address susceptible subpopulations in risk assessment. While the underlying rationale for this approach is that there is a wide range of responsiveness within populations that is dependent on specific mechanisms (Preston 1996), the degree of over- or under-protection that results from this default is largely unknown. Not only do hosts' biological characteristics vary (e.g., Hattis et al. 2001), but so too do their behaviors and exposures, which may be affected by social and other factors (Balbus and Parkin 2000). Some risk assessors have treated susceptibility only as a genetic trait, while others have linked exposures to host characteristics. Defining subgroups on the basis of only a few of these factors introduces an unknown degree of uncertainty into the risk assessment.

Risk assessors acknowledge that susceptibility is a trait that has meaning in risk assessment only when it is evaluated in relationship to the population's mean and/or distribution of susceptibility. Hattis has described interindividual variations in susceptibility across the entire exposure-to-effect pathway (Hattis 1996; Hattis et al. 1999). It is widely understood that there is substantial variation in susceptibility across human populations at any point in time, but intraindividual variations over time are less clear. Few risk assessments to date have indicated that susceptibility is a dynamic state that may change significantly during individuals' lifetimes. In microbial risk assessments, some efforts have been made to model the change in susceptibility status depending on one's age, concurrent health condition, prior exposures, and immune status (Soller, Eisenberg, and Oliveri 1999).

Susceptibility in Microbial Risk Assessment

As it is currently conceived, risk assessment is neither the beginning nor the end of a risk management process (CSA 1997; P/C 1997). The assessment is often begun after risk managers and/or stakeholders have defined the risk problem; the risk assessor may have had no or very little role in the initiation of the process. Although both the Presidential/Congressional Commission and Canadian Standards Association models present lists of subpopulations potentially of concern, neither document notes the importance of clarifying the concept of susceptibility early in the risk management process (CSA 1997; P/C 1997). Their omission mirrors the lack of early attention to susceptibility in the National Research Council (NRC) reports (NRC 1983, 1994). Perhaps because of the wide acceptance and use of the NRC paradigm, the role of susceptibility is not often sufficiently recognized at the beginning of the risk management or even the risk assessment process. The Food Quality Protection Act (1996) and Safe Drinking Water Act Amendments (1996) requirements regarding susceptibility, however, are causing a greater recognition of the need to explicitly address susceptibility in the problem initiation stage. Population issues, including target populations, are mentioned in the problem formulation step of the

ILSI (2000) paradigm, but the text does not explicitly address susceptible subpopulations or susceptibility.

Subpopulations have been considered to varying extents in risk assessment approaches (CSA 1997; NRC 1994; P/C 1997). In the ILSI (2000) model, susceptibility and the factors that affect susceptibility are discussed only in the characterization of human health effects portion of the analytic step. Recognition is given to the need for new methods to incorporate the impacts of susceptibility factors more effectively in microbial risk estimates. However, the case has also been made that susceptibility needs to be systematically and thoroughly addressed in every step of risk assessment in order to develop better estimates of subpopulations' risks (Parkin 2001).

Methods for Assessing Susceptible Subpopulations' Risks

There is no generally accepted risk assessment methodology for characterizing variations in individual and subgroup responses to hazards. There are very few data available on subpopulations' inherent, acquired, behavioral, or other factors that may affect their susceptibility. Monte Carlo (Gilks, Richardson, and Spiegelhalter 1996) or statistical simulation methods have been promoted as effective means to estimate the variability of factors in populations (Haas, Rose, and Gerba 1999). However, the quality of the data available to initiate the calculation determines whether the end results will be suitable for estimating the risks to susceptible subpopulations. There is much yet to be learned about the value of Monte Carlo strategies, the data, and the sensitivity analyses that are needed to produce meaningful risk estimations.

Until there is greater experience with addressing susceptibility in risk assessment, assessors need to carefully consider susceptibility in each step of their risk assessment models (Parkin 2001). Whether the assessor is using a traditional chemical risk assessment paradigm (NRC 1983), a model for radiological risks (NRC 1999), or a more recent microbial risk assessment method (ILSI 2000), the assessor must clearly define the concept of susceptibility and the technical methods to be used to evaluate it in each assessment. When subpopulations' risks are a concern, a definition of susceptibility that is suitable for risk assessment is one that utilizes population-scale concepts. The modeling and predictive goals of risk assessment may be best met by using definitions of susceptibility that incorporate population, variation, distribution, and probability concepts.

However, susceptibility has limited meaning in a risk assessment if the problem formulation does not also clearly state the specific external agent (exposure) and the specific adverse health effect(s) of concern. For a single agent, there may be different health effects to be addressed for different subpopulations, so the effects and subgroups of concern need to be explicitly linked in the problem statement. For example, *Mycobacterium avium* complex (MAC) is a bacterial infection that causes a life-threatening disseminated infection in people with AIDS (Hocqueloux et al. 1998) but a disproportionate incidence of pulmonary problems in women (Kennedy and Weber 1994; Kubo et al. 1998). As a result, a problem formulated to state the need to address infections and pulmonary problems in the AIDS patients and women would be too imprecise. For MAC, the specific health outcome of interest needs to be linked clearly to the subpopulation in which the outcome is known to occur.

Risk assessors must ensure that they do not proceed with an assessment until they are very clear about the risk managers' and stakeholders' definitions of susceptibility and susceptible subpopulations. In order to gain sufficient understanding of others' concerns, risk assessors may need to participate in interactive processes and review their approaches to susceptibility with managers and stakeholders before conducting the technical assessment. The purpose and specific questions to be addressed are important issues in defining the scope and scale of an assessment. Identification of the susceptible subpopulations and health effects of concern needs to be done by building on risk managers' and stakeholders' insights, perceptions, and prior experiences with the issues and populations of concern. The results of this collaborative approach to the assessment's scope and scale, including the definition of susceptible subpopulations, need to be clearly stated in the problem formulation.

When the assessor gathers data to characterize exposures and the people exposed, many factors that may affect susceptibility within the population must be considered. For example, the source and temporality of the pathogen may affect which subpopulations are exposed, how much, and how often. Sometimes population-based surveys have been used to determine the range of human exposures (USDA 1998), but these have been done primarily for foodborne agents and rely on small numbers of people over short periods of time. Although these surveys have provided valuable information on total populations, they have also provided difficult challenges for estimating subgroup exposures (Kahn, Jacobs, and Stralka 1999).

When the assessor works to describe the subpopulations at risk, the inherent and exposure pathway factors that may influence the hosts' probability and severity of response must be thoroughly considered. For example, little information exists on the immune status of populations or how that status changes over time. Few susceptibility or exposure factors have been described for any agents, so assessors frequently must rely on default assumptions. When data are available, Monte Carlo analyses may be used to explore which factors contribute most to the risk estimate. Participants in a workshop on microbial risk assessment recommended that dose–response curves be generated for each subpopulation of concern (ILSI 2000). They also noted that these curves may be affected by secondary transmission, requiring careful evaluation of whether certain subgroups (e.g., residents of nursing homes and other institutions) may be at greater risk of experiencing secondary pathways that would not be considered important in an assessment for the population as a whole. Although it has been US Environmental Protection Agency (USEPA) policy since 1995 to present the range of exposures and to describe subpopulations at risk, the lack of data and strategies to estimate their risks effectively has limited the impact of this policy.

Some recent work has demonstrated approaches that hold promise for productive directions for susceptible subpopulation issues (e.g., Krewski et al. 1999; Soller, Eisenberg, and Oliveri 1999). However, the need remains to develop methods to incorporate more factors that can affect susceptibility and to conduct sensitivity analyses to understand which factors are most important for specific subpopulations' risk estimates. More refined risk assessment methods for susceptible subpopulations can be expected to more effectively improve strategies to protect subgroups' health.

Summary of Susceptible Subpopulations

Few risk assessments to date have indicated that susceptibility is a dynamic state that may change significantly during individuals' lifetimes. Even if we could define people's susceptibility at one point in time, its nature is dynamic, so each person's susceptibilities are constantly in flux. Therefore, concepts of susceptibility have focused on (1) the probability that an individual will be exposed to a questionable agent and then react to it, (2) the comparison of an individual's susceptibility to that of the majority of the population, or (3) the variation of individual states of vulnerability within a population (Parkin and Balbus 2000). Important factors in overall susceptibility include immune status, pregnancy, underlying illness, and lifestyle (e.g., smoking and drinking habits). Genetic factors may either result in lifelong or periodic susceptibility.

Typically, a list of at-risk subpopulations, based on readily observed characteristics, is used as a default option to address susceptible subpopulations in risk assessment. While the underlying rationale for this approach is that there is a wide range of responsiveness within populations that is dependent on specific mechanisms (Preston 1996), the degree of over- or under-protection that results from this default is largely unknown. Not only do hosts' biological characteristics vary, but so too do their behaviors and exposures, which may be affected by social and other factors (Balbus and Parkin 2000).

Groups that usually comprise high-risk subpopulations are infants and children, pregnant women, immunocompromised people, and the elderly. Table 4-1 includes a brief summary of how microbial contaminants may affect these groups more than the general population.

Neonates

Many studies have compared immunologic function of newborn infants with that in adults. The results often show lower concentrations of immunologic factors and decreased function. Despite these limitations in immunity in premature and full-term infants, the rate of invasive infectious diseases is low in the absence of other risk factors. After birth, neonates may be exposed to infectious diseases in the nursery or in the community. Newborn infants may be less capable of responding to infection due to one or more immunologic limitations.

Neonates (birth to 1 month old) are usually afforded some protection from microbial risk through passive immunity from their mothers. Maternal infection is a necessary prerequisite for transplacental infections. For some etiologic agents, maternal antibody is protective in the woman and her fetus. For other agents, maternal antibody may improve the outcome of infection or have no effect. Even without maternal antibody, transplacental transmission of infection to a fetus is variable, and the placenta often functions as an effective barrier. There are diverse modes of transmission of

Table 4-1: Categories of patients at higher risk than the general population for specific exposures*

Contaminant	Neonates	Infants–children	Pregnant women	Immunocompromised	Elderly
E. coli O157:H7		XXX			XX
Enteroviruses	XXX				
Cryptosporidia				XXX	XX
Hepatitis E			XXX		
Adenovirus				XXX	
MAC				XXX	XX

*The number of Xs is a qualitative judgment by the authors of the relative degree of susceptibility, weighing both frequency and severity of documented adverse health effects.

infectious agents from mother to fetus or newborn infant. Vertical transmission of infection from mother to infant may take place *in utero*, just before delivery, or during the process of delivery. Prenatal infections that are known to be transmitted transplacentally include syphilis, *Borrelia burgdorferi*, rubella, cytomegalovirus, parvovirus B19, hepatitis B virus, herpes simplex virus, HIV, varicella zoster, *Listeria monocytogenes*, toxoplasmosis, tuberculosis, and *Trypanosoma cruzi* (Stoll and Kliegman 2000).

Examples of pathogens that are dangerous to newborns are enteroviruses, such as echoviruses and coxsackieviruses. The most serious complications associated with enteroviral infections in neonates are hepatitis and myocarditis, with a resulting fatality rate reported as high as 83% (Modlin 1986). It should be noted that exposure of the neonate is primarily through the mother, either transplacentally or through direct contact, rather than through neonates ingesting water themselves. Also, pregnant women infected with *Listeria* or *Campylobacter* spp. may experience only a mild, flu-like illness. But *Listeria* infections during pregnancy can lead to premature delivery, infection of the newborn, stillbirth, or neonatal infection with sepsis and meningitis (Gribble et al. 1981; Lorber 2000).

Infants/Children

Children are not necessarily more susceptible to microbial infection or severe illness than adults. With some pathogens, such as coxsackievirus and echovirus, the rate of developing clinical disease can be low in children under 5 years and high in adults. Many enterovirus outbreaks in infants (beyond the neonatal state) result in asymptomatic infections (Abzug 1995; Dagan 1996). Also, hepatitis A and E are less serious diseases in children than adults. On the other hand, children are more likely to develop significant disease due to rotavirus infection than adults, and children under 5 years are more susceptible to serious outcomes from *Escherichia coli* O157:H7, including hemolytic uremic syndrome, which is the most common cause of kidney failure in children (CDC 2000a).

Most immunity functions are near adult levels in small children, but children are susceptible to different exposure risks, which can make them more susceptible. Recreational water outbreaks are more commonly reported than drinking water outbreaks in the United States, and children are most likely to be swimming in lakes, ponds, streams, or pools. The CDC identified *Cryptosporidium* as the most commonly reported cause of recreational water outbreaks in 1997 and 1998 (Barwick et al. 2000). In one study that measured enteroviral levels in community swimming pools, 100% of the wading pools that were tested contained measurable enterovirus (Keswick, Gerba, and Goyal 1981). Enrollment in day-care and either wearing or being around children in diapers have shown to be major risk factors in outbreak situations.

Pregnant Women

To prevent rejection of the fetus by the mother's immune system, cell-mediated immunity must be down-regulated during pregnancy. Certain intracellular pathogens are controlled by cell-mediated immunity, such as *Salmonella, Listeria, Shigella,* hepatitis A, and hepatitis E. Based on this lessening of cell-mediated immunity in the mother, one would assume that pregnant women would be highly susceptible to these pathogens. But in most cases, waterborne diseases do not seriously threaten the health of pregnant women; rather, the risk is borne by their children (see section on Neonates). However, there are exceptions; for example, hepatitis E. With the hepatitis E virus, the case fatality rate for pregnant women is extremely high—possibly up to 40% (Craske 1992). Although outbreaks of hepatitis E have not been reported in this country, waterborne outbreaks affecting thousands of people have been reported in other, mainly developing countries, and travelers have introduced cases to the United States (Gerba, Rose, and Haas 1996).

Elderly

The elderly are at increased risk of infection and disease from microbial contamination because of many factors: reduced immunity; a high incidence of frailty from malnutrition or existing chronic illness; and institutional exposure (e.g., hospitals and nursing homes). Primary immunity (also called passive immunity) consists of phagocytosis, complement, and natural killer cells. This form of immunity does not depend on prior exposure to pathogens to be effective, so age has little effect on it. However, acute and chronic diseases, and especially malnutrition, which is a problem for the elderly, may compromise these defense mechanisms. Secondary or acquired immunity is activated only after exposure to a pathogen and is also compromised by chronic diseases, malnutrition, and immunosuppressive agents (Castle 1994). Acquired immunity is also affected with advanced aging due to defects in helper T-cell function and possibly B-cell function, which also explains the elderly's diminished antibody response to vaccines (Cantrell and Norman 1998).

Because they are more likely to be hospitalized or in nursing homes, and because fecal incontinence facilitates the spread of pathogens, the elderly suffer a lot of gastrointestinal illness, and diarrhea is a serious complication. In addition, as we age, the gut-associated lymphoid tissue becomes senescent, which can increase susceptibility to infection. There is also a decrease in acid secretion, and a lowered pH in the stomach inhibits the defense from enteric pathogens. A 1991 study reviewed death certificate data listing diarrhea as a cause between 1979 and 1987 (Lew et al. 1991). The majority of these deaths occurred among the elderly (>74 years) with risk factors of being white, female, and residing in a long-term care facility. The case fatality rates in nursing homes for certain waterborne pathogens, such as rotavirus and *E. coli* O157:H7, can be two orders of magnitude greater than that in the general population (Gerba, Rose, and Haas 1996). Outbreaks of Norwalk virus and other caliciviruses have been frequently reported in nursing homes. There is also evidence to suggest that the elderly are more susceptible to the effects of cryptosporidiosis (Bannister and Mountford 1989; Neill et al. 1996).

Immunocompromised

The immunosuppressed population includes not only people with AIDS but also transplant patients, persons undergoing chemotherapy, and those suffering from less common congenital or acquired immune system dysfunction. Cryptosporidiosis is deadly for the immunocompromised. During the Milwaukee outbreak in 1993, almost all the deaths occurred in those with AIDS.

AIDS Patients

In their 2000 year-end surveillance report, the CDC estimated that more than 127,000 people in this country were living with HIV, and another 322,000 were living with AIDS (CDC 2000b). AIDS is the most advanced stage of HIV infection, and is defined partly by the body's inability to fight off opportunistic infections. There are a number of pathogens that can be transmitted by water to which AIDS patients are especially susceptible: *Cryptosporidium*, MAC, and microsporidia, for example. AIDS patients are at high risk for developing chronic diarrhea from many sources, including infectious pathogens. On the other hand, AIDS patients do not appear to be any more susceptible to infection or severe outcome from some infections associated with water transmission such as *Helicobacter pylori* and *Aeromonas* spp. The advent of potent antiretroviral drugs for people with HIV has also dramatically decreased the number of patients who are affected by opportunistic infections of all kinds (Griffith 1999; Havlir et al. 2000).

Chemotherapy Patients

The number of new cancer patients has increased steadily over the past 20 years (Morris and Potter 1997), which has probably corresponded to an increase in the number of chemotherapies administered. In addition to the immunosuppressive features of cancer drug treatment, there are other factors that contribute to these patients' increased susceptibility. For example, some drugs are antimicrobials, which can disturb the ecology of the gut flora, making patients susceptible to enteric pathogens. Chemotherapy patients often suffer from neutropenia (a decrease in the number of neutrophils—a type of white blood cell), which also contributes to their susceptibility to more severe disease if infected (Kallianpur 1999).

Transplant Patients

Medical advances have resulted in increased numbers of transplant patients, who must be on immunosuppressing drugs for the rest of their lives. In 2000, there were almost 23,000 transplants in this country—up almost 8,000 from 10 years before (United Network for Organ Sharing 2001). These patients become part of the population of chronically immunosuppressed because of their drug regimens.

Transplant patients are especially susceptible to many kinds of infections, including those that can be considered waterborne. As well, the outcomes of infections may be much more severe in transplant patients than in immunocompetent people. For example, adenovirus infections that could easily be subclinical in most people cause disseminated disease in people who have undergone transplants. Liver and bone marrow transplant patients are susceptible to adenovirus hepatitis (Bertheau et al. 1996; Saad et al. 1997), and kidney transplant patients are susceptible to hemorrhagic cystitis (Foy 1997). One review estimated that 11% of transplant patients become infected with adenovirus (Hierholzer 1992). The case fatality rate for bone marrow and kidney transplant patients from adenovirus was 60% and 18%, respectively.

Patients undergoing allogeneic bone marrow transplants have developed severe complications from enterovirus infections, including the report of an outbreak of coxsackievirus A1 resulting in increased mortality in this population (Schwarer et al. 1997; Townsend et al. 1982). These patients may also be at greater risk for microsporidiosis (Kelkar et al. 1997).

Proposed Framework to Assess Microbial Risk in Drinking Water to Sensitive Subpopulations

Risk assessments are generally used to guide regulatory decisions regarding a level of exposure resulting in a population level of "acceptable" risks. When expected exposures exceed the defined risk level, action is taken to reduce the concentration of pollutants to an acceptable level. Defining what kind of vulnerability merits protective actions and where to draw the regulatory line on who to protect and how to best protect them are the issues with which USEPA and other agencies struggle. The Presidential/Congressional Commission on Risk Assessment and Risk Management (1997) states that "Genetic, nutritional, metabolic, and other differences make some segments of a population more susceptible than others to the effects of a given exposure to a given chemical; however, current regulatory approaches for reducing risks associated with chemical exposures generally do not include information on differences in individual susceptibility or encourage information gathering evidence to identify them." The same statement can be made about microbial risk assessment.

The goal of our research is to develop a framework to evaluate microbial pathogens and their importance as waterborne contaminants, specifically to sensitive subpopulations. We then apply this proposed framework to each microbial pathogen on the contaminant candidate list (CCL): adenovirus, *Aeromonas hydrophila*, calicivirus, coxsackievirus, cyanobacteria, echovirus, *H. pylori*, microsporidia, and MAC. This process will help characterize the pathogens' comparative risk, define data gaps for future research, and offer assessment criteria to policymakers for use in the development of regulations.

The framework comprises host health issues and exposure issues that are summarized as follows:

Health effects: Includes factors such as the prevalence of illness in the population and degree of morbidity and mortality.

Dose response: What is known about pathogenesis, including infectious dose and the probability of developing symptoms or disease after infection.

Efficacy of medical treatment: The availability of an efficacious medical treatment for a pathogenic infection or disease. If available, the length and expense of the course of treatment.

Secondary spread: Once someone is infected, the likelihood that they will transmit the infection to others.

Chronic sequelae: The long-term health outcomes that may follow an acute infection.

Routes of exposure: How pathogens spread from person to person.

Concentrations at intake: The occurrence of the pathogen in various source waters.

Efficacy of water treatment: How well conventional drinking water treatment (settling, filtration, flocculation/coagulation, and disinfection) removes the pathogen of interest.

Survival/amplification in the distribution system: One step beyond treatment—how well the pathogen is inactivated with a standard chlorine residual in case of a breech in the distribution system. Also, the pathogen's ability to colonize water distribution systems and under what conditions.

In evaluating a pathogen, certain elements of the framework are most relevant to the question of susceptible populations.

Variation in Infection Rate and Degree of Morbidity and Mortality

There are different ways a subpopulation can be categorized as susceptible, in terms of the frequency and severity of infection, which is evident from data on the CCL microbes. A waterborne pathogen may affect a certain subpopulation by (1) having a higher rate of infection than the general population, but the same disease response; (2) having a higher rate of infection than the general population, but a lessened disease response; (3) having the same incidence of infection as the general population; or (4) both—a higher rate of infection and an increased rate of morbidity/mortality or level of disease severity. Table 4-2 exemplifies these concepts with four CCL microbes.

For example, children can be more susceptible to infection from certain pathogens than adults because of their inherently reduced immunity, which is a natural state. Though they are more likely to become infected, many children's outcomes are mild or asymptomatic, and infection usually confers some level of protection from later exposures. In this sense, children fall into the category of having a higher rate of infection but decreased morbidity and/or mortality compared with adults. This is also a transient category; children are not automatically in this state at birth, though they may acquire temporary immunity from their mothers by breastfeeding. Also, once a child has been infected by a particular pathogen and develops immunity, he or she leaves that category and enters either the "general" population or another category based on the presence of some other risk factor for susceptibility. This cycling in and out of categories of susceptibility varies for each pathogen.

Table 4-2: Variation of incidence and morbidity/mortality with four waterborne microbes

Pathogen	Susceptible Population	Incidence	Morbidity/Mortality	References
MAC	AIDS/Elderly	Decreasing in AIDS; increasing in elderly	One of many end-stage infections in AIDS; chronic pulmonary disease in elderly	Havlir et al. 2000; Kennedy and Weber 1994; Palella et al. 1998; Prince et al. 1989; Reich and Johnson 1991
Adenovirus	Immunocompromised (e.g., transplant)	Same as general population	Greater risk of severe/ fatal outcome	Hierholzer 1992; Saad et al. 1997
H. pylori	Immunocompromised (e.g., AIDS)	Same as general population	Same as general population	Battan et al. 1990; Edwards et al. 1991
Calicivirus	Unidentified genetic factor	Higher than general population	Higher rate of illness/ reinfection than general population	Moe et al. 1999

Route of Exposure and Degree of Waterborne Exposure

Once the population has been classified in terms of its infection rate and outcome, we can consider likely routes of exposure in the sensitive populations and the role of waterborne exposure. We unfortunately know little in this area, though it is obviously a key component in the risk assessment process. USEPA addressed this issue when announcing the final CCL by asking: "Is the contaminant known or substantially likely to occur in public water systems with a frequency and at levels posing a threat to public health?" (USEPA 1998). The obstacle lies in defining the term *substantially.* Unfortunately, we lack the analytical techniques to accurately detect most CCL microbes in water samples. Most of the pathogens are shed in the feces of humans and possibly animals, which is the basis for determining plausible waterborne contamination. Traditionally, contaminants that are fecal in origin have been estimated by fecal coliform indicator bacteria. However, as more sophisticated molecular detection methods, such as polymerase chain reaction, have been employed in research, we have learned that many pathogens—particularly viruses and protozoa—can show up in the absence of our long-standing indicator. In addition, fecal coliforms may not reflect the concentrations, viability, or human pathogenicity of potentially harmful agents. So, continuing the assumption that all CCL microbials can infect via a waterborne route, we turn to the possible role of waterborne exposure in the transmission of disease. The fact that we have no more than an assumption about whether or not some pathogens occur in drinking water results in further questioning the role of drinking water ingestion in transmission (Table 4-3). One way to approach the question is to hypothesize: If the microbe of interest were completely removed from drinking water, what could the impact be on the incidence of infection? Where reliable knowledge is available on other routes of transmission, we can estimate the answer to the question with more certainty.

Medical Treatment Efficacy and Chronic Sequelae

For some infections, the efficacy of medical treatment influences the later development of chronic sequelae. In healthy populations, most exposures result in asymptomatic or mild and self-limiting disease. When treatment is required, many times all that is available is palliative care until the acute phase is over. Often, once the patient has survived the acute phase, recovery is complete—even in susceptible patients. The most frequent chronic condition reported in association with waterborne microbial disease is diarrhea.

Table 4-3: Elements of waterborne exposure

Pathogen	Susceptible Population	Possible Routes of Exposure	Role of Waterborne Exposure	References
MAC	AIDS	Inhaling aerosols, ingestion	Unknown	Fry, Meissner, and Falkinham 1986; Peters et al. 1995
Echovirus	Neonates	Transplacental infection or nosocomial	Unlikely	Abzug 1995; Dagan 1996
Calicivirus	Unidentified genetic factor	Water, food, fomite, inhaling aerosols	Waterborne outbreaks have been documented—otherwise, unknown	Green 1997; Moe et al. 1999; Noel et al. 1997

Table 4-4: Relationship between medical treatment and chronic sequelae

Pathogen	Susceptible Population	Efficacy of Medical Treatment	Possible Chronic Sequelae	References
H. pylori	Children with risk factors	If identified, excellent	Gastric ulcer; gastric cancer	Graham 1994; Moayyedi and Axon 1998
Pulmonary MAC	Elderly with risk factors	Long, difficult, marginally effective drug treatment	Chronic pulmonary disease	Iseman 1996; Kennedy and Weber 1994

However, some of the pathogens on the CCL have been tentatively linked with such serious chronic diseases as insulin-dependent diabetes, cancer, lung disease, and neurological sequelae. *H. pylori* and pulmonary MAC are examples of this scenario, which are seen in Table 4-4.

Summary

Risk assessors recognize that challenges still exist in formulating a process to evaluate microbial risks in drinking water, especially for susceptible subpopulations, whose exposures may not be well understood. Each individual is born with an inherent level of susceptibility to pathogens, and that level may increase or decrease a number of times over a person's lifetime. We have developed a process to highlight and compare the knowns and unknowns of the pathogens on USEPA's CCL. This template is intended to help frame and guide important research and decisions regarding regulation of microbial contamination of drinking water.

Bibliography

Abzug, M.J. 1995. Perinatal Enterovirus Infections. In *Human Enterovirus Infections.* Edited by H.A. Rotbart. Washington, DC: ASM Press. pp. 221–238.

Balbus, J.M., and R. Parkin. 2000. Social factors of susceptibility in risk assessment: application to microbial pathogens. *Risk Policy Report,* 7:36–38.

Balbus, J., R. Parkin, and M. Embrey. 2000. Susceptibility in microbial risk assessment: definitions and research needs. *Environ. Health Perspect.,* 108:901–905.

Bannister, P., and R.A. Mountford. 1989. *Cryptosporidium* in the elderly: a cause of life-threatening diarrhea. *Am. J. Med.,* 86(4):507–8.

Barwick, R.S. D.A. Levy, G.F. Craun, M.J. Beach, R.L. Calderon. 2000. Surveillance for waterborne-disease outbreaks—United States, 1997–1998. *MMWR,* 49(4):1–21.

Battan, R., M.C. Raviglione, A. Palagiano, J.F. Boyle, M.T. Sabatini, K. Sayad, and L.J. Ottaviano. 1990. *Helicobacter pylori* infection in patients with acquired immune deficiency syndrome. *Am. J. Gastroenterol.,* 85:1576–1579.

Bertheau, P., N. Parquet, F. Ferchal, E. Gluckman, and C. Brocheriou. 1996. Fulminant adenovirus hepatitis after allogeneic bone marrow transplantation. *Bone Marrow Transplant.,* 17:295–298.

Canadian Standards Association (CSA). 1997. *Risk Management: Guideline for Decision-Makers.* CAN/CSA-Q850-97. Toronto: Canadian Standards Association.

Cantrell, M., and D.C. Norman. 1998. Infections. In *Duthie: Practice of Geriatrics.* 3rd ed. Philadelphia: W.B. Saunders Company.

Castle, S. 1994. Age-related Changes in Host Defenses. In *Antimicrobial Therapy in the Elderly Patient.* Edited by T.T. Yoshikawa and D.C. Norman. New York: Marcel Dekker.

Centers for Disease Control and Prevention (CDC). 2000a. Division of Bacterial and Mycotic Diseases. Disease Information: *Escherichia coli* O157:H7 [Online]. Available: <http://www.cdc.gov/ncidod/dbmd/diseaseinfo/escherichiacoli_g.htm>. [cited January 30, 2002]

Centers for Disease Control and Prevention (CDC). 2000b. HIV/AIDS Surveillance Report. National Center for HIV, STD and TB Prevention. Divisions of HIV/AIDS Prevention. Vol. 12, No. 2.

Centers for Disease Control and Prevention (CDC). 2001. Two fatal cases of adenovirus-related illness in previously healthy young adults — Illinois, 2000. *MMWR*, 50(26):553–555.

Craske, J. 1992. Hepatitis C and non-A non-B hepatitis revisited: hepatitis E, F and G. *J. Infect.*, 25(3):243–50.

Dagan, R. 1996. Nonpolio enteroviruses and the febrile young infant: epidemiologic, clinical and diagnostic aspects. *Pediatr. Infect. Dis. J.*, 15:67–71.

Edwards, P.D., J. Carrick, J. Turner, A. Lee, H. Mitchell, and D.A. Cooper. 1991. *Helicobacter pylori*-associated gastritis is rare in AIDS: antibiotic effect or a consequence of immunodeficiency? *Am. J. Gastroenterol.*, 86:1761–1764.

Food Quality Protection Act (FQPA). Pub. L. 104-170, 110. Food Quality Protection Act of 1996 (Aug. 3, 1996). Stat. 1489.

Foy, H.M. 1997. Adenoviruses. In *Viral Infections of Humans.* 4th ed. Edited by A.S. Evans and R.A. Kaslow. New York: Plenum Press.

Fry, K.L., P.S. Meissner, and J.O. Falkinham III. 1986. Epidemiology of infection by nontuberculous mycobacteria. VI. Identification and use of epidemiologic markers for studies of *Mycobacterium avium, M. intracellulare* and *M. scrofulaceum*. *Am. Rev. Respir. Dis.*, 134:39–43.

Gerba, C.P., J.B. Rose, and C.N. Haas. 1996. Sensitive populations: who is at the greatest risk? *Int. J. Food Microbiol.*, 30:113–123.

Gilks, W.R., S. Richardson, D.J. Spiegelhalter, eds. 1996. *Markov Chain Monte Carlo in Practice.* London: Chapman and Hall.

Graham, D.Y. 1994. Benefits from elimination of *Helicobacter pylori* infection include major reduction in the incidence of peptic ulcer disease, gastric cancer, and primary gastric lymphoma. *Prev. Med.*, 23:712–716.

Green, K.Y. 1997. The role of human calicivirus in epidemic gastroenteritis. *Arch. Virol.*, 13(Suppl):153–165.

Gribble, M.J., I.E. Salit, J. Isaac-Renton, and A.W. Chow. 1981. Campylobacter infections in pregnancy. Case report and literature review. *Am. J. Obstet. Gynecol.*, 140(4):423–426.

Griffith, D.E. 1999. Risk-benefit assessment of therapies for *Mycobacterium avium* complex infections. *Drug Saf.*, 21:137–152.

Haas, C.N., J.B. Rose, and C.P. Gerba. 1999. *Quantitative Microbial Risk Assessment.* New York: John Wiley & Sons.

Hattis, D. 1996. Human interindividual variability in susceptibility to toxic effects: from annoying detail to a central determinant of risk. *Toxicol.*, 111:5–14.

Hattis, D., P. Banati, R. Goble, and D.E. Burmaster. 1999. Human interindividual variability in parameters related to health risks. *Risk Anal.*, 19:711–726.

Hattis, D., A. Russ, R. Goble, P. Banati, and M. Chu. 2001. Human interindividual variability in susceptibility to airborne particles. *Risk Anal.*, 21:585–599.

Havlir, D.V., R.D. Schrier, F.J. Torriani, K. Chervenak, J.Y. Hwang, and W.H. Boom. 2000. Effect of potent antiretroviral therapy on immune responses to *Mycobacterium avium* in human immunodeficiency virus-infected subjects. *J. Infect. Dis.*, 182(6):1658–1663.

Hierholzer, J.C. 1992. Adenovirus in the immunocompromised host. *Clin. Microbiol. Rev.*, 5:262–274.

Hocqueloux, L., P. Lesprit, J.L. Herrmann, A. de La Blanchardiere, A.M. Zagdanski, J.M. Decazes, and J. Modai. 1998. Pulmonary *Mycobacterium avium* complex disease without dissemination in HIV-infected patients. *Chest*, 113:542–548.

International Life Sciences Institute (ILSI). 2000. *Revised Framework for Microbial Risk Assessment.* Washington, DC: ILSI.

Iseman, M.D. 1996. Pulmonary Disease Due to *Mycobacterium avium* Complex. In Mycobacterium avium-*Complex Infection: Progress in Research and Treatment.* Edited by J.A. Korvick and C.A. Benson. New York: Marcel Dekker.

Kahn, H., H. Jacobs, and K. Stralka. 1999. *Preliminary Estimates of Water Consumption in the United States Based on the CSFII.* Washington, DC: USEPA.

Kallianpur, A. 1999. Supportive care in hematologic malignancies. In *Wintrobe's Clinical Hematology.* 10th ed. Baltimore: Lippincott, Williams & Wilkins.

Kelkar, R., P.S. Sastry, S.S. Kulkarni, T.K. Saikia, P.M. Parikh, and S.H. Advani. 1997. Pulmonary microsporidial infection in a patient with CML undergoing allogenic marrow transplant. *Bone Marrow Transplant.*, 19:179–182.

Kennedy, T.P., and D.J. Weber. 1994. Nontuberculous mycobacteria. An underappreciated cause of geriatric lung disease. *Am. J. Respir. Crit. Care Med.*, 149:1654–1658.

Keswick, B.H., C.P. Gerba, and S.M. Goyal. 1981. Occurrence of enteroviruses in community swimming pools. *Am. J. Public Health*, 71:1026–1030.

Krewski, D., S.N. Rai, J.M. Zielinski, and P.K. Hopke. 1999. Characterization of uncertainty and variability in residential radon cancer risks. *Ann. NY Acad. Sci.*, 896:245–271.

Kubo, K., Y. Yamazaki, T. Hachiya, M. Hayasaka, T. Honda, M. Hasegawa, and S. Sone. 1998. *Mycobacterium avium-intracellulare* pulmonary infection in patients without known predisposing lung disease. *Lung*, 176:381–391.

Lew, J.F., R.I. Glass, R.E. Gangarosa, I.P. Cohen, C. Bern, and C.L. Moe. 1991. Diarrheal deaths in the United States, 1979 through 1987: a special problem for the elderly. *JAMA*, 265(24):3280–3284.

Lorber, B. 2000. Lysteria monocytogenes. In *Principles and Practice of Infectious Diseases*. 5th ed. Edited by G.L. Mandell, J.E. Bennet, and R. Dolin. Philadelphia: Churchill Livingstone, Inc.

Moayyedi, P., and A.T.R. Axon. 1998. Is there a rationale for eradication of *Helicobacter pylori*? Cost-benefit: the case for. *Br. Med. Bull.*, 54:243–250.

Modlin, J.F. 1986. Perinatal echovirus infection: Insights from a literature review of 61 cases of serious infection and 16 outbreaks in nurseries. *Rev. Infect. Dis.*, 8(6), 918–926.

Moe, C., D. Rhodes, S. Pusek, F. Tseng, W. Heizer, C. Kapoor, B. Gilliam, P. Stewart, M. Harb, S. Miller, M. Sobsey, J. Herrmann, N. Blacklow, and R. Calderon. 1999. Determination of Norwalk virus dose–response in human volunteers. Presented at *Health Effects Stakeholder Meeting for the Stage 2 DBPR and LT2ESWTR*, February 12, 1999. Washington, DC.

Morris, J.G., and M. Potter. 1997. Emergence of new pathogens as a function of changes in host susceptibility. *Emerg. Infect. Dis.*, 3(4).

National Research Council (NRC). 1983. *Risk Assessment in the Federal Government: Managing the Process*. Washington, DC: National Academy Press.

National Research Council. (NRC). 1994. *Science and Judgment in Risk Assessment*. Washington, DC: National Academy Press.

National Research Council (NRC). 1999. *Health Effects of Exposure to Radon (BEIR VI)*. Washington, DC: National Academy Press.

Neill, M.A., S.K. Rice, N.V. Ahmad, and T.P. Flanigan. 1996. Cryptosporidiosis: an unrecognized cause of diarrhea in elderly hospitalized patients. *Clin. Infect. Dis.*, 22:168–170.

Noel, J.S., T. Ando, J.P. Leite, K.Y. Green, K.E. Dingle, M.K. Estes, Y. Seto, S.S. Monroe, and R.I. Glass. 1997. Correlation of patient immune responses with genetically characterized small round-structured viruses involved in outbreaks of nonbacterial acute gastroenteritis in the United States, 1990 to 1995. *J. Med. Virol.*, 53:372–383.

Palella, F.J., Jr., K.M. Delaney, A.C. Moorman, M.O. Loveless, J. Fuhrer, G.A. Satten, D.J. Aschman, and S.D. Holmberg.1998. Declining morbidity and mortality among patients with advanced human immunodeficiency virus infection. HIV Outpatient Study Investigators. *N. Engl. J. Med.*, 338:853–860.

Parkin, R.T. 2001. Strategies for incorporating susceptibility in risk assessment. In *Proceedings of the Eighth International Inhalation Symposium*. Fraunhofer Institut für Toxikologie und Aerosolforschung, Hannover, Germany.

Parkin, R.T., and J.M. Balbus. 2000. Variations in concepts of "susceptibility" in risk assessment. *Risk Anal.*, 20:603–611.

Peters, M., C. Muller, S. Rusch-Gerdes, C. Seidel, U. Gobel, H.D. Pohle, and B. Ruf. 1995. Isolation of atypical mycobacteria from tap water in hospitals and homes: is this a possible source of disseminated MAC infection in AIDS patients? *J. Infect.*, 31(1):39–44.

The Presidential/Congressional Commission on Risk Assessment and Risk Management. 1997. *Framework for Environmental Health Risk Management: Final Report.* Vol. I. Washington, DC: The Presidential/ Congressional Commission on Risk Assessment and Risk Management.

Preston, R.J. 1996. Genetic susceptibility and sensitivity to cancer. CIIT Activities, 16:1–6. Research Triangle Park, NC: CIIT.

Prince, D.S., D.D. Peterson, R.M. Steiner, J.E. Gottlieb, R. Scott, H.L. Israel, W.G. Figueroa, and J.E. Fish. 1989. Infection with *Mycobacterium avium* complex in patients without predisposing conditions. *N. Engl. J. Med.*, 321:863–868.

Reich, J.M., and R. Johnson. 1991. *Mycobacterium avium* complex pulmonary disease. Incidence, presentation, and response to therapy in a community setting. *Am. Rev. Respir. Dis.*, 143:1381–1385.

Saad, R.S., A.J. Demetris, R.G. Lee, S. Kusne, and P.S. Randhawa. 1997. Adenovirus hepatitis in the adult allograft liver. *Transplantation*, 64:1483–1485.

Safe Drinking Water Act Amendments of 1996, 42 *US Code* 300 (1974).

Schwarer, A.P., S.S. Opat, A.M. Watson, D. Spelman, F. Firkin, and N. Lee. 1997. Disseminated echovirus infection after allogeneic bone marrow transplantation. *Pathology*, 29(4):424–425.

Soller, J.A., J.N. Eisenberg, and A.W. Oliveri. 1999. *Evaluation of Pathogen Risk Assessment Framework*. Oakland, CA: Eisenberg, Oliveri & Associates, Inc.

Stoll, B.J., and R.M. Kliegman. 2000. High-risk pregnancies. In *Nelson Textbook of Pediatrics.* 16th ed. Edited by R.E. Behrman, R.M. Kliegman, and H.B. Jenson. Philadelphia: W.B. Saunders.

Townsend, T.R., E.A. Bolyard, R.H. Yolken, W.E. Beschorner, C.A. Bishop, W.H. Burns, G.W. Santos, and R. Saral.1982. Outbreak of Coxsackie A1 gastroenteritis: a complication of bone-marrow transplantation. *Lancet*, 1(8276):820–823.

United Network for Organ Sharing. 2001 [Online]. Available <http://www.unos.org/Frame_default.asp?Category=Newsdata>. [cited January 31, 2002]

US Department of Agriculture (USDA). 1998. *1994–96 Continuing Survey of Food Intake by Individuals (CSFII) and 1994–96 Diet and Health Knowledge Survey.* Washington, DC: USDA.

US Environmental Protection Agency (USEPA). 1998. Announcement of the Drinking Water Contaminant Candidate List. *Fed. Reg.*, 63:10273–10287.

CHAPTER 5

Adenovirus in Drinking Water

by Martha A. Embrey

Executive Summary

Problem Formulation

Occurrence of Illness (*enteric*)

- In summarized studies of sources of diarrheal illness, enteric adenovirus was identified as a sole cause around 5–10% of the time, representing between 0 and 40% of all serotypes in adenovirus cases; generally, 0 and 20% prevalence of fecal shedding in healthy hosts.

Role of Waterborne Exposure

- Adenovirus has not been positively associated with any outbreaks from drinking water. A number of outbreaks have occurred as a result of exposure to recreational water.

Degree of Morbidity and Mortality (*enteric*)

- Long-lasting (>1 week) diarrheal illness; mild to moderate in nature; self-limiting; does not appear to affect sensitive subpopulations with any more severity or frequency.

Detection Methods in Water/Clinical Specimens (*enteric*)

- Cell culture (only way to show infectivity), electron microscopy, ELISA, EIA, monoclonal antibodies, gene probes, PCR (nested PCR appears to be the most sensitive and specific).

Mechanisms of Water Contamination

- Human fecal matter/sewage.

Concentrations at Intake

- Dependent on level of human fecal contamination in the source water. In published reports, concentrations in river water have ranged from 0–25 pfu/L and from 70–3,200 CPU/L. One study reported average concentrations in raw sewage as 1,950 IU/L and 250 IU/L in secondary effluent. Seasonal trends have not been shown.

Efficacy of Water Treatment

- Indirect removal through the capture of adenovirus particles with coagulation/flocculation is 85% effective and 93% effective for enteroviruses. Viruses can be inactivated with 0.1 mg/L of free chlorine. Membrane filtration technology has the potential of removing virtually all viral pathogens.

Survival/Amplification in Distribution

- Not applicable in viruses.

Routes of Exposure

- Direct (touching), aerosolization; ingestion, nosocomial, genitally (nonenteric);
- Most likely fecal–oral for enteric transmission.

Dose Response

Infectious Dose

- Unknown

Probability of Illness Based on Infection

- About 50% of childhood enteric adenovirus infections result in illness; the probability of illness increases when the infection is respiratory.
- In studies of immunocompromised individuals, about one half to one third of those infected with enteric adenoviruses became ill.
- In another study, 20% of the infected (with mainly subgenus B) patients developed severe, disseminated disease.

Efficacy of Medical Treatment (*enteric*)

- The illness is self-limiting. Rehydration therapy may be necessary in some children.

Secondary Spread (*enteric*)

- Limited evidence of secondary spread except among very young (preschool age) children.

Chronic Sequelae (*enteric*)

- None known in either immunocompetent or immunosuppressed patients.

Introduction

In the early 1950s, a series of epidemics of febrile respiratory disease among military recruits produced a number of research efforts to find the cause. The new pathogen was identified in 1953 through the newly available tool of cell culture for viruses (Foy 1997). Because it was first isolated from adenoid tissue, it was named adenovirus. Physically, adenovirus is made up of a double-stranded DNA genome without an envelope, and it ranges in size from 60 to 90 nm. It is the only virus made up of double-stranded DNA. The adenovirus family is made up of six subgenera labeled A through F (Table 5-1), divided by DNA homology of the viral genomes (Turner et al. 1987). Within the six subgenera are 47 known serotypes or species defined by quantitative neutralization with hyperimmune sera. Clinical manifestations vary among subgenera and even species. Additionally, many variants occur within each species, which could explain why geographic differences in disease occur (Foy 1997). The different species of adenovirus produce a variety of illnesses including upper and lower respiratory illness, conjunctivitis, cystitis, and gastroenteritis. New species are formed by mutations and recombinations, and the newest species (42–47) have been isolated from the fecal samples of acquired immune deficiency syndrome (AIDS) patients (Foy 1997). Diarrhea can be a symptom accompanying any adenovirus infection, but subgenus F specifically causes gastrointestinal illness. This chapter will focus on that subgenus comprising species 40 and 41, also called the enteric adenoviruses.

Table 5-1: Adenovirus classification

Subgenus	Species	Disease Syndromes
A	12,18,31	Nonspecific enteric infection
B	3,7,11,14,16,21,34,35	Upper and lower respiratory illness, acute respiratory disease, cystitis
C	1,2,5,6	Upper and lower respiratory illness
D	8–10,13,15,17,19, 20,22–30,32,33,36–39,42–47	Epidemic keratoconjunctivitis, conjunctivitis
E	4	Acute respiratory disease, pharyngoconjunctival fever, pneumonia
F	40,41	Gastroenteritis (mainly infantile)

Adapted from Foy 1997.

Epidemiology

Adenovirus is not reportable; therefore, no national or population-based morbidity and mortality figures exist. Most of our epidemiological data come from the study of select populations, specifically military recruits, children in institutions (e.g., hospitals), children in day-care centers, and groups of families.

General

In the studies of children, adenovirus 1, 2, 5, and 6 (subgenus C) appear to be endemic, whereas other species, especially 3 and 7, tend to be more epidemic or sporadic in nature. Forty to 60% of children have antibodies to 1, 2, and 5, and few have antibodies to 3, 4, or 7, making adults more susceptible to those serotypes (Singh-Naz and Rodriguez 1996). Because asymptomatic, healthy people can shed virus, it is difficult to confidently link adenovirus with illness when it occurs (Foy 1997). Occurrence studies comparing infection in healthy and ill people have shown from 0 to 20% of asymptomatic people shedding virus (see Table 5-2). Family studies in New York and Seattle (Fox, Hall, and Cooney 1977; Fox et al. 1969) showed large numbers of asymptomatic infections—especially in the lower numbered species. The subjects also demonstrated a high frequency of recrudescent virus shedding. Most illnesses caused by adenoviruses are acute and self-limited, but though the disease phase may be short, the virus can remain in the gastrointestinal tract and continue to be excreted for a long time (all species, not just the enteric). Species within subgenus C especially may continue to be excreted for months or even years after the disease has run its course. Any type of adenovirus infection may be accompanied by diarrhea, but virus can be excreted even if diarrhea is not present (Wadell et al. 1994). Large proportions of infection in subgenera A and D tend to be asymptomatic; the species within B and E tend to result in a higher rate of symptomatic respiratory illness (Foy 1997). Most illness is mild to moderate, but the virus can invade organs and cause more serious diseases like pneumonia and hepatitis that may result in fatality. This is rare in the immunocompetent person, but immunosuppressed people are at higher risk for disseminated disease. Immunity is species specific; volunteers who were challenged with a species for which they already had antibodies rarely developed symptomatic infection (Foy 1997). Because virus can be isolated from tonsils (in 50% of extracted tonsils) and lymphocytes, it appears infection can remain latent for a long time—perhaps for life—only to be reactivated from AIDS or other immune deficiency (Durepaire et al. 1995; Foy 1997; Khoo et al. 1995). Data on the infective dose have not been reported.

Studies have looked at the occurrence of adenovirus, specifically subgenus F, in children with gastroenteritis and diarrhea; most are cross-sectional surveys, though a few have also included well or asymptomatic children as controls (see Table 5-2). An important consideration when comparing these types of studies is the assay and technique used to detect the virus from fecal samples. The quality of the assays is variable, so comparisons have to be general. Many of the studies used electron microscopy to screen for the presence of any adenovirus, then more specific methods to detect the enteric adenovirus species. In the studies summarized in Table 5-2, most researchers used monoclonal antibody techniques, though some used enzyme-linked immunosorbent assays (ELISA) or cell culture. These methods are different in terms of their sensitivity and specificity. As discussed earlier, polymerase chain reaction (PCR) is the newest and most accurate method to detect specific species. The study by Allard and colleagues (1992) was the only one summarized here that used the PCR technique to test their population of 150 children and adults. Interestingly, they identified no samples as adenovirus 40 or 41.

Generally, many species of adenovirus are endemic throughout the world, though at different occurrence rates. In the studies summarized in Table 5-2, the rate of total adenovirus among children with gastroenteritis ranged from 4 to 50%, while the rate of subgenus F ranged from 0 to 43% of the total adenovirus rate. A percentage of asymptomatic children are usually infected, and the study that included sick and well adults showed that 18% of the well adults tested positive for enteric adenovirus (Allard, Albinsson, and Wadell 1992). Geography or a country's state of development did not seem to have any effect on the rates. There is no seasonality associated with rate of infection and disease with any species of adenovirus, like there is with rotavirus (Bates et al. 1993; Lew et al. 1991; Wang and Chen 1997).

Adenovirus is a common cause of acute lower respiratory tract infections in children. Recent reports have shown that subgenus B has frequently been the cause and that genome type 7h has been associated with especially severe symptoms and even death (Larranaga et al. 2000; Videla, Carballal, and Kajon 1999). Another nosocomial outbreak of adenovirus 7d caused serious illness and death in pediatric patients in Chicago (Gerber et al. 2001).

Adenovirus has been known as a frequent cause of outbreaks in military installations until a vaccine came into use in the 1970s. In 1995, the adenovirus vaccine went out of production, and surveillance among military trainees at four centers was instituted to measure the impact of the loss of vaccine (Gray et al. 2000). From October 1996 to June 1998, over half (n = 1,814) of the throat cultures of symptomatic soldiers

Table 5-2: Selected studies on the occurrence of adenovirus in children

Study	Place	N*	% Total AV†	% EAV (40/41)‡	Other
Uhnoo et al. 1984, 1986	Sweden	200 well 416 ill	1.5 13.5	0 7.9	In- and outpatient children with acute gastroenteritis
Rodriguez et al. 1985	Washington state	270: 134 children; 136 adults	6.7	1.1	EAV cases all <24 months with gastroenteritis
Kotloff et al. 1989	Baltimore, Md.	372 well 538 ill	— —	1.3 5.2	2-year prospective study; children <2 years
Cruz et al. 1990	Guatemala	191 well 385 ill	— —	4.7 14.0 of total well/ill samples	Aged 0–3; of 59 hospitalized: 51% rotavirus/31.2% EAV
Kim et al. 1990	Korea	90 well 345 ill	— —	2.0 9.0	6% rotavirus/AV 40/41 combined; 94% with EAV<24 months
Lew et al. 1991	Arizona	129 well 565 ill	8.0 8.0	2.0 2.0	4% ill, 1% well = astrovirus; 100% of AV infected children <35 months
Ruuska and Vesikari 1991	Finland	248 ill	4.0	—	26% rotavirus; 4% bacterial; $^2/_3$ unidentified; sickest children = 75% rotavirus; 0–2.5 years
Allard et al. 1992	Sweden	50 well adults 50 ill children 50 ill adults	18.0 50.0 24.0	0 0 0	PCR methodology with 40/41 primers; mean age 32 years/21 months/31 years, respectively
Mistchenko et al. 1992	Argentina	766 well 180 ill	14.4 13.3	0.8 33.0 of total AV	Family-based study of children <15 years

*N = number of study subjects.
†AV = adenovirus.
‡EAV = enteric adenoviruses.

Table continued on next page.

Table 5-2: Selected studies on the occurrence of adenovirus in children (continued)

Study	Place	N*	% Total AV†	% EAV (40/41)‡	Other
Bates et al. 1993	England	1,426 ill	17.8	16.4 of total AV	78.3% rotavirus; 7.9% astrovirus; hospitalized children <5 years
Donelli et al. 1993	Rome	417 ill	7.0	—	18.2% rotavirus; astrovirus; hospitalized children
Harsi et al. 1995	Brazil	79 well 67 ill	11.4 10.0	— 43.0 of total AV	Hospitalized children <2 years
Grimwood et al. 1995	Australia	4,473 ill	—	3.1(40/41) = 14.0 (40) and 86.0 (41) of total AV	Children hospitalized with acute gastroenteritis
Bryden et al. 1997	England	452 ill	32.0 (non 40/41)	22.0(40); 46.0(41) of total AV	50% of positive samples were from infants <1 year
Wang and Chen 1997	China	44 ill	100	58.0(40); 32.0(41);16.0 (both)	100% AV infected population; EAVs predominated in children <3 years
Bon et al. 1999	France	414 ill	3.1	3.1	Infants and children who presented to outpatient clinics
Lin et al. 2000	Taiwan	64 ill	100	100	No seasonal clustering; 43.5% had diarrhea >7 days; 17.2% acquired the infection nosocomially; 76.6% <2 years

*N = number of study subjects.
†AV = adenovirus.
‡EAV = enteric adenoviruses.

tested positive for adenovirus—mainly 4, 7, 3, and 21.The Centers for Disease Control and Prevention estimates that 10 to 12% of all military recruits became ill with adenovirus since 1999 (CDC 2001). Large epidemics of adenovirus 4 have been reported in army facilities (Hendrix et al. 1999; McNeil et al. 2000), and recently the deaths of two Navy recruits were associated with adenovirus (CDC 2001).

Many outbreaks of pharyngoconjunctival fever (PCF) from nonenteric adenoviruses have come from swimming in pools and lakes. Papers published in the 1920s described a syndrome of pharyngitis, conjunctivitis, and fever (hallmarks of PCF) related to swimming (Bahn 1927; Paderstein 1925 cf. Foy 1997). PCF is caused mainly by adenovirus 3 and 7, but adenovirus 1, 4, and 14 have also been implicated (Foy 1997). After virus-culturing techniques became available in the 1950s, these types of swimming pool outbreaks were traced to adenoviruses. A community swimming pool was associated with an outbreak of 77 people (Turner et al. 1987). Adenovirus 7 was cultured from the throats of some of the ill people, though the pool water was not cultured. A chlorinator had malfunctioned at the pool. In 1977, 105 people developed the symptoms of PCF after visiting a swimming pool; through throat cultures, adenovirus 3 was implicated as the cause of their infections. The outbreak coincided with a decrease in chlorine level, suggesting the pool as the source of the outbreak (Martone et al. 1980). Published accounts have confirmed a link with swimming pool outbreaks through the detection of adenovirus in pool water. Adenovirus 4 was detected in the water of a Georgia pool after 72 people became ill; all had pool contact (D'Angelo et al. 1979). More recently, Greek researchers used PCR to detect adenovirus in pool water after 80 swimmers developed PCF (Papapetropoulou and Vantarakis 1998). Clearly, recreational water is a source of adenovirus infection in swimmers.

Enteric

The viruses in subgenus F, adenovirus 40 and 41, were sometimes called fastidious adenoviruses because they were impossible to culture. The two are now more commonly referred to as enteric adenoviruses because they produce gastrointestinal illness. Though diarrhea can occur during infection with any species of adenovirus, subgenus F specifically causes gastroenteritis and diarrhea. In addition to adenovirus 40/41, adenovirus 31 has been more closely associated with diarrhea (Turner et al. 1987; Foy 1997; LeBaron et al. 1990) than any of the other nonenteric adenoviruses. A Canadian study of the stool samples of ill children from 1983 to 1986 showed 18% of adenovirus infections were from adenovirus 31, 16.9% were from 40, and 38% were from 41 (Brown 1990). The researchers noted that over the 3-year study period, the incidence of adenovirus 40 decreased, 41 increased, and 31 stayed constant. Similarly, a retrospective study in Toronto found that adenovirus 31 represented 17% of their 105 identified adenovirus cases (Krajden et al. 1990). The authors noted that the clinical syndrome caused by adenovirus 31 was indistinguishable from that of the enteric adenoviruses. The majority of adenovirus 31 cases in this study were acquired nosocomially. A molecular comparison was made of adenovirus 31 to different adenovirus subgenera and species (Pring-Akerblom and Adrian 1995). The amino acid sequence of 31 was closer to the enteric adenoviruses than any other species outside subgenus A. This might explain why adenovirus 31 is more associated with diarrheal episodes than other nonenteric species.

Respiratory symptoms can sometimes occur with 40/41 but not always (Lin et al. 2000; Wadell et al. 1994). Some estimate that adenovirus 40/41 contribute from 5 to 20% of hospitalizations for diarrhea in developed countries (LeBaron et al. 1990). Children less than a few years old are the most vulnerable to infection (LeBaron et al. 1990; Lew et al. 1991). Some reports have shown the highest occurrence of infection is in children less than 6 months old (Bates et al. 1993; de Jong et al. 1993). Though rotavirus is overwhelmingly the leading cause of childhood gastroenteritis, enteric adenovirus is very common (Shinozaki et al. 1991; Wadell et al. 1994). The incubation period for adenovirus ranges from 2 to 10 days (Foy 1997; Wadell et al. 1994), and the course of related illness is usually a week or more—longer than rotavirus or other enteric viruses. The primary symptom of an enteric adenovirus infection is diarrhea rather than vomiting or fever, and the symptoms are usually mild to moderate in nature. Rotavirus, on the other hand, is associated with more severe gastrointestinal illness (Cruz et al. 1990; Lew et al. 1991; Wadell et al. 1994), though its course is shorter than adenovirus. Enteric adenovirus infections are usually self-limited and only require supportive therapy in immunocompetent hosts, though oral rehydration may be required for some children. As for antibiotic treatment, enteric adenovirus has been shown to be sensitive to interferon in vitro, but any clinical application is unknown (Wadell et al. 1994). Other specific treatments have not been tested.

A prospective study of children enrolled in day-care centers in Texas has generated data elucidating the role of enteric adenoviruses in group settings (Van et al. 1992). Stool specimens were collected weekly from children 6 to 24 months old, regardless of their symptoms. The monitoring took place during four study periods over five years, enrolling 1,310 children and identifying 131 diarrheal outbreaks. Ten outbreaks affecting 249 children were associated with enteric adenoviruses. Infection rate during the 10 outbreaks ranged from 20 to 60% (mean 38%), and 46% of the infected children remained asymptomatic. Of the 131 total outbreaks, 25% were associated with rotavirus and 48% remained unknown, though the samples were tested for 10 pathogens. In another day-care study, 565 samples were taken from children with diarrhea and 129 samples from well children, primarily under the age of 35 months (Lew et al. 1991). Of these samples, adenovirus of any subtype was identified in 8% of both well and ill children; enteric adenovirus was identified in 2% of both well and ill children. This indicates that about 50% of the infected children developed illness.

A number of researchers around the world who published retrospective or prospective studies looking at occurrence found that the ratio of adenovirus 41 to 40 began to rise around the mid-1980s to early 1990s (Bates et al. 1993; Bryden et al. 1997; Scott-Taylor and Hammond 1995). Bryden and colleagues reported a 40/41 ratio of 1:1.6 for 1991 to 1993; in 1994 the ratio was almost 1:5. The reasons for this shift in infection characteristics may be a result of subtypes within adenovirus 41 becoming more predominant as they evolve (de Jong et al. 1993).

Sensitive Subpopulations

Although nonenteric adenovirus infection produces disease that is usually moderate in people with normal immune systems, the immunocompromised are at higher risk for serious or fatal disseminated disease. Though adenovirus can result in the same mild or asymptomatic infections in the immunocompromised (Cox et al. 1994; Khoo et al. 1995), the infection can disseminate into any body system and cause pneumonitis, meningoencephalitis, hepatitis (especially in liver and bone marrow transplant patients) (Bertheau et al. 1996; Saad et al. 1997), and hemorrhagic cystitis (especially in kidney transplant patients) (Foy 1997). Once the disease becomes fulminant, it usually progresses to death (Hierholzer 1992). The enteric adenoviruses are rarely isolated from immunocompromised patients with gastroenteritis or diarrhea (Durepaire et al. 1995; Khoo et al. 1995) and are generally not associated with serious illness in the immunocompromised.

Hierholzer published a review of more than 300 cases of adenovirus infection on immunocompromised patients: children with severe combined immunodeficiency syndrome (SCIDS), transplant patients, and AIDS patients. The different groups tended to be stricken with certain species of the virus. SCIDS patients were susceptible to types 1, 7, and 31; children with bone marrow transplants were susceptible to the serotypes in subgenera A, B, C, and E; adults with kidney transplants were infected with mostly species from subgenus B, especially 11, 34, and 35; and AIDS patients were seen with all types of species from all subgenera. The newest species from subgenus D (42–47) have only been seen in AIDS patients (Foy 1997).

Hierholzer's case review (1992) noted that 11% of transplant patients became infected with adenovirus. The case fatality rate was 60% for bone marrow transplant patients and 18% for kidney transplant patients. Because AIDS patients are susceptible to many opportunistic pathogens as their immunity decreases, it is difficult to ascertain which infection ultimately causes death. It is estimated that 12% of AIDS patients develop active adenovirus infection, and 45% of these are fatal within two months (Hierholzer 1992).

During a 6-year period, a prospective study tracked the incidence of adenovirus infection in 1,051 bone marrow transplant patients (Shields et al. 1985). Adenovirus was isolated from 51 of the 1,051 (4.9%). Ten of those patients went on to develop invasive infection—mostly in the lungs but also in the kidney and liver; four died. Because there was no common source of infection, the authors speculated that the infections resulted from the reactivation of latent virus. In a retrospective study of bone marrow transplant patients, 42 out of 201 patients had positive adenovirus cultures (Flomenberg et al. 1994). The incidence was significantly higher in pediatric cases than adults (31.3% versus 13.6%, $p = .003$). One third of those infected developed definite or probable adenovirus-related illness. Of 2,899 adult bone marrow transplant patients at the M.D. Anderson Cancer Center in Houston, Texas, 3% were diagnosed with adenovirus infection (La Rosa et al. 2001). Outcomes ranged from asymptomatic viruria to upper respiratory tract infections to disseminated disease. The mortality rate was 26% and associated with pneumonia and disseminated disease. In another population in the United Kingdom, 17% of 572 consecutive bone marrow transplant patients developed adenovirus infection (Baldwin et al. 2000). Significantly more children than adults were affected, and the mortality rate was only 1%.

A recent study reported on adenovirus recovered by culture from the urine of AIDS patients (Echavarria et al. 1998). In 1993 and 1994, the occurrence was 18%, whereas in 1995 and 1996, the occurrence had dropped to 5%. The authors suggested that the increased use of multiple antiretroviral therapies in AIDS patients has kept immunities higher in this population or the drugs had a direct effect on the virus. Another longitudinal study followed 63 HIV-positive people from 5 to 27 months (Khoo et al. 1995). In this population, 29% (18/63) became infected with adenovirus. Seventeen of the 18 had gastrointestinal infection with mostly species of subgenus D. Neither adenovirus 40 nor 41 were detected. Only 8 of the 33 patients who developed diarrhea had adenovirus—compared with 9 of 30 who did not have diarrhea. A number of other enteropathogens were isolated in the patients with diarrhea. Less than 50% of the adenovirus-infected patients developed gastrointestinal symptoms.

Durepaire and colleagues (1995) tested for adenovirus in 103 stool samples of HIV-positive people and 200 matched controls to compare its prevalence and role in diarrhea. Adenovirus was more prevalent in the HIV-positive samples (8.7%) than in the control samples (2.5%). No significant difference was found between the HIV-positive patients with and without diarrhea; however, a difference did exist between HIV-positive patients with diarrhea and control patients with diarrhea. Most of the serotypes isolated from HIV-positive samples were from subgenus D. Adenovirus 40/41 was isolated in only four stools—two each from the HIV and control groups. Three stools were diarrheic. The authors noted that the 8.7% prevalence was within the range of 0 to 15.4% reported in six other studies of HIV-positive patients. Adenovirus was detected in 16% of HIV-positive patients with diarrhea between 1991 and 1995 (Sabin et al. 1999). Those with adenovirus infection had lower CD4 counts and a significantly lower median survival rate than those without adenovirus.

There have also been reports of adenovirus infection in neonates resulting in serious or fatal pneumonia, bronchiolitis, and disseminated disease (Foy 1997; Pichler et al. 2000; Rosenlew et al. 1999). Evidence from case studies suggests that children may acquire infection from their mothers during birth (Abzug and Levin 1991). A recent study suggested that adenovirus infection of the placenta during pregnancy causes poor outcomes, such as fetal growth restriction, oligohydramnios, and nonimmune fetal hydrops (Koi et al. 2001).

Chronic Sequelae

Adenovirus can remain latent in the body (in tonsils, lymphocytes, and adenoidal tissue) for years and be reactivated under certain conditions, such as loss of full immunity. The long-term effect of this latent infection is unknown (Foy 1997).

Certain species of adenovirus are oncogenic (from subgenera A and B) only when human virus is introduced into an in vivo (animal) model (Foy 1997). Investigations into possible human cancer effects through a search for tumor antigens and DNA sequences had negative results (Mackey et al. 1979; Wold et al. 1979), though the research is rather outdated. Durepaire et al. (1995) suggest that these species may, theoretically at least, be associated with the development of tumors in AIDS patients. New investigations using PCR might give data that are more definitive on this subject.

Adenovirus infection can cause hepatitis, which is often fatal in transplant and AIDS patients (Bertheau et al. 1996; Saad et al. 1997). Although data suggest that adenovirus can cause hepatitis in dogs and possibly other animals (Foy 1997), there is no evidence that adenovirus leads to hepatitis in an immunocompetent host, except in the very rare instance of infection progressing to disseminated disease centered in the liver.

Chronic sequelae from adenovirus-related pneumonia are rare but, when reported, have been associated with chronic pulmonary changes, such as chronic airway obstruction (Singh-Naz and Rodriguez 1996).

There are no reports of enteric adenoviruses causing any chronic sequelae in either immunocompetent or immunosuppressed hosts.

Transmission

Adenovirus infects via the conjunctiva, the nasal mucosa, the gastrointestinal tract, and genitally. It can be transmitted by directly touching the eye, touching a fomite then an eye or mouth, or breathing in aerosol containing the virus. The information on contact versus airborne transmission has not yet been characterized. For the enteric adenoviruses, fecal–oral transmission is probably the significant pathway in children (who are affected substantially more often than adults). Some infected children have excreted the enteric adenoviruses from the respiratory tract, but usually the virus is excreted intermittently through the feces (Foy 1997). When an infection is based in the gastrointestinal tract, the amount of fecal shedding is greater (eight times more) than when it is centered in another part of the body (Wadell et al. 1994). Since all types of adenovirus (besides enteric alone) are excreted in feces, contaminated water could be a source of exposure for any type—either through ingestion, inhalation, or direct contact with the eyes. Contact with recreational water has been associated with several adenoviral outbreaks over the years. No outbreaks have been associated with either food or drinking water, and no water-related outbreaks of enteric adenovirus have been reported.

Outbreaks of adenovirus 40 and 41 have occurred in young children's group settings, such as day-care centers (Lew et al. 1991; Van et al. 1992), health care centers (Kotloff et al. 1989; Krajden et al. 1990), and orphanages (Chiba et al. 1983). In these outbreaks, the adult contacts were usually not affected. When general diarrheal outbreaks in children are investigated for multiple pathogens, rotavirus is almost always the primary infective agent, though the majority of samples usually remains unidentified (Donelli et al. 1993; Kim et al. 1990; Van et al. 1992).

In a community-based study in Argentina, 49 families were tracked prospectively for two years to look at their incidence of enteric adenovirus (Mistchenko et al. 1992). One hundred-eighty fecal samples were taken from children (0–14 years) who became sick during the study period. Of those samples, 14.4% were positive for any adenovirus, and 3.3% of them tested positive for 40 or 41. In samples from asymptomatic subjects, adenovirus was found in 13.3% of the 766 samples, and 0.8% of those were 40/41. The study protocol called for all family members to be tested when one child came up positive for adenovirus 40/41. Though 41% of the siblings developed diarrhea after the index child in their family, none of their fecal samples tested positive for enteric adenoviruses. The detection methods for the clinical samples were ELISA for adenoviruses generally, then a dot-blot hybridization technique to identify adenovirus 40 or 41. These techniques are not as sensitive as PCR, but PCR was not in widespread use for detecting enteric adenovirus at that time.

Another longitudinal study of adenovirus incidence in 70 families in the Washington, D.C., area was published in 1985 (Rodriguez et al. 1985). Of the 18 ill children who had adenovirus in their stool, 3 were identified as enteric (adenovirus 41). None of the contacts or family members of either the nonenteric infections or the adenovirus 41 infections appeared to shed adenovirus in their stool. In comparison, the authors noted that in a similar family study, rotavirus spread easily to adults (25%) and children (56%). A 1978 adenovirus outbreak in young children seemed to be transmitted from child to child, except in one case where neither older siblings nor parents got sick (Richmond et al. 1979). Another study prospectively tracking adenoviral infection in AIDS patients and their partners showed no indication of secondary spread among partners, though 29% of the AIDS patients became infected (Khoo et al. 1995).

Long-term immunity is thought to be acquired in early childhood, and generally, the majority of children have been infected by the time they are 4 years old (LeBaron et al. 1990). In two different day-care outbreaks, the likelihood of infection with the enteric viruses decreased as the age of the child increased (Lew et al. 1991; Van et al. 1992). The attack rate in one day-care center was 38% for exposed children; 46% of them remained asymptomatic (Van et al. 1992). It appears that enteric adenoviruses strike mainly the very young—under a few years of age—and can be spread to other very young children. However, older children and adults are less affected by secondary spread, presumably because of acquired immunity.

Detection Methods

With any virus, cell culture is the referent detection method because it is the only way to prove infectivity. With adenovirus, subgenera A through E can generally be cultured in human cell lines, though it can be time consuming. For many years, it was thought that the enteric adenoviruses could not be cultured at all; they were detectable only through electron microscope and classified as "unculturable." Researchers have since found unusual cell lines in which adenovirus 40/41 can be isolated to a limited extent (e.g., Graham 293, Chang conjunctival cells, Buffalo Green monkey kidney, and CaCo-2 cells) (Irving, deLouvois, and Nichols 1996; Pinto et al. 1995; Singh-Naz and Rodriguez 1996; Wadell et al. 1994). Other, quicker detection methods are standardly used, though they only detect the presence of the virus, not its infectivity. Frequently in clinical samples, electron microscopy is used to identify adenovirus; however, serotyping must then be done through immunological techniques, such as immunoelectron microscopy, ELISA, and monoclonal antibodies. DNA analyses have included restriction endonuclease analysis and dot-blot hybridization (Rousell et al. 1993; Foy 1997). de Jong et al. (1993) demonstrated greater detection sensitivity using virus isolation rather than ELISA methods and found that ELISA and electron microscopy with virus neutralization were about the same. It is difficult to compare the sensitivity and specificity of different assays because any differences in technique performing the same assay may produce different results.

The advent of PCR techniques has provided a faster, more sensitive, and specific way to detect distinct serotypes in both clinical and environmental samples. Puig (1994) compared cell culture, one-step PCR, and nested PCR on sewage and river water samples. Nested PCR was the most accurate, allowing for the detection of <10 particles, which is 100 to 1,000 times more sensitive than cell culturing. Similarly, Allard et al. (1990) compared latex agglutination, polyacrylamide gel electrophoresis, and PCR on 60 stool samples. The positive specimens found using each test were 20/60 using latex agglutination, 16/60 using polyacrylamide gel electrophoresis, and 51/60 using PCR. When they looked specifically for enteric adenoviruses, the latex agglutination method detected two positives; PCR with enteric adenovirus primers detected 16 positives. PCR is recognized as the most sensitive and specific way to detect enteric adenoviruses, though again, the technique does not demonstrate infectivity. Avellon and colleagues (2001) found PCR to have a higher sensitivity than immunofluorescence in clinical samples.

Environmental Occurrence

Many types of adenovirus have been detected in sewage, rivers, oceans, swimming pools, and shellfish. Positive samples are found throughout all seasons of the year. No association with fecal coliform levels has been documented. Following is a brief summary of selected water occurrence studies:

- 29 surface water samples were tested for a number of viruses, including adenovirus 40 and 41 (Chapron et al. 2000), using a combination of cell culture and PCR methods. Fourteen (48.3%) of the samples were positive for enteric adenoviruses and 78.6% of those were deemed infective via cell culture (US).
- Researchers used a nested PCR method to detect adenoviruses in coastal California waters (Jiang, Noble, and Chu 2001). They found that 4 of 12 samples tested positive with a most probable number (MPN) of 880 to 7,500 genomes per liter of water.
- Monthly samples of raw sewage, effluent, river water, and seawater were tested using nested PCR amplification; adenovirus was detected in 14/15 sewage, 2/3 effluent, 15/23 river water, 7/9 seawater samples; any sample positive for an enterovirus or hepatitis A was also positive for adenovirus; no correlation between fecal coliform level and virus occurrence was reported (Spain) (Pina et al. 1998).
- Thirty-six samples of effluent over 15 months were tested using cell culture; adenovirus was detected in all samples; concentration ranged from 70 to 3,200 cytopathic units (CPU)/L; types 1, 2, 3, 5, 7, and 15 were detected (Greece) (Krikelis et al. 1985).
- Weekly testing of urban river water over 5 years used cell culture for detecting reovirus, enteroviruses, and adenoviruses tested; levels of adenovirus were low compared to other viruses but detected consistently over the study period (samples ranged from 0 to 25 pfu/L); types 2, 3, 5, and 6 were most prevalent (Japan) (Tani et al. 1995).
- Beach area samples, including open water and sewage effluent, were tested using cell culture and ELISA; 4/66 samples were positive for adenovirus, 2 in effluent, 2 in open water samples; fecal coliforms were in the normal range (Australia) (Kueh and Grohmann 1989).

- Raw sewage, primary effluent, and secondary effluent were sampled over 12 months using cell culture; 25/26 raw sewage, 23/26 primary effluent, 23/26 secondary effluent samples were positive for adenovirus; mean concentrations in sewage, primary effluent, and secondary effluent were 1,950, 1,350, and 250 international units (IU)/L, respectively (Australia) (Irving and Smith 1981).

Clearly, adenoviruses are consistently found in all kinds of water sources in geographically diverse areas. Although no study focused specifically on the occurrence of enteric adenoviruses in water, the fact that they are shed in feces implies they should be present in these sources as well. Since adenovirus can be transmitted through aerosolization, it seems possible that contaminated water (perhaps unchlorinated groundwater) could infect someone through showering.

Like other viruses, adenovirus does not grow outside the body of their host, so growth in the environment is not an issue. However, adenovirus is able to resist inactivation under fairly rigorous environmental conditions. It is able to endure pH from 5.0 to 9.0, temperatures from 4° to 36°C, and being frozen. It remains undiminished after 70 days at 4°C (Irving, deLouvois, and Nichols 1996), but it is destroyed by 56°C heat after 30 minutes (Foy 1997).

Abad and colleagues (1994) investigated the survival of several viruses—hepatitis A, rotavirus, adenovirus, and poliovirus—on object surfaces. Testing parameters included surface type (porous or nonporous), humidity, and temperature. Generally, hepatitis A and rotavirus were more resistant to inactivation under any condition. Adenovirus and poliovirus were susceptible to inactivation after drying on a surface. Adenovirus survived longer at 4° C than at 20° C and was unaffected by humidity.

Water Treatment

Authors report that adenovirus is inactivated by the levels of chlorine used in drinking water treatment, though there are not many supportive data. Irving, deLouvois, and Nichols (1996) reports 0.1 mg/L of chlorine is adequate for inactivation; to put that number in perspective, the maximum residual disinfection goal for chlorine under the Stage 1 Disinfection By-product/Microbial Rule of the Safe Drinking Water Act is 4.0 mg/L (USEPA 1998). Fuchs and Wigand (cf. Foy 1997) concluded that adenovirus 3 appears to be as susceptible to chlorination as *Escherichia coli* in vitro. However, Irving and Smith (1981) tested raw sewage and primary and secondary effluent for adenovirus using cell culture methods (so infectivity is assured). They found that normal sewage treatment using coagulation, flocculation, and disinfectant did not remove adenovirus as efficiently as the enteroviruses, either before or after chlorination. The treatment process included chlorination of the raw sewage at a concentration of 8.0 mg/L, followed by primary sedimentation, aeration with activated sludge, secondary sedimentation, and finally, chlorination at 2.0 mg/L. After 30 minutes of contact time, the chlorine level in the secondary effluent had fallen to 0 to 0.1 mg/L. In the final effluent, enteroviruses were reduced by 93%, and adenovirus was reduced by 85%. Adenovirus was found in almost twice as many secondary effluent samples as enterovirus, and the average concentrations were higher than for enterovirus. After the secondary effluent was chlorinated again, five samples out of seven were positive for adenovirus with an average concentration of 300 IU/L and a range of 0 to 1,150 IU/L. The authors stated that the calculations were based on crude averages and that additional research indicated a possible effect of the timing of the sampling on the results of the removal figures. They did, however, conclude from their data that adenovirus was more resistant to sewage treatment than the enteroviruses.

Using nested PCR, Pina and colleagues (Pina et al. 1998) tested raw sewage and found adenoviruses present in 93% of the samples. Three samples were taken from the effluent after treatment; two of those samples still tested positive, supporting the data that sewage treatment does not completely remove viruses. The infectivity of the virus in the effluent was unknown, since the samples were not cultured. Because treatment included only settling and not disinfection, one may assume their infectivity was unaffected.

Abad and colleagues (1994) compared inactivation of four human enteric viruses in the presence of free chlorine and copper and silver ions. Rotavirus and hepatitis A showed little inactivation under all the conditions tested. Adenovirus (type 2) lasted longer than poliovirus but significantly less than hepatitis and rotavirus. After 30 minutes of 0.5 mg/L free chlorine treatment, the adenovirus showed almost a 3 $\log_{10}$ titer reduction; whereas, poliovirus showed a 4 $\log_{10}$ reduction. Meng and Gerba (1996) compared the ability of ultraviolet (UV) radiation to inactivate adenovirus 40 and 41, coliphages MS-2 and PRD-1, and poliovirus type 1. The UV doses required to achieve 90% inactivation in these agents were 30, 23.6, 14, 8.7, and 4.1 mW s/cm^2, respectively. The significant increase in dosage necessary for the enteric adenoviruses compared with the others might result from their double-stranded DNA makeup. Adenovirus is the only virus comprised of DNA, not RNA, which may affect its resistance to inactivation. Though these reports on the resistance of adenovirus to sewage treatment and disinfection exist, the effect of drinking water treatment on adenovirus is unknown.

Membrane filtration technology, such as micro-, ultra-, and nanofiltration as well as reverse osmosis, has the capacity to remove water contaminants down to the ion level—greater than conventional water treatment. Ultrafiltration with a nominal pore size of 0.01 µm can achieve 4-log virus removal or greater (Najm and Trussell 1999; Taylor and Wiesner 1999). This technology will continue to gain in popularity as costs become more competitive and regulatory approval is assured.

Conclusions

Many species of adenovirus, including the enteric types, occur worldwide, and adenovirus is frequently cited as the second leading cause of childhood diarrhea after rotavirus. Certain species of adenovirus can cause severe and often fatal infections for the immunocompromised; however, types 40/41 are not recognized as problematic types for that group. The group most commonly infected by adenovirus 40/41 is very young children, especially those under 1 or 2 years old. The course of illness is longer (>1 week) than other enteric viruses but not as severe as rotavirus. The infection generally runs its course with no treatment beyond the palliative. Considering most children become infected very early in life and outbreaks have been reported in group settings, the primary route of transmission is most likely fecal–oral.

Adenovirus is found in sewage and all sorts of other water bodies. Its potential for disinfection by chlorine appears to be average compared with other viruses, though studies have shown that sewage treatment does not remove all adenovirus from the effluent. Therefore, it is plausible that enteric adenoviruses can contaminate drinking water, but no studies have investigated that source.

Bibliography

Abad, F.X., R.M. Pinto, and A. Bosch. 1994. Survival of enteric viruses on environmental fomites. *Appl. Envirol. Microbiol.*, 60:3704–3710.

Abzug, M.J., and M.J. Levin. 1991. Neonatal adenovirus infection: four patients and review of the literature. *Pediatrics*, 87:890–896.

Allard, A., B. Albinsson, and G. Wadell. 1992. Detection of adenoviruses in stools from healthy persons and patients with diarrhea by two-step polymerase chain reaction. *J. Med. Virol.*, 37:149–157.

Allard, A., R. Girones, P. Juto, and G. Wadell. 1990. Polymerase chain reaction for detection of adenoviruses in stool samples. *J. Clin. Microbiol.*, 28:2659–2667.

Avellon, A., P. Perez, J.C. Aguilar, R. Lejarazu, and J.E. Echevarria. 2001. Rapid and sensitive diagnosis of human adenovirus infections by a generic polymerase chain reaction. *J. Virol. Methods*, 92:113–120.

Baldwin, A., H. Kingman, M. Darville, A.B. Foot, D. Grier, J.M. Cornish, N. Goulden, A. Oakhill, D.H. Pamphilon, C.G. Steward, and D.I. Marks. 2000. Outcome and clinical course of 100 patients with adenovirus infection following bone marrow transplantation. *Bone Marrow Transplant.*, 26:1333–1338.

Bates, P.R., A.S. Bailey, D.J. Wood, D.J. Morris, and J.M. Couriel. 1993. Comparative epidemiology of rotavirus, subgenus F (types 40 and 41) adenovirus and astrovirus gastroenteritis in children. *J. Med. Virol.*, 39:224–228.

Bertheau, P., N. Parquet, F. Ferchal, E. Gluckman, and C. Brocheriou. 1996. Fulminant adenovirus hepatitis after allogeneic bone marrow transplantation. *Bone Marrow Transplant.*, 17:295–298.

Bon, F., P. Fascia, M. Dauvergne, D. Tenenbaum, H. Planson, A.M. Petion, P. Pothier, and E. Kohli. 1999. Prevalence of group A rotavirus, human calicivirus, astrovirus, and adenovirus type 40 and 41 infections among children with acute gastroenteritis in Dijon, France. *J. Clin. Microbiol.*, 37:3055–3058.

Brown, M. 1990. Laboratory identification of adenoviruses associated with gastroenteritis in Canada from 1983 to 1986. *J. Clin. Microbiol.*, 28:1525–1529.

Bryden, A.S., A. Curry, H. Cotterill, C. Chesworth, I. Sharp, and S.R. Wood. 1997. Adenovirus-associated gastro-enteritis in the north-west of England: 1991–1994. *Br. J. Biomed. Sci.*, 54:273–277.

Center for Disease Control and Prevention (CDC). 2001. Two fatal cases of adenovirus-related illness in previously healthy young adults—Illinois, 2000. *MMWR*, 50(26):553-55.

Chapron, C.D., N.A. Ballester, J.H. Fontaine, C.N. Frades, and A.B. Margolin. 2000. Detection of astroviruses, enteroviruses, and adenovirus types 40 and 41 in surface waters collected and evaluated by the information collection rule and an integrated cell culture-nested PCR procedure. *Appl. Environ. Microbiol.*, 66:2520–2525.

Chiba, S., S. Nakata, I. Nakamura, K. Taniguchi, S. Urasawa, K. Fujinaga, and T. Nakao. 1983. Outbreak of infantile gastroenteritis due to type 40 adenovirus. *Lancet*, 2:954–957.

Cox, G.J., S.M. Matsui, R.S. Lo, M. Hinds, R.A. Bowden, R.C. Hackman, W.G. Meyer, M. Mori, P.I. Tarr, and L.S. Oshiro. 1994. Etiology and outcome of diarrhea after marrow transplantation: a prospective study. *Gastroenterology*, 107:1398–1407.

Cruz, J.R., P. Caceres, F. Cano, J. Flores, A. Bartlett, and B. Torun. 1990. Adenovirus types 40 and 41 and rotaviruses associated with diarrhea in children from Guatemala. *J. Clin. Microbiol.*, 28:1780–1784.

D'Angelo, L.J., J.C. Hierholzer, R.A. Keenlyside, L.J. Anderson, and W.J. Martone. 1979. Pharyngoconjunctival fever caused by adenovirus type 4: report of a swimming pool-related outbreak with recovery of virus from pool water. *J. Infect. Dis.*, 140:42–47.

de Jong, J.C., K. Bijlsma, A.G. Wermenbol, M.W. Verweij-Uijterwaal, H.G. van der Avoort, D.J. Wood, A.S. Bailey, and A.D. Osterhaus. 1993. Detection, typing, and subtyping of enteric adenoviruses 40 and 41 from fecal samples and observation of changing incidences of infections with these types and subtypes. *J. Clin. Microbiol.*, 31:1562–1569.

Donelli, G., F. Superti, A. Tinari, M.L. Marziano, D. Caione, C. Concato, and D. Menichella. 1993. Viral childhood diarrhoea in Rome: a diagnostic and epidemiological study. *New Microbiol.*, 16:215–225.

Durepaire, N., S. Ranger-Rogez, J.A. Gandji, P. Weinbreck, J.P. Rogez, and F. Denis. 1995. Enteric prevalence of adenovirus in human immunodeficiency virus seropositive patients. *J. Med. Virol.*, 45:56–60.

Echavarria, M., M. Forman, J. Ticehurst, J.S. Dumler, and P. Charache. 1998. PCR method for detection of adenovirus in urine of healthy and human immunodeficiency virus-infected individuals. *J. Clin. Microbiol.*, 36:3323–3326.

Flomenberg, P., J. Babbitt, W.R. Drobyski, R.C. Ash, D.R. Carrigan, G.V. Sedmak, T. McAuliffe, B. Camitta, M.M. Horowitz, and N. Bunin. 1994. Increasing incidence of adenovirus disease in bone marrow transplant recipients. *J. Infect. Dis.*, 169:775–781.

Fox, J.P., C.D. Brandt, F.E. Wassermann, C.E. Hall, I. Spigland, A. Kogon, L.R. Elveback. 1969. The virus watch program: a continuing surveillance of viral infections in metropolitan New York families. VI. Observations of adenovirus infections: virus excretion patterns, antibody response, efficiency of surveillance, patterns of infections, and relation to illness. *Am. J. Epidemiol.*, 89(1):25-50.

Fox, J.P., C.E. Hall, and M.K. Cooney. 1977. The Seattle Virus Watch. VII. Observations of adenovirus infections. *Am. J. Epidemiol.*, 105:362–386.

Foy, H.M. 1997. Adenoviruses. In *Viral Infections of Humans.* 4th ed. Edited by A.S. Evans and R.A. Kaslow. New York: Plenum Press.

Gerber, S.I., D.D. Erdman, S.L. Pur, P.S. Diaz, J. Segreti, A.E. Kajon, R.P. Belkengren, and R.C. Jones. 2001. Outbreak of adenovirus genome type 7d2 infection in a pediatric chronic-care facility and tertiary-care hospital. *Clin. Infect. Dis.*, 32:694–700.

Gray, G.C., P.R. Goswami, M.D. Malasig, A.W. Hawksworth, D.H. Trump, M.A. Ryan, and D.P. Schnurr. 2000. Adult adenovirus infections: loss of orphaned vaccines precipitates military respiratory disease epidemics. For the Adenovirus Surveillance Group. *Clin. Infect. Dis.*, 31:663–670.

Grimwood, K., R. Carzino, G.L. Barnes, and R.F. Bishop. 1995. Patients with enteric adenovirus gastroenteritis admitted to an Australian pediatric teaching hospital from 1981 to 1992. *J. Clin. Microbiol.*, 33:131–136.

Harsi, C.M., D.P. Rolim, S.A. Gomes, A.E. Gilio, K.E. Stewien, E.R. Baldacci, and J.A. Candeias. 1995. Adenovirus genome types isolated from stools of children with gastroenteritis in Sao Paulo, Brazil. *J. Med. Virol.*, 45:127–134.

Hendrix, R.M., J.L. Lindner, F.R. Benton, S.C. Monteith, M.A. Tuchscherer, G.C. Gray, and J.C. Gaydos. 1999. Large, persistent epidemic of adenovirus type 4-associated acute respiratory disease in US army trainees. *Emerg. Infect. Dis.*, 5:798–801.

Hierholzer, J.C. 1992. Adenoviruses in the immunocompromised host. *Clin. Microbiol. Rev.*, 5:262–274.

Irving, T.E., J. deLouvois, and G.L. Nichols. 1996. Fact Sheets on Emerging Waterborne Pathogens: Final Report to the Department of the Environment. WRC and Public Health Laboratory Service. DWI 4248/1.

Irving, L.G., and F.A. Smith. 1981. One-year survey of enteroviruses, adenoviruses, and reoviruses isolated from effluent at an activated-sludge purification plant. *Appl. Environ. Microbiol.*, 41:51–59.

Jiang, S., R. Noble, and W. Chu. 2001. Human adenoviruses and coliphages in urban runoff-impacted coastal waters of Southern California. *Appl. Environ. Microbiol.*, 67:179–184.

Khoo, S.H., A.S. Bailey, J.C. de Jong, and B.K. Mandal. 1995. Adenovirus infections in human immunodeficiency virus-positive patients: clinical features and molecular epidemiology. *J. Infect. Dis.*, 172:629–637.

Kim, K.H., J.M. Yang, S.I. Joo, Y.G. Cho, R.I. Glass, and Y.J. Cho. 1990. Importance of rotavirus and adenovirus types 40 and 41 in acute gastroenteritis in Korean children. *J. Clin. Microbiol.*, 28:2279–2284.

Koi, H., J. Zhang, A. Makrigiannakis, S. Getsios, C.D. MacCalman, G.S. Kopf, J.F. Strauss, III, S. Parry. 2001. Differential expression of the coxsackievirus and adenovirus receptor regulates adenovirus infection of the placenta. *Biol. Reprod.*, 64(3):1001–1009.

Kotloff, K.L., G.A. Losonsky, J.G. Morris Jr., S.S. Wasserman, N. Singh-Naz, and M.M. Levine. 1989. Enteric adenovirus infection and childhood diarrhea: an epidemiologic study in three clinical settings. *Pediatrics*, 84:219–225.

Krajden, M., M. Brown, A. Petrasek, and P.J. Middleton. 1990. Clinical features of adenovirus enteritis: a review of 127 cases. *Pediatr. Infect. Dis. J.*, 9:636–641.

Krikelis, V., N. Spyrou, P. Markoulatos, and C. Serie. 1985. Seasonal distribution of enteroviruses and adenoviruses in domestic sewage. *Can. J. Microbiol.*, 31:24–25.

Kueh, C.S., and G.S. Grohmann. 1989. Recovery of viruses and bacteria in waters off Bondi beach: a pilot study. *Med. J. Aust.*, 151:632–638.

Kukkula, M., P. Arstila, M.L. Klossner, L. Maunula, C.H. Bonsdorff, and P. Jaatinen. 1997. Waterborne outbreak of viral gastroenteritis. *Scand. J. Infect. Dis.*, 29:415–418.

La Rosa, A.M., R.E. Champlin, N. Mirza, J. Gajewski, S. Giralt, K.V. Rolston, I. Raad, K. Jacobson, D. Kontoyiannis, L. Elting, and E. Whimbey. 2001. Adenovirus infections in adult recipients of blood and marrow transplants. *Clin. Infect. Dis.*, 32:871–876.

Larranaga, C., A. Kajon, E. Villagra, and L.F. Avendano. 2000. Adenovirus surveillance on children hospitalized for acute lower respiratory infections in Chile (1988–1996). *J. Med. Virol.*, 60:342–346.

LeBaron, C.W., N.P. Furutan, J.F. Lew, J.R. Allen, V. Gouvea, C. Moe, and S.S. Monroe. 1990. Viral agents of gastroenteritis. Public health importance and outbreak management. *MMWR*, 39:1–24.

Lew, J.F., C.L. Moe, S.S. Monroe, J.R. Allen, B.M. Harrison, B.D. Forrester, S.E. Stine, P.A. Woods, J.C. Hierholzer, and J.E. Herrmann. 1991. Astrovirus and adenovirus associated with diarrhea in children in day care settings. *J. Infect. Dis.*, 164:673–678.

Lin, H.C., C.L. Kao, C.Y. Lu, C.N. Lee, T.F. Chiu, P.I. Lee, H.Y. Tseng, H.L. Hsu, C.Y. Lee, and L.M. Huang. 2000. Enteric adenovirus infection in children in Taipei. *J. Microbiol. Immunol. Infect.*, 33:176–180.

Mackey, J.K., M. Green, W.S. Wold, and P. Rigden. 1979. Analysis of human cancer DNA for DNA sequences of human adenovirus type 4. *J. Natl. Cancer Inst.*, 62:23–26.

Martone, W.J., J.C. Hierholzer, R.A. Keenlyside, D.W. Fraser, L.J. D'Angelo, and W.G. Winkler. 1980. An outbreak of adenovirus type 3 disease at a private recreation center swimming pool. *Am. J. Epidemiol.*, 111:229–237.

McNeill K.M., B.F. Ridgely, S.C. Monteith, M.A. Tuchscherer, J.C. Gaydos. 2000. Epidemic spread of adenovirus type 4-associated acute respiratory disease between U.S. Army installations. *Emerg. Infect. Dis.*, 6(4):415–419.

Meng, Q.S., and C. Gerba. 1996. Comparative inactivation of enteric adenoviruses, poliovirus and coliphages by ultraviolet radiation. *Water Res.*, 30:2665–2668.

Mistchenko, A.S., K.H. Huberman, J.A. Gomez, and S. Grinstein. 1992. Epidemiology of enteric adenovirus infection in prospectively monitored Argentine families. *Epidemiol. Infect.*, 109:539–546.

Najm, I., and R.R. Trussell. 1999. New and Emerging Drinking Water Treatment Technologies. In *Identifying Future Drinking Water Contaminants.* Washington, DC: National Academy Press.

Papapetropoulou, M., and A.C. Vantarakis. 1998. Detection of adenovirus outbreak at a municipal swimming pool by nested PCR amplification. *J. Infect.*, 36:101–103.

Pichler, M.N., J. Reichenbach, H. Schmidt, G. Herrmann, and S. Zielen. 2000. Severe adenovirus bronchiolitis in children. *Acta. Paediatr.*, 89:1387–1389.

Pina, S., M. Puig, F. Lucena, J. Jofre, and R. Girones. 1998. Viral pollution in the environment and in shellfish: human adenovirus detection by PCR as an index of human viruses. *Appl. Environ. Microbiol.*, 64:3376–3382.

Pinto, R.M., R. Gajardo, F.X. Abad, and A. Bosch. 1995. Detection of fastidious infectious enteric viruses in water. *Environ. Sci. Technol.*, 29:2636–2638.

Pring-Akerblom, P., and T. Adrian. 1995. Sequence characterization of the adenovirus 31 fibre and comparison with serotypes of subgenera A to F. *Res. Virol.*, 146:343–354.

Puig, M., J. Jofre, F. Lucena, A. Allard, G. Wadell, and R. Girones. 1994. Detection of adenoviruses and enteroviruses in polluted waters by nested PCR amplification. *Appl. Environ. Microbiol.*, 60:2963–2970.

Richmond, S.J., E.O. Caul, S.M. Dunn, C.R. Ashley, S.K. Clarke, and N.R. Seymour. 1979. An outbreak of gastroenteritis in young children caused by adenoviruses. *Lancet*, 1:1178–1181.

Rodriguez, W.J., H.W. Kim, C.D. Brandt, R.H. Schwartz, M.K. Gardner, B. Jeffries, R.H. Parrott, R.A. Kaslow, J.I. Smith, and H. Takiff. 1985. Fecal adenoviruses from a longitudinal study of families in metropolitan Washington, D.C.: laboratory, clinical, and epidemiologic observations. *J. Pediatr.*, 107:514–520.

Rosenlew, M., M. Stenvik, M. Roivainen, A.L. Jarvenpaa, and T. Hovi. 1999. A population-based prospective survey of newborn infants with suspected systemic infection: occurrence of sporadic enterovirus and adenovirus infections. *J. Clin. Virol.*, 12:211–219.

Rousell, J., M.E. Zajdel, P.D. Howdle, and G.E. Blair. 1993. Rapid detection of enteric adenoviruses by means of the polymerase chain reaction. *J. Infect.*, 27:271–275.

Ruuska, T., and T. Vesikari. 1991. A prospective study of acute diarrhoea in Finnish children from birth to $2^1/_2$ years of age. *Acta. Paediatr. Scand.*, 80:500–507.

Saad, R.S., A.J. Demetris, R.G. Lee, S. Kusne, and P.S. Randhawa. 1997. Adenovirus hepatitis in the adult allograft liver. *Transplantation*, 64:1483–1485.

Sabin, C.A., G.S. Clewley, J.R. Deayton, A. Mocroft, M.A. Johnson, C.A. Lee, J.E. McLaughlin, and P.D. Griffiths. 1999. Shorter survival in HIV-positive patients with diarrhoea who excrete adenovirus from the GI tract. *J. Med. Virol.*, 58:280–285.

Scott-Taylor, T.H., and G.W. Hammond. 1995. Local succession of adenovirus strains in pediatric gastroenteritis. *J. Med. Virol.*, 45:331–338.

Shields, A.F., R.C. Hackman, K.H. Fife, L. Corey, and J.D. Meyers. 1985. Adenovirus infections in patients undergoing bone-marrow transplantation. *N. Engl. J. Med.*, 312:529–533.

Shinozaki, T., K. Araki, Y. Fujita, M. Kobayashi, T. Tajima, and T. Abe. 1991. Epidemiology of enteric adenoviruses 40 and 41 in acute gastroenteritis in infants and young children in the Tokyo area. *Scand. J. Infect. Dis.*, 23:543–547.

Singh-Naz, N., and W. Rodriguez. 1996. Adenoviral infections in children. *Adv. Pediatr. Infect. Dis.*, 11:365–388.

Tani, N., Y. Dohi, N. Kurumatani, and K. Yonemasu. 1995. Seasonal distribution of adenoviruses, enteroviruses and reoviruses in urban river water. *Microbiol. Immunol.*, 39:577–580.

Taylor, J.S., and M. Wiesner. 1999. Membranes. In *Water Quality and Treatment: A Handbook of Community Water Supplies.* 5th ed. Edited by American Water Works Association. New York: McGraw-Hill.

Turner, M., G.R. Istre, H. Beauchamp, M. Baum, and S. Arnold. 1987. Community outbreak of adenovirus type 7a infections associated with a swimming pool. *South. Med. J.*, 80:712–715.

Uhnoo, I., G. Wadell, L. Svensson, and M.E. Johansson. 1984. Importance of enteric adenoviruses 40 and 41 in acute gastroenteritis in infants and young children. *J. Clin. Microbiol.*, 20:365–372.

Uhnoo, I., G. Wadell, L. Svensson, E. Olding-Stenkvist, E. Ekwall, and R. Molby. 1986. Aetiology and epidemiology of acute gastro-enteritis in Swedish children. *J. Infect.*, 13:73–89.

US Environmental Protection Agency (USEPA). 1998. National primary drinking water regulations: stage 1 disinfection by-product/microbial rule. *Fed. Reg.*, 63:69389–69476.

Van, R., C.C. Wun, M.L. O'Ryan, D.O. Matson, L. Jackson, and L.K. Pickering. 1992. Outbreaks of human enteric adenovirus types 40 and 41 in Houston day care centers. *J. Pediatr.*, 120:516–521.

Videla, C., G. Carballal, and A. Kajon. 1999. Genomic analysis of adenovirus isolated from Argentinian children with acute lower respiratory infections. *J. Clin. Virol.*, 14:67–71.

Wadell, G., A. Allard, M. Johansson, L. Svensson, and I. Uhnoo. 1994. Enteric Adenoviruses. In *Viral Infections of the Gastrointestinal Tract.* Edited by A.Z. Kapikian. New York: Marcel Dekker.

Wang, B., and X. Chen. 1997. The molecular epidemiological study on enteric adenovirus in stool specimens collected from Wuhan area by using digoxigenin labeled DNA probes. *J. Tongji Med. Univ.*, 17:79–82.

Wold, W.S., J.K. Mackey, P. Rigden, and M. Green. 1979. Analysis of human cancer DNAs for DNA sequence of human adenovirus serotypes 3, 7, 11, 14, 16, and 21 in group B1. *Cancer Res.*, 39:3479–3484.

CHAPTER 6

Aeromonas hydrophila (and other pathogenic species) in Drinking Water

by Martha A. Embrey

Executive Summary

Problem Formulation

Occurrence of Illness

- Prevalence studies of adults and children with gastroenteritis show an isolation rate of between 0 and 10% in most study populations, independent of common risk factors in waterborne illness, such as geography, socioeconomic status, and, to some degree, age. Young children are most likely to be identified as infected with *Aeromonas*, but that could be a result of reporting bias.

Role of Waterborne Exposure

- *Aeromonas* spp. are found in all types of water: lakes, rivers, marine environments, wastewater, and chlorinated drinking water. Because *Aeromonas* is also found frequently in all types of food and is ubiquitous in the environment, the role of water ingestion in the development of gastroenteritis is unknown. Exposure to fresh water is the primary risk factor associated with wound infection.

Degree of Morbidity and Mortality

- *Aeromonas*-related gastroenteritis is generally self-limited, acute, watery diarrhea of a few days' to a few weeks' duration. Chronic diarrhea of more than a few weeks and more serious sequelae, such as sepsis, usually occur only in immunocompromised people.

Disseminated, systemic disease in the immunocompromised (especially those with underlying liver disease or cancer) causes a high fatality rate.

Detection Methods in Water/Clinical Specimens

- Clinical and environmental specimens can be cultured for *Aeromonas*. The procedure for collecting *Aeromonas* in water samples is the same as for how coliforms and *E. coli* are collected—passed through a membrane filter, then grown in culture. *Aeromonas* will grow in most media, though ampicillin-dextrin is reported to be most effective for water samples.

Mechanisms of Water Contamination

- *Aeromonas* spp. are common and ubiquitous in the environment: water, food, soil, and cold- and warm-blooded animals. They are also shed through feces.

Concentrations at Intake

- The aeromonads occur at generally high concentrations in all types of source water. River and marine water concentrations have been reported up to 10^6/100 mL; lake concentrations have been reported up to 10^5/100 mL, and well water up to 10^2/100 mL.

Efficacy of Water Treatment

- Standard water treatment with chlorine disinfectant decreases concentrations significantly (<10/100 mL) but not necessarily completely. Water temperature, contact time, and level of organic material in the source water are all related to disinfection efficacy. *Aeromonas* appears to be as susceptible to chlorine disinfection as *E. coli*.

Survival/Amplification in Distribution

- Significant. *Aeromonas* can colonize wells and distribution systems for months and years. Substantial regrowth can occur after disinfection, and the aeromonads colonize biofilm, which makes them resistant to disinfectant residuals. Several studies have demonstrated long-term colonization of single water sources with *Aeromonas*.

Routes of Exposure

- Ingestion of contaminated water and food; dermal contact with water or soil. Most infections are community-acquired; some are nosocomial (especially in the immunocompromised).

Dose Response

Infectious Dose

- Studies in which volunteers have been fed up to 10^{10} organisms have resulted in few patients becoming infected. It appears that very high doses of select strains are needed to colonize and infect a susceptible host.

Probability of Illness Based on Infection

- Unknown for gastroenteritis. Symptomatic and asymptomatic people shed *Aeromonas* in their feces.

Efficacy of Medical Treatment

- *Aeromonas* is resistant to many antibiotics, including penicillin; however, many other antibiotics, including third generation cephalosporins, are efficacious in treating all kinds of infections, including chronic gastroenteritis.

Secondary Spread

- None.

Chronic Sequelae

- Case reports have been published detailing chronic diarrhea from undiagnosed *Aeromonas* infections. A course of antibiotics effectively eradicates chronic infection, once diagnosed.

Introduction

The *Aeromonas* species comprise gram-negative, anaerobic rods that measure 0.3 to 1.0 μm × 1.0 to 3.5 μm. Morphologically, they are indistinguishable from other enteric organisms like *Escherichia coli* (Smith and Cheasty 1999). Aeromonads are not recognized as part of the normal human gut flora, and researchers estimate that in developed countries, less than 1% of people—perhaps higher during warm months—carry *Aeromonas* in their gastrointestinal tract (Janda 1999; Smith and Cheasty 1999). Surveillance of asymptomatic populations has shown prevalences of up to 9% in healthy people (see Table 6-1).

There have been tumultuous changes in *Aeromonas* taxonomy in recent years. The two original classifications were *A. hydrophila,* which included all the mesophilic strains that infect fish and people, and *A. salmonicida,* which primarily affect fish. Now the mesophilic species (what used to be just *A. hydrophila*) are divided into three genetically distinct groups: *A. caviae, A. hydrophila,* and *A. veronii* (with biotypes *sobria* and *veronii*) (Janda 1999; Smith and Cheasty 1999). Within these species are 14 named hybridization groups: HG1–HG14. In the past, clinical specimens were classified only to the *Aeromonas* genus. Now, laboratories generally classify them to one of the three main species: *hydrophila, caviae*, or *sobria* (Hanninen and Siitonen 1995; Smith and Cheasty 1999). Currently, more complex detection methods are being used to biotype and serogroup samples—primarily for research purposes (Janda 1999).

Many new *Aeromonas* species are still being discovered, but only a few have been established as major human pathogens thus far. *A. hydrophila, A. caviae,* and *A. veronii* (biotype *sobria*) are considered major and more likely to be clinically significant (see Table 6-2). Newly identified species, like *A. jandaei* and *A. schubertii,* have been established through isolation from extraintestinal (wound) infections (Janda and Abbott 1998); very little information on occurrence and pathogenicity is available for them. *A. eucrenophila* and *A. trota* frequently have been isolated from clinical samples (Albert et al. 2000; Demarta et al. 2000; Overman and Janda 1999).

The Aeromonas genus is antigenically diverse, made up of more than 96 serogroups. Janda and colleagues (1996) identified the serological properties of 273 strains and classified them by serogroup. They found that the serogroups are not species specific, and that serogroups O:11, O:16, O:18, and O:34 predominate in the United States. Different serogroups also appear to be associated with different clinical effects—thus, the serogroup may determine virulence or some other unique property related to pathogenicity.

Table 6-1: ***Aeromonas*** **species prevalence in symptomatic and asymptomatic populations**

Reference	Location	Isolated-Ill	Isolated-Well	Notes
Gracey, Burke, and Robinson 1982*	Australia	10% (n = 1,156)	0.6% (n = 1,156)	Illness most common in children <2 years; mainly mild diarrhea of short duration; some >2 weeks; some "dysentery-like"
Figura et al. 1986	Italy	3.7% (n = 561)	2.1% (n = 576)	Children, ages unknown; *A. caviae* most frequent
Challapalli et al. 1988	USA	7.0%	2.0%	Ill children = 1 to 27 months old; 90% = self-limiting illness, ≤10 days; *A. caviae* most frequent
San Joaquin and Pickett 1988	USA	2.5% (n = 2,120)	0.5% (n = 380)	91% <3 years old; acute watery diarrhea; eight patients with chronic diarrhea. *A. caviae* most frequent
Golik et al. 1990	Israel	1.7% (n = 1,005)	0 (n = 500)	Adult hospital inpatients with acute diarrhea; *Aeromonas* spp. identified as sole-source pathogen
Deodhar, Saraswathi, and Varudkar 1991	India	1.8% (n = 2,480)	0 (n = 512)	Mixed age study population with acute gastroenteritis; most patients <5 years old; *A. hydrophila* most frequent
Pazzaglia et al. 1991	Peru	52.4% (n = 391)	8.7% (n = 138)	Children <18 months old; 58% of cases were polymicrobic
Pazzaglia Escamilla, and Batchelor 1991	Peru	9.2% (n = 655)	3.5% (n = 287)	Adult US citizens living in Lima
Utsalo et al. 1995	Nigeria	3.0% (n = 296)	2.0% (n = 616)	Infected cases: 100% well children = 6 years or more; 100% ill children = 5 years or less
Samonis et al. 1997	Greece	.05% (n = 3,600)	—	Mixed-age population with diarrhea; *A. hydrophila* most frequent
Komathi, Ananthan, and Alavandi 1998	India	6.5% (n = 200)	0% (n = 52)	Study group = <10 years old; *A. hydrophila* most frequent
Teka et al. 1999	Bangladesh	5.5% (n = 7,398)	—	Study group = <5 years old; *A. caviae* most frequent at 32%; 83% of children <3 years old
Albert et al. 2000	Bangladesh	7.2% (n = 1,735)	3.3% (n = 830)	Study group = children ≤5 years old
Juan et al. 2000	Taiwan	2.3% (n = 2,150)	—	Study group = children 5 months to 16 years; positive culture for *A. hydrophila* only
Essers et al. 2000	Switzerland	4.8% (n = 312)	—	Children admitted to hospital for acute diarrhea

*Shading indicates study population comprised of children only.

Table 6-2: Currently named *Aeromonas* species

Major Pathogens	Minor Pathogens	Environmental Species*
A. hydrophila (HG1)	*A. veronii/veronii* (HG10)	*A. salmonicida* (HG3)
A. caviae (HG4)	*A. jandaei* (HG9)	*A. sobria* (HG7)
A. veronii/sobria (HG8)	*A. schubertii* (HG12)	*A. media* (HG5)
	A. trota	*A. eucrenophila* (HG6)
		A. allosaccharophila
		A. encheleia (HG11)
		A. bestiarum (HG2)
		A. popoffii

Adapted from Janda and Abbott 1998.

*Most commonly seen in water, animals, or soil, though some have been isolated from clinical samples

Epidemiology

The aeromonads are ubiquitous in all environments. This global prevalence has made it difficult for surveillance efforts to trace sources of infection in *Aeromonas*-related gastrointestinal illness; additionally, no animal models have been found for *Aeromonas*-related diarrhea, which has hampered more extensive research. The relationship between gastrointestinal disease and the aeromonads is still under debate to some extent (Albert et al. 2000; Janda and Abbott 1998; Schiavano et al. 1998), mainly because of its inability to fulfill Koch's postulates. Most current scientific literature and, therefore, this paper assumes a cause-and-effect association.

The population prevalence of any type of *Aeromonas* infection varies. Infection occurs in both children and adults—in symptomatic and asymptomatic people—and in developed and developing countries. All 14 hybridization groups have been recovered in human feces, which is evidence of their abundance in the environment (most notably water). However, the presence of *Aeromonas* in the gastrointestinal tract does not prove its pathogenicity—it may just reflect a transient colonization in the body (Janda and Abbott 1998). Generally, at least 85% of isolates from human feces are either *A. hydrophila* HG1, *A. caviae* HG4, or *A. veronii* biotype *sobria* HG8/10 (Hanninen and Siitonen 1995; Janda and Abbott 1998; Janda et al. 1996). Epidemiological studies that are more than 10 to 15 years old do not identify *Aeromonas* down to the species level, much less to the hybridization group. Therefore, comparing prevalence studies using various levels of species identification can be somewhat misleading. The most recent studies use technically advanced, molecular detection methods.

Table 6-2 summarizes a selection of studies comparing infection between cases with diarrhea with controls who are asymptomatic. Generally, there is no predictable pattern of prevalence in either group or in any population, except for an increased rate of infection in the warm, summer months.

To date, no single-source outbreaks of *Aeromonas*-related illness have been identified in the literature. Case reports of "outbreaks" have been questionable because of the lack of species characterization or the variety of species that are identified in outbreak patients (de la Morena et al. 1993). Without the sophisticated molecular detection methodology that has recently been developed, studies could only make estimates on relative risk between exposure to water and infection.

For example, a 1986 study linked patients who had large amounts of *A. hydrophila* in their stool with drinking untreated water (Holmberg et al. 1986). The same study demonstrated further evidence of a causal link with gastroenteritis: Nine patients took antimicrobials to which their strains were resistant, and they remained ill; five others took antibiotics to which their strains were susceptible, and their symptoms resolved. The study, however, did not sample the drinking water to compare with the patient samples.

Using newly developed DNA detection techniques, Davin-Regli et al. (1998) were able to make more confident predictions regarding a cluster of *A. hydrophila* infections in seven hospitalized patients over two months. They sampled the patients and took 26 tap water samples from the hospital, assuming the source was water. They used random amplification of polymorphic DNA (RAPD) and enterobacterial repetitive intergenic consensus-polymerase chain reaction (ERIC-PCR) to test the association between the patients' infections and hospital water. Between 1 and 30,000 colonies of *A. hydrophila* were isolated in 13/26 (50%) of the water samples; however, no genetically identical strains were common to both clinical and environmental samples, and the researchers found wide heterogeneity among the samples. Two identical strains were isolated from water in three separate hospital units, suggesting that the water distribution system had been colonized. Although two of the patients were infected by one identical strain, it did not match up with any of the water isolates. Therefore, the two patients shared one source of infection, probably other than water (they were in the same unit together for 32 days), but the other cases appeared sporadic and unconnected.

In Switzerland, Demarta and colleagues (2000) used ribotyping to compare *Aeromonas* spp. isolated from children with gastroenteritis to isolates from their contacts and environment. Fifteen of all isolated strains (47%) from the children with gastroenteritis had the same riboprofile as those from asymptomatic contacts and environments.

In May 1988, California became the first state to make *Aeromonas* infection reportable, allowing for a population-based study of this condition (King, Werner, and Kizer 1992). In the first year, the state health department completed 219 investigations. The incidence rate for all infections combined was 10.6 cases per 1 million population; 81% of those infections were enteric; 9% were wound related. Investigators did not identify any common-source outbreaks among the gastrointestinal cases. Though 5 of the 219 cases subsequently died, they all had serious underlying medical conditions that primarily led to their deaths; however, medical records were not reviewed. Officials did speculate that there was significant underreporting of this infection during the year of surveillance. However, even with that consideration, they estimated that the true incidence rate was probably much lower than that of other enteric pathogens (e.g., *Salmonella* and *Shigella* at 189 and 266 cases per 1 million population, respectively). The authors concluded,

> . . . much of the morbidity and mortality among patients from whom *Aeromonas* isolates are obtained is associated with serious underlying health problems and that infection is more of a personal health problem than a public health problem. The lack of evidence of common-source enteric disease outbreaks, the lack of evidence that *Aeromonas* by itself causes severe disease, and the lack of interventional strategies to prevent these infections all suggest that ongoing surveillance may contribute little to the prevention and control of *Aeromonas* infection.

Consequently, officials discontinued the surveillance efforts and mandatory reporting in California.

Clinical

The pathogenic mechanisms of these bacteria are complex and not well characterized (Chang et al. 1997). The effects of infection with *Aeromonas* species can include gastroenteritis, wound infection, bacteremia, sepsis, peritonitis, meningitis, pneumonia, and conjunctivitis (Forbes, Sahm, and Weissfeld 1998). Both immunocompetent and immunocompromised people experience illness from *Aeromonas* infection, though serious morbidity and mortality occur mainly in the immunocompromised.

Gastrointestinal Illness

Aeromonas has been implicated as a cause of gastroenteritis for some time, though the exact mechanisms are still unresolved. Isolation rates in patients with gastroenteritis are not consistently high, and healthy people also shed the organism (Chang et al. 1997).

Evidence linking *Aeromonas* with gastrointestinal illness include case reports, case-control studies, and the production of various enterotoxins by many strains; however, a lack of recognized outbreaks or animal models for *Aeromonas*-related gastroenteritis is contradictory to the evidence (Janda 1999; Janda and Abbott 1998; Smith and Cheasty 1999). The weight of increasing clinical and epidemiological data indicate that mesophilic aeromonads should be regarded as enteric pathogens, even with the lack of documented outbreaks or animal models (Janda and Abbott 1998; Smith and Cheasty 1999).

Symptoms in both children and adults range from mild diarrhea to a febrile dysentery-type illness. The most common response is a mild to moderate, self-limiting, watery diarrhea of short duration (a few days to a few weeks) (Bloom and Bottone 1990; Smith and Cheasty 1999; Wilcox et al. 1992). Chronic (up to 3 months) watery diarrhea has also been reported (Gracey, Burke, and Robinson 1982). Occasionally, diarrhea will contain blood and mucus—this can occur in either an acute or chronic expression of illness.

In one study of 80 infected adult patients, symptoms of *Aeromonas*-related gastroenteritis ranged from acute, self-limited diarrhea to chronic, indolent diarrhea, with 16% confirmed cases of colitis (George et al. 1985). A study of 36 patients of all ages showed that adults tended to suffer chronic symptoms, while children's illness was more severe, but acute (Holmberg et al. 1986).

Skin and Soft-Tissue Infection

Wounds are the second most common location of *Aeromonas* infection after the gastrointestinal tract (Janda and Abbott 1998). These can range from mild infections kept to the outer skin layers to infection involving muscle, joints, and bone. Exposure can result in asymptomatic colonization of the skin (Gold and Salit 1993). Localized soft-tissue trauma and freshwater exposure are the principle risk factors in skin and soft-tissue infections; this is the primary route of infection in a healthy host. Eleven cases of soft-tissue infection from *Aeromonas* were diagnosed between 1979 and 1992 in Canada: 78% of the patients were exposed to fresh water and 18% (2) were nosocomial. Ten patients required surgical intervention for their wounds (Gold and Salit 1993).

Most wound infections associated with *Aeromonas* are polymicrobic—they are frequently isolated with other microbes (e.g., *Clostridium perfringens*)—resulting in possibly synergistic infections (Gold and Salit 1993; Janda and Abbott 1998; Pazzaglia et al. 1991). *A. hydrophila* is the species predominantly associated with wound infections (Janda and Abbott 1998). Soft-tissue infections can progress to a systemic disease and death in people with underlying illness (Gold and Salit 1993).

Sepsis

When sepsis strikes people who have other medical conditions or severe wound infections, the mortality rate can range from 25–90% (Janda and Abbott 1998). *A. hydrophila, A. veronii* (both biotypes), *A. caviae, A. jandaei,* and possibly *A. schubertii* have all been implicated in sepsis (Janda and Abbott 1998). Patients who have underlying diseases like cancer or cirrhosis are more susceptible to sepsis, and their prognoses are poor (Ko et al. 2000; Martino et al. 2000). A Taiwanese study found patients with cirrhosis or neoplasms had a higher mortality rate from *Aeromonas*-related sepsis than other patients (89% versus 11%) (Lau et al. 2000).

Meningitis

Meningitis is a very rare outcome. Only a handful of cases have been reported in the literature (Janda and Abbott 1998).

Peritonitis

Peritonitis is also rare; more than 45 cases have been reported, mostly associated with bacteremia in people with significant underlying conditions. *A. hydrophila* is the predominant source of this outcome, which can result in roughly a 60% fatality rate (Janda and Abbott 1998; Muñoz et al. 1994).

Respiratory Tract

Respiratory infections range from pneumonia and empyema to fatal complications involving soft tissues. The immunocompetent tend to develop respiratory illness after water exposure through accident (near drowning) or swimming. Respiratory infections in the immunocompromised are possibly caused from *Aeromonas* disseminating from their gastrointestinal tracts. These cases can have a mortality rate up to 50% (Janda and Abbott 1998). However, adverse outcomes are associated less with the presence or severity of a patient's underlying condition than with factors like the incidence of pneumonia, hemoptysis, or concomitant sepsis.

Conjunctivitis

This rare sequela can be superficial or serious—including the development of corneal ulcers. Water-related sources are commonly reported, but many *Aeromonas* eye infections have no known exposure source (Janda and Abbott 1998).

Treatment

No definitive guidelines on medical treatment for *Aeromonas* infection exist. For gastroenteritis, treatment is probably unnecessary, as it is usually a self-limiting illness. Antibiotics may be required for severe, bloody, or chronic diarrhea or in the immunocompromised (Smith and Cheasty 1999). For wound infection and sepsis, a number of antibiotics can be considered, but aeromonads are resistant to penicillins, erythromycin, tetracycline, and some cephalosporins (Alavandi, Subashini, and Ananthan 1999; Overman and Janda 1999; Forbes, Sahm, and Weissfeld 1998). About 90% of clinical strains are susceptible to a range of third generation cephalosporins (Gold and Salit 1993). Frequently, *A. hydrophila* is misdiagnosed as a streptococcal or staphylococcal infection, which results in inappropriate therapy with drugs to which *A. hydrophila* is resistant (Gold and Salit 1993; Weber, Wertheimer, and Ognjan 1995).

Susceptible Subpopulations

Immunocompromised

Many clinical manifestations of *Aeromonas* infection are either more common or more severe in the immunocompromised than in the immunocompetent. However, healthy people can be more susceptible to certain types of infection, notably, wound infection and gastroenteritis. The gastrointestinal tract is the most common location for infection, and people of all ages, both immunocompromised and healthy, are at risk. A case series suggested that bone marrow transplant recipients may have a higher colonization rate than other patient groups (Sherlock, Burdge, and Smith 1987), but little other evidence supports a higher enteric infection rate in the immunocompromised. However, the immunocompromised are far more likely to develop bacteremia and sepsis, which can come from a disseminated gastrointestinal infection and are often serious or fatal in this population (Bloom and Bottone 1990; Janda and Abbott 1998; Lau, Peng, and Chang 2000). The most susceptible to disseminated disease are those with cancer or liver disease (Janda and Abbott 1998; Muñoz et al. 1994). These patients are also most likely to develop rarer, critical secondary sequelae such as peritonitis, meningitis, and respiratory disease (Janda and Abbott 1998). In a case series (Funada and Matsuda 1997), patients with blood diseases who developed *Aeromonas* bacteremia in the hospital were mostly male, elderly, and leukemic. In these 17 patients, the overall mortality was 35%. *A. caviae* has been more associated with sepsis in cancer patients (Janda and Abbott 1998).

Though most wound infections are community acquired and occur in healthy individuals, some are nosocomially acquired (Gold and Salit 1993). Other types of *Aeromonas* infections can also be nosocomial; therefore, patients who spend more time in hospitals, such as those with chronic diseases, are more susceptible to hospital-based infections (Davin-Regli et al. 1998; Janda and Abbott 1998). Additionally, people in long-term institutional settings, like nursing homes, would also be more susceptible because of their immune status and group setting. Bloom and Bottone (1990) reported on 17 nursing home patients with diarrhea associated with *A. hydrophila* (though only four samples were positively cultured). Seventy-six percent of the patients had mild illness lasting less than 48 hours; one patient died after developing profuse diarrhea, fever, and tachycardia.

Children

Many of the studies published on the prevalence of *Aeromonas*-related gastroenteritis have focused on children. This is not necessarily because it is seen more frequently in children but because children's diarrhea receives more study. Also, in hospital- and clinic-based study populations, children are more likely to be given health care for diarrhea than adults. Table 6-3 summarizes selected studies of the prevalence of *Aeromonas* species in cases of gastroenteritis and diarrhea. The majority of studies had a prevalence of less than 5% attributable to *Aeromonas*; the rest were 10% or below, except for one study (Pazzaglia et al. 1991), in which *Aeromonas* was isolated from the stools of 52% of 391 Peruvian children with gastroenteritis. Compared with other studies, this seemed like an aberration; however, the authors pointed out that 58% of those cases were polymicrobic—other enteric pathogens were isolated from the same patients. Therefore, the majority of the diarrhea cases could not be attributed solely to *Aeromonas*. This reported rate of isolation still appears to be unusually large compared with other rates. Studies of adult patients or mixed-age populations have shown similar rates of 0–10% (see Table 6-2). Both asymptomatic children and adults shed *Aeromonas* in their feces. In published reports, 0–2% of asymptomatic children shed the organism; the highest reported rate of 8.7% was in the Peruvian study noted above for its unusually high isolation rate in sick children. No differences in prevalence between developed or developing countries are apparent.

The clinical course of infection in children is variously reported. Illness seems to occur most frequently in the youngest children—a few years old or less. It is unknown if that is an effect of host immune response or a young child's increased exposure because of poorer hygiene (e.g., diapers, etc.). As detailed in Table 6-3, the majority of studies report children's illness as an acute, but self-limited, watery diarrhea that lasts from a few days to a week or two. Some reviewers conclude that children's illnesses tended to be acute, while adults' tended to be chronic (Merino et al. 1995; Smith and Cheasty 1999). However, biased reporting may result in flawed assumptions (e.g., adults less likely than children to be treated for acute diarrhea). Several investigators have described cases of children whose illness progressed to chronic diarrhea, dysentery-like illness, and dehydration (Gluskin et al. 1992; Gracey, Burke, and Robinson 1982; San Joaquin and Pickett 1988). Vomiting and/or fever sometimes accompany the diarrhea but not necessarily. Bloody stools or fecal leukocytes occur from time to time (Ashdown and Koehler 1993; Challapalli et al. 1988; Gluskin et al. 1992). Severe complications, such as bacteremia, are rarely reported but do occur (San Joaquin and Pickett 1988). Cases of hemolytic uremic syndrome, which has most recently been associated with *E. coli* O157:H7, have also been associated with *Aeromonas* (Bogdanovic et al. 1991; Robson, Leung, and Trevenen 1992; San Joaquin and Pickett 1988).

A. caviae has been isolated most frequently in children with gastroenteritis. However, *A. caviae* isolates associated with illness do not always show typical virulence properties, like cytotoxin production (Havelaar et al. 1992; Janda 1999; San Joaquin and Pickett 1988; Smith and Cheasty 1999), or their virulence properties do not differ substantially from other isolates' virulence properties (Challapalli et al. 1988; Figura et al. 1986). Challapalli et al. (1988) did not correlate different isolates with the level of disease severity, though *A. caviae* has been associated with more serious sequelae in children, such as chronic diarrhea (San Joaquin and Pickett 1988).

Chronic Sequelae

Aeromonas infections include such serious clinical consequences as bacteremia, pneumonia, and peritonitis, and these syndromes cause a high fatality rate in certain people whose immune systems are compromised. However, these diseases are acute in nature. The only chronic sequelae reported in association with *Aeromonas* infection have been chronic diarrhea

Table 6-3: *Aeromonas* species isolation in children with gastroenteritis

Reference	Location	%/n = Ill	Notes
Gracey, Burke, and Robinson 1982	Australia	10% (n = 1,156)	Illness most common in children <2 years; mainly mild diarrhea of short duration; some >2 weeks; some "dysentery-like"
Challapalli et al. 1988	United States	7% (n = 32)	Ill children = 1 to 27 months old; 90% = self-limiting illness, ≤10 days; *A. caviae* most frequent
San Joaquin and Pickett 1988	United States	2.5% (n = 2,120)	92% <3 years old; acute watery diarrhea; eight patients with chronic diarrhea. *A. caviae* most frequent
Pazzaglia et al. 1991	Peru	52.4% (n = 391)	Children <18 months old; 58% of cases were polymicrobic
Gluskin et al. 1992	Israel	4% (n = 32,810)	Children to age 13; 94% <3 years; peak incidence and morbidity = 2 to 6 months
Wilcox et al. 1992	United Kingdom	2.5% (n = 1,026)	Mainly mild diarrhea lasting 2 to 3 days; *A. caviae* most frequent
Ashdown and Koehler 1993	Australia	4.9% (n = 912)	Mixed-age study group; 60% of ill patients <5 years old; 38% of ill patients >25 years old; *A. hydrophila* most frequent
de la Morena et al. 1993	United States	2.9% (n = 381)	Study group from day-care centers; *A. caviae* most frequent
Utsalo et al. 1995	Nigeria	3% (n = 296)	Infected cases: 100% well children = 6 years or more; 100% ill children = 5 years or less
Notario et al. 1996	Argentina	1.9% (n = 570)	Surveillance of acute diarrhea in children between 1985 and 1993
Ahmed et al. 1997	Pakistan	4.2% (n = 1,003)	Most prevalent in children <10 years
Figura et al. 1997	Italy	1.4% (n = 6,403)	10-year surveillance of children with diarrhea
Komathi, Ananthan, and Alavandi 1998	India	6.5% (n = 200)	Study group = <10 years old; *A. hydrophila* most frequent
Teka et al. 1999	Bangladesh	5.5% (n = 7,398)	Children <5 years old; *Aeromonas* spp. identified as sole pathogen; *A. caviae* accounted for 32%; majority of symptoms were acute vomiting and watery diarrhea
Essers et al. 2000	Switzerland	4.8% (n = 312)	Children admitted to hospital for diarrhea
Juan et al. 2000	Taiwan	2.3% (n = 2,150)	Outpatients ranging from 5 months to 16 years; *A. hydrophila* was the only type under study

and/or gastroenteritis. Chronic, persistent diarrhea lasting months or more than a year has been attributed mainly to the *A. hydrophila* and *A. caviae* species (Janda and Abbott 1998; Rautelin et al. 1995; San Joaquin and Pickett 1988; Smith and Cheasty 1999) and has occurred in both adults and children. Several case reports also attributed stubborn cases of colitis to *Aeromonas* infection (Deutsch and Wedzina 1997; Farraye et al. 1989; Marsik and Werlin 1984). Though considered chronic (lasting more than a few weeks), all of these reported episodes were resolved after a course of an appropriate antibiotic. The chronic nature of infection may result from misdiagnosis or inappropriate treatment of an acute infection.

One recent case series suggested that chronic prostatitis in several men was associated with bacteria phylogenically similar to *Aeromonas* species (Riley et al. 1998).

Transmission

Transmission is generally limited to sporadic, individual infection from ingestion of food or water or exposure of skin and mucus to water and soil. No person-to-person transmission has been reported; however, because *Aeromonas* is shed in human feces, fecal–oral transmission is plausible. No single point-source outbreaks have been confirmed to date (Janda and Abbott 1998; Nichols 1996). Though the presence of *Aeromonas* spp. in drinking water and food creates the opportunity for outbreaks (Krovacek et al. 1995; Massa, Altieri, and D'Angela 2001; Soriano et al. 2000), most infections in healthy people involve an injury exposed to an environmental source—usually recreational or occupational water exposure or gastroenteritis from ingestion. Most of these cases are associated with freshwater contact, though *Aeromonas* has been isolated from marine and estuarine environments as well; cases of anglers being inoculated by fish fins or hooks have also been reported. Soil can also be a source of infection (Forbes et al. 1998; Smith and Cheasty 1999). Pathogenic *Aeromonas* spp. have been recovered from the feces of cats and dogs, so zoonotic transmission is plausible (Ghenghesh et al. 1999). Most infections are community acquired, but some (especially among the immunocompromised) are nosocomial in nature.

Obi et al. (1997) published a cross-sectional study of persons with diarrhea and asymptomatic controls categorized by whether they lived in an urban or rural setting in Nigeria. In the sick patients, *Aeromonas* was isolated from 5% of the urban residents compared with 15% of the rural residents. Lower rates with a similar distribution between residences occurred in asymptomatic people. These results suggest that the rural population may have significantly different exposures (e.g., more contact with soil or untreated water, more time spent outside) that contribute to the difference in prevalence. Drinking water could also be a factor if the rural water supplies had inferior treatment.

Pazzaglia and colleagues (1990) measured the rate of intestinal colonization of *Aeromonas* spp. in 52 Peruvian newborns. In the first week, *Aeromonas* spp. were found in the stools of 23% of the infants, none of whom became clinically ill. A hospital survey suggested that the hospital water supply may have been the source of infection.

Dose Response

Little is known about the host immune response to *Aeromonas* by exposure through ingestion (Smith and Cheasty 1999). Most published information involves dermal exposure resulting in a skin infection. Of concern to drinking water, however, is ingestion of the microbe and the disease outcome.

Infectious Dose

Limited information on dose response is available primarily from animal and volunteer feeding studies. Pitarangsi et al. (1982) fed *Aeromonas* strains isolated from ill people to five rhesus monkeys at a dose of 10^9/mL. They did not observe any effects. In a 1964 human volunteer study, Rosner (cf. Janda and Abbott 1998) found that only certain *Aeromonas* isolates at doses greater than 10^9 CFU caused colonization/shedding in human volunteers. In a more recent challenge study, only 2 of 57 volunteers who were fed virulent doses ranging from 10^4 to 10^{10}/mL became infected with *Aeromonas* (Morgan et al. 1985). A Swedish outbreak of food poisoning was associated with *A. hydrophila* isolated from leftover food (Krovacek et al. 1995). Researchers found levels of 10^6–10^7/g in the suspect food; 24 of 32 people who ate the food became sick. However, clinical samples from patients were not tested to confirm the source. Limited data suggest that prior treatment with antibiotics may be a risk factor for infection by decreasing the necessary infectious dose (Rusin et al. 1997; Moyer 1987); this has been suggested as a risk factor for other bacterial infections as well (Rusin et al. 1997). The patient's depressed immune status may also decrease the dose needed for colonization and infection (Rusin et al. 1997).

Some conclude that most adults are more resistant to infection than children (Janda and Abbott 1998), though prevalence studies of adults with gastroenteritis do not necessarily support that notion (see Table 6-2). It appears that only very high doses of certain strains will cause gastroenteritis in select hosts and that true outbreaks (single strain, high concentration in foods, clustering in susceptible populations) are unlikely to occur (Janda and Abbott 1998; Janda et al. 1996; Nichols 1996).

Virulence

The virulence factors related to the *Aeromonas* species have not been well characterized (Forbes et al. 1998). The lack of animal models that validate *Aeromonas*-associated diarrhea and a still-emerging serotyping regime have hindered more data on dose response (Janda 1999; Smith and Cheasty 1999). However, potential virulence factors, such as adhesions, enterotoxins, hemolysins, and protease, are produced in some form by all three major pathogenic species (Gibotti et al. 2000; Smith and Cheasty 1999; Trower et al. 2000).

Pathogenicity and virulence factors on diarrhea-related strains include pili that adhere to intestinal cells, tissue, and enterocytes (Bondi et al. 2000; Kirov, O'Donovan, and Sanderson 1999), as well as cytotoxicity, hemolytic activity, and enterotoxin production (Albert et al. 2000). Some of the enterotoxins reacted with cholera toxins (Trower et al. 2000). Albert et al. (2000) have studied the genetics of three enterotoxins produced by *Aeromonas* spp. and found that concentrations of two of three enterotoxins were significantly higher in samples from ill children versus samples from healthy children and samples from the environment, and that there was a synergistic effect between those two toxins that resulted in more severe diarrhea.

Handfield and coworkers (1996) evaluated the pathogenicity of *Aeromonas* isolates from food and drinking water for factors related to hemolysis, hemagglutination, and cytotoxicity. All three processes were common in isolates from both food and drinking water. In another study, the feces of 292 diarrheic patients and 195 healthy people for eight *Aeromonas* strains were analyzed for toxin production and adhesion/invasive ability—both virulence properties (Schiavano et al. 1998). Some strains isolated from the people with gastroenteritis demonstrated one property or the other, or both. One strain that produced cytotoxin and displayed adherence was cultured from an asymptomatic subject. Other researchers (Haque et al. 1996) have found environmental and diarrheal isolates that produce a Shiga-like toxin similar to that produced by *E. coli* O157:H7. Although the amount of toxin produced by the *Aeromonas* species was 5- to 1,000-fold lower than that from *E. coli*, the levels were sufficient to produce gastroenteritis, though it is not known how the toxin causes illness. Trower and colleagues (2000) found an enterotoxin produced by an isolate of *A. veronii* biovar *sobria* cross-reacted to cholera toxin antibodies. Others found hemolytic and cytotoxic *Aeromonas* spp. in potable water supplies in India (Alavandi, Subashini, and Ananthan 1999). Clearly, strains found in both food and drinking water have the potential to be pathogenic to humans, but multiple virulence factors, as well as variation in the host, seem to strongly influence the organism's infectivity.

In a recent review of the risk of opportunistic bacterial pathogens, Rusin and colleagues (1997) performed a risk assessment of *Aeromonas* based on available data on water occurrence and infective dose. Though their modeling process was not detailed, it was based on the traditional National Academy of Sciences four-part paradigm: hazard assessment, exposure assessment, dose response, and risk characterization (NRC 1983). In drinking water assumed to have 200 CFU/2 L and 38,000 CFU/2 L, they estimated a daily risk of infection between 7.3×10^{-9} to 1.4×10^{-6}, respectively. Sensitivity analysis showed the main source of variability in the model to be the concentration of organisms in the water, contributing 95.5% of the uncertainty. They concluded that the risk is generally below 1/10,000 for a single exposure but that accurate occurrence data are critical to the success of the model.

Detection Methods

Because of the confusing and ever-changing taxonomy, *Aeromonas* species have been difficult to separate and classify. The aeromonads can be easily cultured, but commercial kits to identify them down to the species level are limited and expensive. Another problem has been commercial assays that cross-react with *Vibrio* species (Abbott et al. 1998; Forbes et al. 1998). Work is progressing on the development of a fast, accurate technique to detect *Aeromonas* from clinical and environmental sources. Molecular techniques like ribotyping and multilocus enzyme electrophoresis and polymerase chain reaction (PCR) (Cascon et al. 1996) are more successful but are expensive and technically demanding (Janda and Abbott 1998). In addition to being used as a detection method, PCR has been used to characterize different strains' production of virulence factors such as hemolysin, protease, and cytotoxin, which may indicate their pathogenicity (Baloda et al. 1995). Ribotyping and pulsed-field gel electrophoresis have been used to differentiate particular strains from clinical and environmental samples to try to link specific food or water isolates to a specific illness (Demarta et al. 2000; Moyer et al. 1992; Smith and Cheasty 1999; Talon et al. 1996). Overman and Janda (1999) tested 56 isolates from clinical specimens for antimicrobial resistance. The isolates belonged to four species: *A. jandaei*, *A. schubertii*, *A. trota*, and *A. veronii* biotype *veronii*. Most isolates were resistant to ampicillin, but the level of susceptibility and resistance beyond that varied greatly over the four species and the 56 isolates. These data may be useful in differentiating these four species from human specimens.

No specific culture medium is recommended for isolating *Aeromonas* species from feces; they grow on most enteric media (Smith and Cheasty 1999). However, ampicillin dextrin agar is recommended as a culture media for isolation from water (Handfield, Simard, and Letarte 1996; Moyer et al. 1992; Nichols 1996). Researchers found that using this medium with a standard membrane filtration procedure (like the method for coliform detection) worked well for recovery and differentiation in water samples mixed with common bacterial contaminants (Handfield, Simard, and Letarte 1996; Havelaar, During, and Versteegh 1987). This approach is the best available for detecting *Aeromonas* in water and probably better than any other method currently available for detecting other microbes on the US Environmental Protection Agency's Contaminant Candidate List (March 2, 1998; 63 FR 10273).

Environmental Occurrence

Aeromonas species are found in all types of water: wastewater, surface water, groundwater, marine and estuarine environments, even chlorinated water supplies. They have been shown to inhabit biofilms (Gavriel et al, 1998) and they experience regrowth in distribution systems (Gavriel, Landre, and Lamb 1998; Smith and Cheasty 1999). Factors that contribute to the growth of aeromonads in the water supply include a high level of organic carbon, reduced chlorine residuals, and warm water and air temperatures (Burke et al. 1984; Smith and Cheasty 1999). Like *Mycobacterium avium* complex, *A. hydrophila* has been detected in hospital water supplies (Picard and Goullet 1987). Environmental occurrence varies by season, with increases occurring in the warmer summer months (Burke et al. 1984; Gavriel, Landre, and Lamb 1998; Smith and Cheasty 1999). Infection rates correspond to the rate of occurrence in the environment. Aeromonads, however, can be isolated in cold weather environments—just at reduced numbers (Janda 1999; Smith and Cheasty 1999).

Different *Aeromonas* species have been found in all kinds of animals, including fish, birds, cats, dogs, pigs, cattle, horses, and monkeys (Smith and Cheasty 1999). They are found in a variety of foods: dairy products, meat, produce, and shellfish (Forbes et al. 1998; Nishikawa and Kishi

1988) (though they are not as common in shellfish as are the vibrios) (Janda 1999). The psychrophilic nature of some aeromonads allows them to grow well at refrigerator temperatures, which can contribute to food-borne spread (Janda and Abbott 1998; Nichols 1996).

Occurrence in Water Sources

The studies investigating the occurrence of *Aeromonas* species in water environments have produced a variety of results depending on factors such as the specificity of the isolates under study, the season, and the water quality. The quantities of *Aeromonas* found in all water sources are consistently high. Drinking water treatment decreases levels substantially, but they frequently undergo regrowth in the distribution system. Table 6-4 summarizes some reports of the occurrence of *Aeromonas* in water sources.

A summary of reports on environmental occurrence follows:

- Two studies recovered more aeromonads in surface water compared with groundwater (Burke et al. 1984; Legnani et al. 1998).
- Hanninen, Oivanen, and Hirvela-Koski (1997) found *Aeromonas* species in almost 100% of fish, roe, and freshwater samples in Finland. However, the strains that have been most commonly associated with diarrhea (*A. hydrophila* HG1, *A. caviae* HG4, and *A. veronii* biotype *sobria* or biotype *veronii* HB 8/10) were uncommon in all of the environmental samples (n = 81).
- Sampling was done at recreational marine areas affected by sewage discharge in Spain (Alonso, Amoros, and Botella 1991). Mesophilic aeromonads were detected in every sample (nine samples over 9 months). Aeromonads and total coliforms showed similar growth and occurrence patterns, though aeromonad concentrations consistently exceeded fecal coliform levels. No seasonal fluctuation occurred, which is unusual. The most frequent species identified were *A. caviae,* then *A. hydrophila,* and then *A. sobria. A. hydrophila* occurred in higher concentrations in the less polluted samples, and *A. caviae* was found most in the samples with the most pollution.
- A cluster of *Aeromonas*-related gastrointestinal infections in Scotland, as reported by Gavriel and colleagues (1998), suggested that the area's drinking water supply might have been the source. Consequently, Gavriel and colleagues sampled 31 drinking water reservoirs for the presence of both *Aeromonas* and residual chlorine on a weekly basis for a year. Infections in the epidemiological study

Table 6-4: *Aeromonas* concentrations in water sources

Reference	Location	Water Type	Minimum per 100/mL	Maximum per 100/mL
Alonso, Amoros, and Botella 1991	Spain	Marine water	1.4×10^2	9.0×10^6
Aulicino et al. 2001	Italy	Marine water	0	10×10^6
Borrell, Figueras, and Guarro 1998	Spain	Marine water	1.0	10×10^5
		River	1.0×10^5	1.0×10^6
		Lake	1.0×10^3	10×10^5
		Treated drinking water	1.0	6.0×10^2
		Untreated drinking water	1.0	1.0×10^5
Faude and Hofle 1997	Germany	Lake	—	8.0×10^1
Gavriel, Landre, and Lamb 1998	Scotland	Reservoir	0	6.0×10^2
Havelaar, Versteegh, and During 1990	Netherlands	River	0	4.7×10^4
		Treated drinking water	0	4.7×10^2
		Distribution system	0	3.3×10^3
Legnani et al. 1998	Italy	Reservoir	1.0	2.4×10^2
Massa, Altieri, and D'Angela 2001	Italy	Well water	1.0×10^1	6.4×10^2
Pettibone 1998	USA	River (urban)	1.8×10^4	4.0×10^6
		River (rural)	3.0	1.6×10^4

peaked during the summer months. The environmental results also showed a significant seasonal pattern—95% of all the isolates were recovered between mid-June and the end of September, during which time the water temperature remained above 12° C. Higher concentrations were measured after heavy rains, which can increase the amount of organic matter in the water from run-off. Organic matter decreases the effectiveness of posttreatment chlorine and promotes after-growth activity in the distribution system. Generally among these samples, decreased levels of residual chlorine resulted in increased levels of *Aeromonas*; however, reservoirs that had continuously high levels of disinfectant occasionally yielded positive samples. Previous authors (Holmes and Niccolls cf. Gavriel, Landre, and Lamb 1998; Nichols 1996) have maintained that 0.2 mg/L of residual should control *Aeromonas* regrowth in drinking water. This study recovered *Aeromonas* in reservoirs with 0.45 mg/L of chlorine residual. These researchers suggested that the organisms might have been protected from chlorine by biofilms or that they had adapted by becoming more resistant to disinfection. Increases in temperature and organic matter may also have had an effect.

- Burke and colleagues (1984) monitored the Perth, Australia, metropolitan water supply over a year, collecting 3,224 samples. Water temperature and residual chlorine were recorded simultaneously with each sample. Sources of the samples were raw surface water or groundwater, treated surface water or groundwater, service reservoirs before and after chlorination, and the reservoirs' distribution systems. *Aeromonas* species were present in raw water, and though their occurrence decreased after chlorination, it increased to raw water levels again by the time it was in the distribution system. Growth in chlorinated samples was associated with the water temperature, the amount of free chlorine, and the interaction between the two. The residuals measured in the distribution system were typically below 0.3 mg/L. Like other studies, *Aeromonas* species occurred more frequently and in greater numbers in summer, and the incidence of *Aeromonas*-associated gastroenteritis in the Perth area followed the occurrence trends in the water samples.
- From June 1992 to June 1993, Pettibone (1998) sampled one site on the urban Buffalo River in New York and four sites upstream on its tributaries. Aeromonad concentrations at the Buffalo River

site in the summer ranged from 18 to 4,000/mL, which was consistently higher than that site's fecal coliform levels, fecal streptococci levels, and heterotrophic plate counts taken at that site in the summer (though they were correlated in the winter months). Conversely, there was a strong correlation between the occurrence of indicator bacteria and *Aeromonas* in the upstream tributary sites in both winter and summer. After summer rainstorms, the *Aeromonas* concentration increased by one log at the downstream site, whereas, it increased only by a factor of two at the tributary sites. This study demonstrated that in terms of microbial loading, the Buffalo River's conditions differed substantially from its upstream tributaries.

- Researchers tested for *Aeromonas* in the domestic water supplies (e.g., chlorinated municipal water, wells, bottled water) in an Indian city and found that 38% of the samples contained *Aeromonas* spp. (Alavandi, Subashini, and Ananthan 1999). The majority were *A. sobria, A. caviae,* and *A. hydrophila*. The 11/37 positive samples all were deemed pathogenic.
- Another *A. hydrophila* isolate capable of producing cytotoxins and enterotoxins was found in chlorinated municipal water in Argentina (Fernandez et al. 2000).

Survival in the Environment

Aeromonads can be persistent in the environment, especially in water sources where the conditions are favorable. Kersters et al. (1996) tracked the survival of a genetically marked strain of *A. hydrophila* in fresh water and nutrient-poor water. The strain survived up to 10 days in filtered, autoclaved, freshwater samples and filtered, autoclaved, nutrient-poor water (spring water, bottled water, tap water). In unfiltered waters, the viability declined after only 1 to 5 days. The authors hypothesized that the natural biota in unfiltered environmental water may have affected *Aeromonas'* ability to survive. On the other hand, another study tracking a particular isolate with strain-specific monoclonal antibodies showed the aeromonads' tenacity in natural water. Faude and Hofle (1997) isolated *A. hydrophila* PU7718 from a freshwater lake in 1990. During testing three years later, the same strain, *A. hydrophila* PU7718, was detected in its same lake habitat, demonstrating genetic stability over time in this natural environment.

Chung and Yu (1990) compared survival of *A. hydrophila* and *E. coli* in distilled water, pond water, and effluent from anaerobic digesters.

Both organisms survived well at equivalent rates of viability in all but the effluent, where *A. hydrophila* survived significantly longer than the *E. coli* (6 days).

Kuhn and colleagues in Sweden (1997) tested a drinking water well for *Aeromonas* isolates over a 4-year period. Of 170 identified phenotypes isolated in the well, only one was found in more than three samples: *A. hydrophila* HG3. This phenotype was found in most of the 40 samples, including the first sample in 1990 and the last sample in 1994. In a previous work, the authors showed that certain *Aeromonas* strains could persist in drinking water systems for several months. These results suggest that a genetically stable strain colonized the well for at least 4 years.

Brandi and coworkers (1999) found that the survival of *Aeromonas* was dependent on the species and the type of water (marine water, mineral water, or treated municipal water). They used strains of *A. hydrophila, A. caviae,* and *A. sobria* that had been isolated from clinical samples. All three species survived longest in mineral water; viable counts were made after 100 days. Isolates from tap water also showed marked survival; whereas, the bacteria died off quickly in seawater. The authors concluded that salinity had a greater effect on viability than the residual chlorination in the tap water.

Finally, Moyer et al. (1992) used ribotyping to differentiate aeromonads from patients with gastroenteritis and from multiple sites throughout a municipal water supply. No overlap existed between any of the clinical or water system samples; however, identical ribotypes of *A. hydrophila* and *A. sobria* were isolated from several sites throughout the distribution system, indicating a system-wide colonization. This colonization presented an opportunity for posttreatment contamination in the distribution system.

Comparison Among Environmental and Clinical Strains

A number of studies, which are described here, have attempted to find associations between environmental and clinical strains of *Aeromonas.*

- Havelaar et al. (1992) isolated 48 different biotypes in 187 human diarrheal stools and 263 drinking water samples; however, only 10 occurred relatively frequently in either source. Of these 10 more common biotypes, they distinguished 380 different strains: 231 from drinking water and 149 from feces. Only seven biotypes overlapped between water and feces—all from *A. hydrophila, A. sobria,* or *A. caviae. A. caviae* was found most frequently in both feces and drinking water; *A. hydrophila* was the second most common strain in drinking water, but appeared infrequently in feces; *A. sobria* strains that appeared in feces and drinking water showed some similarity, but *A. sobria* occurred infrequently in drinking water. The fecal samples came from ill people, so they were biased toward detecting pathogenic strains. The pathogenic strains found in ill people were rarely found in drinking water.

- In a study of environmental and clinical sources, 983 *Aeromonas* isolates were identified to the genomospecies level (Borrell, Figueras, and Guarro 1998). Of the clinical samples, predominating was *A. veronii* biotype *sobria*, then *A. caviae*, then *A. hydrophila*. Overall, *A. veronii* biotype *sobria* was seen most frequently in both clinical (31.8%) and environmental samples (21.8%). In treated drinking water specifically, 12.5% of samples had *A. hydrophila*, 25% had *A. veronii* biotype *sobria*, and none had *A. caviae*; however, *A. caviae* was found in high levels of sampled milk and seawater. The average concentration of all species in disinfected water was 260 CFU/100 mL, which was probably a consequence of the low or nonexistent level of residual chlorine (0–0.1 ppm), which allowed regrowth in the distribution.
- Janda et al. (1996) isolated 268 *Aeromonas* isolates from clinical, animal, and environmental sources and analyzed them down to serogroup designations. More than 96 distinct serogroups were recovered from the combined samples. Certain serotypes, like O:11, O:34, and O:16, seemed to predominate in the clinical samples, indicating their importance in human infection. *A. hydrophila* was the most common species from any source, accounting for between 30 and 40% of all strains that were identified. *A. hydrophila*, *A. caviae*, and *A. veronii* biotype *sobria* accounted for more than 85% of the clinical isolates, 56% of the animal isolates, and 69% of the environmental (freshwater) isolates.
- A Finnish study looked at 332 isolates from drinking water, fresh water, chicken and beef, and human fecal samples (Hanninen and Siitonen 1995). As in other studies, *A. hydrophila*, *A. caviae*, and *A. veronii* biotype *sobria* predominated in the fecal samples (80–90%). Only three isolates from the drinking water samples matched any from the fecal samples and none from the food; however, isolates from the fecal samples and the food samples substantially overlapped. The authors concluded that in these samples, food, rather than water, was the source of more pathogenic strains.

- In a study by Havelaar et al. 1992, researchers compared *Aeromonas* strains from 187 diarrhea and 263 drinking water samples. The three major species were dominant in both the feces and the water, with *A. caviae* being the most prevalent in both types of samples. More extensive bio- and serotyping showed that though species overlapped, the individual strains did not. The authors concluded that there was little similarity between strains isolated from clinical samples and those from drinking water samples.
- In a more recent study looking at ribotyping patterns, Demarta et al. (2000) compared 29 strains of *Aeromonas* isolated from sick children with 104 strains isolated from the children's homes and contacts. About 47% of the isolates from the patients showed the same riboprofile as found in the overall group of contacts and environments. Three patients had the same riboprofiles as their own domestic environment.

Though some similarity exists between *Aeromonas* strains from the human gastrointestinal tract and those in the environment, especially at the species level, more detailed serotyping often reveals little overlap. Since many types of isolates are found in the environment, only certain pathogenic strains have the virulence properties that allow them to colonize and infect people (Janda et al. 1996). More sophisticated methodology may show more coherence between clinical and environmental isolates.

Water Treatment

Most drinking water treatment processes appear to reduce numbers of aeromonads to levels below 10 CFU/100 mL (Havelaar, Versteegh, and During 1990; Nichols 1996). However, as seen in examples cited above, treated water can experience regrowth in storage reservoirs or distribution systems. The concentrations and rate of regrowth depend on temperature, organic content of the water, residence time in the distribution system, and the amount of residual chlorine. Some report that free chlorine residuals of 0.2–0.5 mg/L should control aeromonad concentrations in distribution systems (Holmes and Niccolls cf. Gavriel, Landre, and Lamb 1998; Nichols 1996). However, in the study by Gavriel and coworkers (1998), regrowth occurred in reservoirs with residual levels of up to 0.45 mg/L. Elements such as temperature and amount of organic matter in the water play a large role in how effective the residual chlorine level is on *Aeromonas.*

French scientists (Chamorey, Forel, and Drancourt 1999) tested *A. hydrophila* strains isolated from a hospital water supply and from the hospital's patients. They tested the effectiveness of chlorine as a disinfectant using interfering substances and a range of exposure times (1–5 minutes), concentrations, and temperatures. They determined a median minimal concentration of 0.95 mg/L without interfering substances to 297 mg/L with interfering substances, which are high concentrations.

An extensive evaluation looked at chlorine resistance and inactivation in 46 clinical isolates and 41 environmental isolates of *Aeromonas* compared with a control group of typical coliform bacteria (i.e., *E. coli, P. fluorenscens, Klebsiella* spp., P. *aeruginosa, Acinetobacter*) (Knochel 1991). The study used a culture disc assay with hypochlorite concentrations of 11, 22, 44, and 88 mg per disc. Though variations existed between the different *Aeromonas* isolates, as a group, they were significantly more susceptible to hypochlorite than the control group at every concentration. In the inactivation part of the study, two strains of *Aeromonas* were more quickly inactivated by monochloramine than an *E. coli* strain, and one strain was slower than *E. coli.*

Sisti, Albano, and Brandi (1998) investigated the susceptibility of pathogenic *Aeromonas* species to disinfection by free chlorine. They found that the bactericidal efficiency of chlorine was strongly influenced by water temperature. At summer temperatures of 20°C, the efficacy was reduced two to three times compared with winter temperatures of 5°C. *A. hydrophila* was "moderately but significantly" more resistant than *A. caviae* or *A. sobria*. All the aeromonads were more susceptible to inactivation than *E. coli.*

If contaminated water does make its way into the distribution system or if a failure in the system allows contamination there, a recent study demonstrated the ease with which *Aeromonas* can attach itself to distribution system pipes. Assanta, Roy, Montpetit (1998) tested the ability of *A. hydrophila* to adhere to different materials used in water distribution systems: stainless steel, copper, and polybutylene. They also measured contact times and water temperature. The results indicated that *A. hydrophila* could attach easily on all materials after exposures of only 1 or 4 hours at either 4° C or 20° C. Polybutylene was most hospitable to colonization, followed by stainless steel, then copper.

Several studies have evaluated the ability of different disinfectants to eliminate *Aeromonas.* Medema et al. (1991) evaluated the disinfecting ability of chlorine dioxide in drinking water systems. A chlorine dioxide dose of 0.2 mg/L is depleted in most drinking water within 10 minutes. *Aeromonas* is sensitive to chlorine dioxide relative to other gram-negative bacteria, and at pH 8, it is as sensitive to chlorine dioxide as it is to free chlorine. The exception is in water that has been filtered through activated carbon to remove organic material, which makes free chlorine more effective. This study concluded that a dose of 0.5 to 1.0 mg/L is necessary to protect against growth in all drinking water, except water that is filtered through activated carbon or that is naturally low in organic matter.

Membrane filtration technology, such as micro-, ultra-, and nanofiltration, as well as reverse osmosis, has the capacity to remove water contaminants down to the ion level. Microfiltration, with a nominal pore size of 0.2 mm, can remove bacterial pathogens better than conventional water treatment, and ultrafiltration, with a nominal pore size of 0.01 mm, can achieve up to a 6-log pathogen removal (Najm and Trussell 1999; Taylor and Wiesner 1999). This technology will continue to gain in popularity as costs become more competitive and regulatory approval is assured.

Conclusions

The pathogenic *Aeromonas* species have the capacity to cause gastroenteritis and wound infections, among other sequelae. These bacteria are ubiquitous in the global environment, and though no single-source outbreaks have been confirmed, transmission is assumed to be mainly through water and food. Though disinfection appears to be relatively effective, *Aeromonas* typically regrows in water distribution systems under the right conditions. Drinking water should be considered a possible source of transmission; however, how much, if any, infection is attributable to drinking water is unknown.

Studies looking at infectious dose suggest that it takes a very large dose ($>10^8$–10^{10}) of particular strains to infect predisposed hosts. Having a serious underlying condition may make patients more susceptible to smaller doses or they could be infected at the same rate and just suffer more frequent and serious illness. The organism's virulence and the host immune response are not well characterized.

Aeromonas-related gastroenteritis occurs in all populations, both children and adults. Asymptomatic people also shed the bacteria in their feces. The normal course includes self-limiting diarrhea; however, chronic diarrhea and colitis have been reported. *Aeromonas* species are resistant to many antibiotics, but many types of antibiotics are still effective. One course of the appropriate antibiotic provides reliable treatment. People with compromised immune systems, especially those with cancer and liver disease, are susceptible to systemic infections, like bacteremia, which can begin with intestinal colonization. These sequelae often result in high case-fatality rates. Though children are more frequently reported with *Aeromonas*-related gastroenteritis, the weight of evidence does not necessarily support that children are more susceptible to either infection or more serious illness than adults.

Overall, the risk from *Aeromonas* is viewed as transient—requiring high doses of particular strains—and the probability of infection is low.

Bibliography

Abbott, S.L., L.S. Seli, M. Catino Jr., M.A. Hartley, and J.M. Janda. 1998. Misidentification of unusual *Aeromonas* species as members of the genus *Vibrio*: a continuing problem. *J. Clin. Microbiol.*, 36:1103–1104.

Ahmed, A., S. Hafiz, A. Zafar, T. Shamsi, J. Rizvi, and S. Syed. 1997. Isolation and identification of *Aeromonas* species from human stools. *J. Pak. Med. Assoc.*, 47:305–308.

Alavandi, S.V., M.S. Subashini, and S. Ananthan. 1999. Occurrence of haemolytic & cytotoxic *Aeromonas* species in domestic water supplies in Chennai. *Indian J. Med. Res.*, 110:50–55.

Albert, M.J., M. Ansaruzzaman, K.A. Talukder, A.K. Chopra, I. Kuhn, M. Rahman, A.S. Faruque, M.S. Islam, R.B. Sack, and R. Mollby. 2000. Prevalence of enterotoxin genes in *Aeromonas* spp. isolated from children with diarrhea, healthy controls, and the environment. *J. Clin. Microbiol.*, 38:3785–3790.

Alonso, J.L., I. Amoros, and M.S. Botella. 1991. Enumeration of motile *Aeromonas* in Valencia coastal waters by membrane filtration. *Water Sci. Technol.*, 24:125–128.

Ashdown, L.R., and J.M. Koehler. 1993. The spectrum of *Aeromonas*-associated diarrhea in tropical Queensland, Australia. *Southeast Asian J. Trop. Med. Public Health*, 24:347–353.

Assanta, M.A., D. Roy, and D. Montpetit. 1998. Adhesion of *Aeromonas hydrophila* to water distribution system pipes after different contact times. *J. Food Prot.*, 61:1321–1329.

Aulicino, F.A., P. Orsini, M. Carere, and A. Mastrantonio. 2001. Bacteriological and virological quality of seawater bathing areas along the Tyrrhenian coast. *Int. J. Environ. Health Res.*, 11:5–11.

Baloda, S.B., K. Krovacek, L. Eriksson, T. Linne, and I. Mansson. 1995. Detection of aerolysin gene in *Aeromonas* strains isolated from drinking water, fish and foods by the polymerase chain reaction. *Comp. Immunol. Microbiol. Infect. Dis.*, 18:17–26.

Bloom, H.G., and E.J. Bottone. 1990. *Aeromonas hydrophila* diarrhea in a long-term care setting. *J. Am. Geriatr. Soc.*, 38:804–806.

Bogdanovic, R., M. Cobeljic, M. Markovic, V. Nikolic, M. Ognjanovic, L. Sarjanovic, and D. Makic. 1991. Haemolytic-uraemic syndrome associated with *Aeromonas hydrophila* enterocolitis. *Pediatr. Nephrol.*, 5:293–295.

Bondi, M., P. Messi, E. Guerrieri, and F. Bitonte. 2000. Virulence profiles and other biological characters in water isolated *Aeromonas hydrophila*. *New Microbiol.*, 23:347–356.

Borrell, N., M.J. Figueras, and J. Guarro. 1998. Phenotypic identification of *Aeromonas* genomospecies from clinical and environmental sources. *Can. J. Microbiol.*, 44:103–108.

Brandi, G., M. Sisti, F. Giardini, G.F. Schiavano, and A. Albano. 1999. Survival ability of cytotoxic strains of motile *Aeromonas* spp. in different types of water. *Lett. Appl. Microbiol.*, 29:211–215.

Burke, V., J. Robinson, M. Gracey, D. Peterson, and K. Partridge. 1984. Isolation of *Aeromonas hydrophila* from a metropolitan water supply: seasonal correlation with clinical isolates. *Appl. Environ. Microbiol.*, 48:361–366.

Cascon, A., J. Anguita, C. Hernanz, M. Sanchez, M. Fernandez, and G. Naharro. 1996. Identification of *Aeromonas hydrophila* hybridization group 1 by PCR assays. *Appl. Environ. Microbiol.*, 62:1167–1170.

Challapalli, M., B.R. Tess, D.G. Cunningham, A.K. Chopra, and C.W. Houston. 1988. *Aeromonas*-associated diarrhea in children. *Pediatr. Infect. Dis. J.*, 7:693–698.

Chamorey, E., M. Forel, and M. Drancourt. 1999. An in-vitro evaluation of the activity of chlorine against environmental and nosocomial isolates of *Aeromonas hydrophila*. *J. Hosp. Infect.*, 41:45–49.

Chang, C.Y., H. Thompson, N. Rodman, J. Bylander, and J. Thomas. 1997. Pathogenic analysis of *Aeromonas hydrophila* septicemia. *Ann. Clin. Lab. Sci.*, 27:254–259.

Chung, K.T., and F.P. Yu. 1990. Survival of *Aeromonas hydrophila* and *Escherichia coli* in aquatic environments. *Chung Wei Shen Wu Chi Mien I Hsueh Tsa Chih*, 23:181–188.

Davin-Regli, A., C. Bollet, E. Chamorey, V. Colonna D'istria, and A. Cremieux. 1998. A cluster of cases of infections due to *Aeromonas hydrophila* revealed by combined RAPD and ERIC-PCR. *J. Med. Microbiol.*, 47:499–504.

de la Morena, M.L., R. Van, K. Singh, M. Brian, M.E. Murray, and L.K. Pickering. 1993. Diarrhea associated with *Aeromonas* species in children in day care centers. *J. Infect. Dis.*, 168:215–218.

Demarta, A., M. Tonolla, A. Caminada, M. Beretta, and R. Peduzzi. 2000. Epidemiological relationships between *Aeromonas* strains isolated from symptomatic children and household environments as determined by ribotyping. *Eur. J. Epidemiol.*, 16:447–453.

Deodhar, L.P., K. Saraswathi, and A. Varudkar. 1991. *Aeromonas* spp. and their association with human diarrheal disease. *J. Clin. Microbiol.*, 29:853–856.

Deutsch, S.F., and W. Wedzina. 1997. *Aeromonas sobria*-associated left-sided segmental colitis. *Am. J. Gastroenterol.*, 92:2104–2106.

Essers, B., A.P. Burnens, F.M. Lanfranchini, S.G. Somaruga, R.O. von Vigier, U.B. Schaad, C. Aebi, and M.G. Bianchetti. 2000. Acute community-acquired diarrhea requiring hospital admission in Swiss children. *Clin. Infect. Dis.*, 31:192–196.

Farraye, F.A., M.A. Peppercorn, P.S. Ciano, and W.N. Kavesh. 1989. Segmental colitis associated with *Aeromonas hydrophila*. *Am. J. Gastroenterol.*, 84:436–438.

Faude, U.C., and M.G. Hofle. 1997. Development and application of monoclonal antibodies for in situ detection of indigenous bacterial strains in aquatic ecosystems. *Appl. Environ. Microbiol.*, 63:4534–4542.

Fernandez, M.C., B.N. Giampaolo, S.B. Ibanez, M.V. Guagliardo, M.M. Esnaola, L. Conca, P. Valdivia, S.M. Stagnaro, C. Chiale, and H. Frade. 2000. *Aeromonas hydrophila* and its relation with drinking water indicators of microbiological quality in Argentine. *Genetica*, 108:35–40.

Figura, N., P. Guglielmetti, A. Zanchi, R. Signori, A. Rossolini, H. Lior, M. Russi, and R.A. Musmanno. 1997. Species, biotype and serogroup of *Campylobacter* spp. isolated from children with diarrhoea over a ten-year period. *New Microbiol.*, 20:303–310.

Figura, N., L. Marri, S. Verdiani, C. Ceccherini, and A. Barberi. 1986. Prevalence, species differentiation, and toxigenicity of *Aeromonas* strains in cases of childhood gastroenteritis and in controls. *J. Clin. Microbiol.*, 23:595–599.

Forbes, B.A., D.F. Sahm, and A.S. Weissfeld. 1998. *Vibrio, Aeromonas, Plesiomonas shigelloides*, and *Chromobacterium violaceum*. In *Bailey & Scott's Diagnostic Microbiology*. Edited by B.A. Forbes, D.F. Sahm, and A.S. Weissfeld. Houston, TX: Mosby.

Funada, H., and T. Matsuda. 1997. *Aeromonas* bacteremia in patients with hematologic diseases. *Intern. Med.*, 36:171–174.

Gavriel, A.A., J.P.B. Landre, and A.J. Lamb. 1998. Incidence of mesophilic *Aeromonas* within a public drinking water supply in north-east Scotland. *J. Appl. Bacteriol.*, 84:383–392.

George, W.L., M.M. Nakata, J. Thompson, and M.L. White. 1985. *Aeromonas*-related diarrhea in adults. *Arch. Intern. Med.*, 145:2207–2211.

Ghenghesh, K.S., S.S. Abeid, M.M. Jaber, and S.A. Ben-Taher. 1999. Isolation and haemolytic activity of *Aeromonas* species from domestic dogs and cats. *Comp. Immunol. Microbiol. Infect. Dis.*, 22:175–179.

Gibotti, A., H.O. Saridakis, J.S. Pelayo, K.C. Tagliari, and D.P. Falcao. 2000. Prevalence and virulence properties of *Vibrio cholerae* non-O1, *Aeromonas* spp. and *Plesiomonas shigelloides* isolated from Cambe Stream (State of Parana, Brazil). *J. Appl. Microbiol.*, 89:70–75.

Gluskin, I., D. Batash, D. Shoseyov, A. Mor, R. Kazak, E. Azizi, and I. Boldur. 1992. A 15-year study of the role of *Aeromonas* spp. in gastroenteritis in hospitalised children. *J. Med. Microbiol.*, 37:315–318.

Gold, W.L., and I.E. Salit. 1993. *Aeromonas hydrophila* infections of skin and soft tissue: report of 11 cases and review. *Clin. Infect. Dis.*, 16:69–74.

Golik, A., D. Modai, I. Gluskin, I. Schechter, N. Cohen, and J. Eshchar. 1990. *Aeromonas* in adult diarrhea: an enteropathogen or an innocent bystander? *J. Clin. Gastroenterol.*, 12:148–152.

Gracey, M., V. Burke, and J. Robinson. 1982. *Aeromonas*-associated gastroenteritis. *Lancet*, 2:1304–1306.

Handfield, M., P. Simard, M. Couillard, and R. Letarte. 1996. *Aeromonas hydrophila* isolated from food and drinking water: hemagglutination, hemolysis, and cytotoxicity for a human intestinal cell line (HT-29). *Appl. Environ. Microbiol.*, 62:3459–3461.

Handfield, M., P. Simard, and R. Letarte. 1996. Differential media for quantitative recovery of waterborne *Aeromonas hydrophila*. *Appl. Environ. Microbiol.*, 62:3544–3547.

Hanninen, M.L., P. Oivanen, V. Hirvela-Koski. 1997. *Aeromonas* species in fish, fish-eggs, shrimp and freshwater. *Int. J. Food Microbiol.*, 34:17-26.

Hanninen, M.L., and A. Siitonen. 1995. Distribution of *Aeromonas* phenospecies and genospecies among strains isolated from water, foods or from human clinical samples. *Epidemiol. Infect.*, 115:39–50.

Haque, Q.M., A. Sugiyama, Y. Iwade, Y. Midorikawa, and T. Yamauchi. 1996. Diarrheal and environmental isolates of *Aeromonas* spp. produce a toxin similar to Shiga-like toxin 1. *Curr. Microbiol.*, 32:239–245.

Havelaar, A.H., M. During, and J.F. Versteegh. 1987. Ampicillin-dextrin agar medium for the enumeration of *Aeromonas* species in water by membrane filtration. *J. Appl. Bacteriol.*, 62:279–287.

Havelaar, A.H., F.M. Schets, A. van Silfhout, W.H. Jansen, G. Wieten, and D. van der Kooij. 1992. Typing of *Aeromonas* strains from patients with diarrhoea and from drinking water. *J. Appl. Bacteriol.*, 72:435–444.

Havelaar, A.H., J.F.M. Versteegh, and M. During. 1990. The presence of *Aeromonas* in drinking water supplies in the Netherlands. *Zentralbl. Hyg. Umweltmed.*, 190:236–256.

Holmberg, S.D., W.L. Schell, G.R. Fanning, I.K. Wachsmuth, F.W. Hickman-Brenner, P.A. Blake, D.J. Brenner, and J.J. Farmer III. 1986. *Aeromonas* intestinal infections in the United States. *Ann. Intern. Med.*, 105:683–689.

Janda, J.M. 1999. *Vibrio, Aeromonas* and *Plesiomonas*. In *Topley & Wilson's Microbiology and Microbial Infections*. Edited by W.J. Hausler and M. Sussman. New York: Oxford University Press.

Janda, J.M., and S.L. Abbott. 1998. Evolving concepts regarding the genus *Aeromonas*: an expanding panorama of species, disease presentations, and unanswered questions. *Clin. Infect. Dis.*, 27:332–344.

Janda, J.M., S.L. Abbott, S. Khashe, G.H. Kellogg, and T. Shimada. 1996. Further studies on biochemical characteristics and serologic properties of the genus *Aeromonas*. *J. Clin. Microbiol.*, 34:1930–1933.

Juan, H.J., R.B. Tang, T.C. Wu, and K.W. Yu. 2000. Isolation of *Aeromonas hydrophila* in children with diarrhea. *J. Microbiol. Immunol. Infect.*, 33:115–117.

Kersters, I., G. Huys, H. Van Duffel, M. Vancanneyt, K. Kersters, and W. Verstraete. 1996. Survival potential of *Aeromonas hydrophila* in freshwaters and nutrient-poor waters in comparison to other bacteria. *J. Appl. Bacteriol.*, 60:266–276.

King, G.E., S.B. Werner, and K.W. Kizer. 1992. Epidemiology of *Aeromonas* infections in California. *Clin. Infect. Dis.*, 15:449–452.

Kirov, S.M., L.A. O'Donovan, and K. Sanderson. 1999. Functional characterization of type IV pili expressed on diarrhea-associated isolates of *Aeromonas* species. *Infect. Immun.*, 67:5447–5454.

Knochel, S. 1991. Chlorine resistance of motile *Aeromonas* spp. *Water Sci. Technol.*, 24:327–330.

Ko, W.C., H.C. Lee, Y.C. Chuang, C.C. Liu, and J.J. Wu. 2000. Clinical features and therapeutic implications of 104 episodes of monomicrobial *Aeromonas* bacteraemia. *J. Infect.*, 40:267–273.

Komathi, A.G., S. Ananthan, and S.V. Alavandi. 1998. Incidence & enteropathogenicity of *Aeromonas* spp in children suffering from acute diarrhoea in Chennai. *Indian J. Med. Res.*, 107:252–256.

Krovacek, K., S. Dumontet, E. Eriksson, and S.B. Baloda. 1995. Isolation, and virulence profiles, of *Aeromonas hydrophila* implicated in an outbreak of food poisoning in Sweden. *Microbiol. Immunol.*, 39:655–661.

Kuhn, I., G. Huys, R. Coopman, K. Kersters, and P. Janssen. 1997. A 4-year study of the diversity and persistence of coliforms and *Aeromonas* in the water of a Swedish drinking water well. *Can. J. Microbiol.*, 43:9–16.

Lau, S.M., M.Y. Peng, and F.Y. Chang. 2000. Outcomes of *Aeromonas* bacteremia in patients with different types of underlying disease. *J. Microbiol. Immunol. Infect.*, 33:241–247.

Legnani, P., E. Leoni, F. Soppelsa, and R. Burigo. 1998. The occurrence of *Aeromonas* species in drinking water supplies of an area of the Dolomite Mountains, Italy. *J. Appl. Microbiol.*, 85:271–276.

Marsik, F., and S.L. Werlin. 1984. *Aeromonas hydrophila* colitis in a child. *J. Pediatr. Gastroenterol. Nutr.*, 3:808–811.

Martino, R., L. Gomez, R. Pericas, R. Salazar, C. Sola, J. Sierra, and J. Garau. 2000. Bacteraemia caused by non-glucose-fermenting gram-negative bacilli and *Aeromonas* species in patients with haematological malignancies and solid tumours. *Eur. J. Clin. Microbiol. Infect. Dis.*, 19:320–323.

Massa, S., C. Altieri, and A. D'Angela. 2001. The occurrence of *Aeromonas* spp. in natural mineral water and well water. *Int. J. Food Microbiol.*, 63:169–173.

Medema, G.J., E. Wondergem, A.M. van Dijk-Looyaard, and A.H. Havelaar. 1991. Effectivity of chlorine dioxide to control *Aeromonas* in drinking water distribution systems. *Water Sci. Technol.*, 24:325–326.

Merino, S., X. Rubires, S. Knochel, and J.M. Tomas. 1995. Emerging pathogens: *Aeromonas* spp. *Int. J. Food Microbiol.*, 28:157–168.

Morgan, D.R., P.C. Johnson, H.L. DuPont, T.K. Satterwhite, and L.V. Wood. 1985. Lack of correlation between known virulence properties of *Aeromonas hydrophila* and enteropathogenicity for humans. *Infect. Immun.*, 50:62–65.

Moyer, N.P. 1987. Clinical significance of *Aeromonas* species isolated from patients with diarrhea. *J. Clin. Microbiol.*, 25:2044–2048.

Moyer, N.P., G.M. Luccini, L.A. Holcomb, N.H. Hall, and M. Altwegg. 1992. Application of ribotyping for differentiating aeromonads isolated from clinical and environmental sources. *Appl. Environ. Microbiol.*, 58:1940–1944.

Muñoz, P., V. Fernandez-Baca, T. Pelaez, R. Sanchez, M. Rodriguez-Creixems, and E. Bouza. 1994. *Aeromonas* peritonitis. *Clin. Infect. Dis.*, 18:32–37.

Najm, I., and R.R. Trussell. 1999. New and Emerging Drinking Water Treatment Technologies. In *Identifying Future Drinking Water Contaminants.* Washington, DC: National Academy Press.

National Research Council (NRC). 1983. *Risk Assessment in the Federal Government.* Washington, DC: National Academy Press.

Nichols, G.L. 1996. Fact Sheets on Emerging Waterborne Pathogens: Final Report to the Department of the Environment: *Aeromonas hydrophila.* WRc and Public Health Laboratory Service. DWI4248/1 [Online]. Available: <www.awwarf.com/newprojects/factshts.html>.

Nishikawa, Y., and T. Kishi. 1988. Isolation and characterization of motile *Aeromonas* from human, food and environmental specimens. *Epidemiol. Infect.*, 101:213–223.

Notario, R., N. Borda, T. Gambande, and E. Sutich. 1996. Species and serovars of enteropathogenic agents associated with acute diarrheal disease in Rosario, Argentina. Rev. Inst. Med. Trop. Sao Paulo, 38:5–7.

Obi, C.L., A.O. Coker, J. Epoke, and R.N. Ndip. 1997. Enteric bacterial pathogens in stools of residents of urban and rural regions in Nigeria: a comparison of patients with and without diarrhoea and controls without diarrhoea. *J. Diarrhoeal Dis. Res.*, 15:241–247.

Overman, T.L., and J.M. Janda. 1999. Antimicrobial susceptibility patterns of *Aeromonas jandaei, A. schubertii, A. trota,* and *A. veronii* biotype *veronii. J. Clin. Microbiol.*, 37:706–708.

Pazzaglia, G., J.R. Escalante, R.B. Sack, C. Rocca, and V. Benavides. 1990. Transient intestinal colonization by multiple phenotypes of *Aeromonas* species during the first week of life. *J. Clin. Microbiol.*, 28:1842–1846.

Pazzaglia, G., J. Escamilla, and R. Batchelor. 1991. The etiology of diarrhea among American adults living in Peru. *Mil. Med.*, 156:484–487.

Pazzaglia, G., R.B. Sack, E. Salazar, A. Yi, E. Chea, R. Leon-Barua, C.E. Guerrero, and J. Palomino. 1991. High frequency of coinfecting enteropathogens in *Aeromonas*-associated diarrhea of hospitalized Peruvian infants. *J. Clin. Microbiol.*, 29:1151–1156.

Pettibone, G.W. 1998. Population dynamics of *Aeromonas* spp. in an urban river watershed. *J. Appl. Microbiol.*, 85:723–730.

Picard, B., and P. Goullet. 1987. Seasonal prevalence of nosocomial *Aeromonas hydrophila* infection related to *Aeromonas* in hospital water. *J. Hosp. Infect.*, 10:152–155.

Pitarangsi, C., P. Echeverria, R. Whitmire, C. Tirapat, S. Formal, G.J. Dammin, and M. Tingtalapong. 1982. Enteropathogenicity of *Aeromonas hydrophila* and *Plesiomonas shigelloides*: prevalence among individuals with and without diarrhea in Thailand. *Infect. Immun.*, 35:666–673.

Rautelin, H., M.L. Hanninen, A. Sivonen, U. Turunen, and V. Valtonen. 1995. Chronic diarrhea due to a single strain of *Aeromonas caviae. Eur. J. Clin. Microbiol. Infect. Dis.*, 14:51–53.

Riley, D.E., R.E. Berger, D.C. Miner, and J.N. Krieger. 1998. Diverse and related 16S rRNA-encoding DNA sequences in prostate tissues of men with chronic prostatitis. *J. Clin. Microbiol.*, 36:1646–1652.

Robson, W.L., A.K. Leung, and C.L. Trevenen. 1992. Haemolytic-uraemic syndrome associated with *Aeromonas hydrophila* enterocolitis. *Pediatr. Nephrol.*, 6:221.

Rusin, P.A., J.B. Rose, C.N. Haas, and C.P. Gerba. 1997. Risk assessment of opportunistic bacterial pathogens in drinking water. *Rev. Environ. Contam. Toxicol.*, 152:57–83:57–83.

Samonis, G., S. Maraki, A. Christidou, A. Georgiladakis, and Y. Tselentis. 1997. Bacterial pathogens associated with diarrhoea on the island of Crete. *Eur. J. Epidemiol.*, 13:831–836.

San Joaquin, V.H., and D.A. Pickett. 1988. *Aeromonas*-associated gastroenteritis in children. *Pediatr. Infect. Dis. J.*, 7:53–57.

Schiavano, G.F., F. Bruscolini, A. Albano, and G. Brandi. 1998. Virulence factors in *Aeromonas* spp. and their association with gastrointestinal disease. *New Microbiol.*, 21:23–30.

Sherlock, C.H., D.R. Burdge, and J.A. Smith. 1987. Does *Aeromonas hydrophila* preferentially colonize the bowels of patients with hematologic malignancies? *Diagn. Microbiol. Infect. Dis.*, 7:63–68.

Sisti, M., A. Albano, and G. Brandi. 1998. Bactericidal effect of chlorine on motile *Aeromonas* spp. in drinking water supplies and influence of temperature on disinfection efficacy. *Lett. Appl. Microbiol.*, 26:347–351.

Smith, H.R., and T. Cheasty. 1999. Diarrhoeal Disease Due to *Escherichia coli* and *Aeromonas*. In *Bacterial Infections*. Edited by W.J. Hausler and M. Sussman. New York: Oxford University Press.

Soriano, J.M., H. Rico, J.C. Molto, and J. Manes. 2000. Assessment of the microbiological quality and wash treatments of lettuce served in University restaurants. *Int. J. Food Microbiol.*, 58:123–128.

Talon, D., M.J. Dupont, J. Lesne, M. Thouverez, and Y. Michel-Briand. 1996. Pulsed-field gel electrophoresis as an epidemiological tool for clonal identification of *Aeromonas hydrophila*. *J. Appl. Bacteriol.*, 80:277–282.

Taylor, J.S., and M. Wiesner. 1999. Membranes. In *Water Quality and Treatment: A Handbook of Community Water Supplies*. 5th ed. Edited by American Water Works Association. New York: McGraw-Hill.

Teka, T., A.S. Faruque, M.I. Hossain, and G.J. Fuchs. 1999. *Aeromonas*-associated diarrhoea in Bangladeshi children: clinical and epidemiological characteristics. *Ann. Trop. Paediatr.*, 19:15–20.

Trower, C.J., S. Abo, K.N. Majeed, and M. von Itzstein. 2000. Production of an enterotoxin by a gastro-enteritis-associated *Aeromonas* strain. *J. Med. Microbiol.*, 49:121–126.

Utsalo, S.J., F.O. Eko, O.E. Antia-Obong, and C.U. Nwaigwe. 1995. Aeromonads in acute diarrhoea and asymptomatic infections in Nigerian children. *Eur. J. Epidemiol.*, 11:171–175.

Weber, C.A., S.J. Wertheimer, and A. Ognjan. 1995. *Aeromonas hydrophila*—its implications in freshwater injuries. *J. Foot Ankle Surg.*, 34:442–446.

Wilcox, M.H., A.M. Cook, A. Eley, and R.C. Spencer. 1992. *Aeromonas* spp. as a potential cause of diarrhoea in children. *J. Clin. Pathol.*, 45:959–963.

CHAPTER 7

Caliciviridae in Drinking Water

by Martha A. Embrey

Executive Summary

Problem Formulation

Occurrence of Illness

- Seroprevalence of identified strains is usually high—approaching 100%—by adulthood. Seropositivity does not affect the susceptibility of reinfection, except perhaps in Sapporo-like viruses. Caliciviruses are estimated to be the number one cause of viral gastroenteritis outbreaks in the United States.

Role of Waterborne Exposure

- Waterborne outbreaks of calicivirus have been identified. Researchers estimated that 23% of US waterborne outbreaks between 1975 and 1981 were due to Norwalk-like viruses. Food has been identified as a common outbreak source.

Degree of Morbidity and Mortality

- Clinical symptoms are generally mild and self-limiting: nausea, vomiting, diarrhea, and fever that last for 24–48 hours. Mortality is rare and has occurred mainly in people with preexisting conditions.

Detection Methods in Water/Clinical Specimens

- For environmental samples: RT-PCR, conventional electron microscopy, or immune electron microscopy. For clinical testing: electron microscopy, rEIA, ELISA, and RT-PCR. Calicivirus cannot be cultured in vitro or passaged in animal models, so infectivity cannot be evaluated through available sampling methods.

Mechanisms of Water Contamination

- Human fecal matter/sewage.

Concentrations at Intake

- Unknown, but dependent on level of human fecal contamination in the source water. Seasonal trends have not been shown. Outbreaks have been associated with wet weather events.

Efficacy of Water Treatment

- Indirect virus removal through the capture of host particles with coagulation/flocculation is 90% effective, increasing to levels up to 99% when followed by sand filtration. Generally, virus removal is most effective with a chlorine residual of 0.4 mg/L × 30 minutes. Most recent data indicate that caliciviruses are less resistant to chlorine than other enteric viruses.

Survival/Amplification in Distribution

- Viruses do not multiply outside their human hosts, and the length of time the virus can survive in the environment is unknown.

Routes of Exposure

- Fecal–oral either person-to-person or through a common source such as water or food. Aerosolization (from vomitus) is probable.

Dose Response

Infectious Dose

- An infectious dose of 10 PCR-detectable units has been shown through oral ingestion.

Probability of Illness Based on Infection

- Attack rates in drinking water outbreaks have ranged from 31–87%. Three studies have shown 46%, 68%, and 85% of people developing symptoms after becoming infected. The probability of developing illness may relate to genetic susceptibility in the host.

Efficacy of Medical Treatment

- The illness is self-limiting. Rehydration therapy may be necessary in rare cases.

Secondary Spread

- Can be high in Norwalk-like viruses but apparently uncommon in Sapporo-like viruses.

Chronic Sequelae

- None.

Introduction

The *Caliciviridae* comprises several viruses, many of which are capable of causing outbreaks of viral gastroenteritis in humans. The human caliciviruses are small, circular, single-stranded RNA, approximately 7.2–7.5 kb long and 23–35 nm in diameter. The taxonomy of the *Caliciviridae* family has been evolving. Of the four genera in the family, two are human pathogens—provisionally named Norwalk-like viruses (NLV) and Sapporo-like viruses (SLV). The other two genera are animal viruses. Within the genus NLV, two different genogroups with 5 and 10 genetic clusters are pathogenic to humans (Ando, Noel, and Fankhauser 2000). Genogroup I has Norwalk virus as the prototype, and genogroup II is comprised of agents like Snow Mountain, Camberwell, and Hawaii (Glass et al. 2000). Previously, the viruses were grouped into three genotypes, and many of the epidemiological studies are reported using this classification: genotype 1: Norwalk, Southampton, Desert Shield, and other Norwalk-like viruses; genotype 2: Snow Mountain, Hawaii, Mexico, Toronto, Camberwell, other Norwalk-like viruses; and genotype 3: Sapporo and Sapporo-like viruses. Genotypes 1 and 2 have been combined into the one NLV genus, and genotype 3 is now the SLV genus. The terminology "small round structured viruses" was also used frequently to describe NLVs. More genetic groups will undoubtedly be uncovered as molecular detection methods improve (Farkas et al. 2000; Han et al. 2000).

Epidemiology

Failure to isolate agents from apparently infectious outbreaks of diarrhea and vomiting led to the widely held assumption that undetected viruses were responsible for such disease. In 1972, this virus was initially detected in diarrheal stools obtained from people during an outbreak of gastroenteritis in Norwalk, Ohio, that involved students in an elementary school and family contacts. We now know that this class of viruses commonly causes outbreaks of gastroenteritis all over the world.

General

Data on occurrence in different populations have varied because of the difficulty in distinguishing between the different genogroups and antigenically different strains using the available assays. However, human caliciviruses are believed to be the most common source of nonbacterial gastroenteritis outbreaks in this and other developed countries, accounting for an estimated 5–17% of cases of diarrhea in the community and 5–7% of cases requiring a doctor's treatment (CDC 2001; Hedlund, Rublilar-Abreu, and Svensson 2000; Maguire et al. 1999). Norwalk virus (the originally identified strain) incidence and prevalence has been the best characterized because reliable assays were developed for it before any other strain. Morphological and epidemiological differences exist between the different pathogenic strains (Caul 1996b; Cubitt 1994).

Outbreaks have occurred in families, schools, nursing homes, institutions, the military, and community-wide. No differences in risk have been detected among groups of different sex, race, occupation, or socioeconomic status (Kapikian, Estes, and Chanock 1996). Infection from NLV can occur in all age groups. In developed countries, Norwalk virus typically has been less prevalent in infants and young children, with seroprevalence increasing dramatically in school-aged children (Parker, Cubitt, and Jiang 1995). Children in developing countries tend to become infected earlier in their lives (Jing et al. 2000). Nursing homes have been the source of several calicivirus outbreaks, though it is difficult to know if the elderly are actually more susceptible or if the group setting increases their exposure in an outbreak situation. Green et al. (2002) estimate that NLVs are the number one etiologic agent of gastroenteritis outbreaks in nursing homes.

A French hospital reported a 10-day-long gastrointestinal outbreak in which polymerase chain reaction (PCR) showed the source of the infection most likely to be tap water (Schvoerer et al. 1999). Of 74 nonbacterial gastroenteritis outbreaks reported to the Centers for Disease Control and Prevention (CDC) in the late 1970s, 42% were associated with Norwalk virus, and in another 23%, a provisional association with Norwalk virus could be made (Kaplan et al. 1982b). Fecal samples from nonbacterial gastroenteritis outbreaks in the United States between 1996 and 1997 showed (by reverse transcriptase-polymerase chain reaction [RT-PCR]) NLVs in 86 of the 90 outbreaks (Fankhauser et al. 1998). The outbreaks occurred in nursing homes and hospitals 43% of the time and were traced to contaminated food in 37% of the cases. Between 1991 and 1996, 95 drinking water outbreaks were reported to the CDC (Kramer et al. 1996; Levy et al. 1998; Moore et al. 1993). The etiology of 37 of those remained unknown; however, 20% (20/95) were consistent with a viral cause. Only three outbreaks were confirmed to be viral—two hepatitis A and one small round structured virus. In 96 waterborne outbreaks reported in the United States between 1975 and 1981, Kaplan et al. (1982a) estimated that, based on descriptive epidemiology, 23% could be associated with Norwalk virus. Among 1,041 victims of calicivirus outbreaks in Sweden, the majority were elderly and either hospitalized or living in a nursing home (Hedlund, Rubilar-Abreu, and Svensson 2000). Another drinking water outbreak in Switzerland affected 1,750 residents and was linked through RT-PCR to NLV strains (Häfliger, Hübner, and Lüthy 2000). The village water source had been contaminated with sewage.

Norwalk virus was once prevalent in the United States and elsewhere (Cubitt et al. 1994), but the prevailing antigenic type of calicivirus has shifted. Today, NLV of genogroup II, like the Camberwell strains, have become the most predominant type of human calicivirus seen in the United States and other developed countries, but circulating strains vary each year (Caul 1996b; Green 1997; Green et al. 2002; Jiang et al. 1996; Wright et al. 1998).

SLV (what had been labeled genotype 3 or sometimes classic human calicivirus) has a different epidemiological profile than NLV; it is also more closely related to animal caliciviruses than other known types (Nakata et al. 1996; Dinulos and Matson 1994). These viruses are almost always seen in infants, very small children (< age 5), and the elderly and not as often in older children or adults. This pattern suggests long-term immunity after this virus is contracted in early childhood that may occasionally wane in the aged (Matsui and Greenberg 2000). In addition, data from rotavirus vaccine trials suggest that some mechanism present in the rotavirus vaccine reduces the severity of SLV infection but has no effect on NLV infection (Pang et al. 2001). Though SLVs have been circulating in certain populations for up to 20 years, occurrence data are less commonly found in the literature. Data from the United Kingdom indicate that 6.5% of all positive diarrhea samples between 1990 and 1995 (n = 90,405) were due to calicivirus infection and that 0.8% of the overall samples were due to classic calicivirus (i.e., SLV) (Caul 1996b). The Sapporo strain was isolated from the feces of 2.9% of sick children in day-care facilities in Houston (n = 375). No Sapporo virus was found in 86 samples from asymptomatic children in that population (Matson et al. 1989).

Clinical

Clinical symptoms of calicivirus infection are generally mild and self-limiting. Symptoms include nausea, vomiting (often explosive), diarrhea, abdominal cramps, fever, chills, and lethargy (Kapikian, Estes, and Chanock 1996). Vomiting occurs more often in children; adults suffer more diarrhea (CDC 2001). Mean onset occurs between 15 and 50 hours, with a mean incubation period of 24 to 48 hours. The duration of the illness is approximately 12 to 24 hours. Viral shedding has generally been thought to follow the course of infection, but recent studies have shown prolonged shedding up to 20 days past infection (Moe et al. 1999). It appears that NLV-related illness may be more severe, with vomiting as the primary symptom, compared to SLV, which is a milder, diarrheic disease (de Wit et al. 2001; Pang et al. 2001). Rehydration management rarely includes the need for intravenous fluids, though hospitalizations for severe dehydration have been reported in both the middle aged and the elderly. In a Finnish study of 148 patients with calicivirus infection (age 2 months to 2 years), 28% needed oral rehydration and 1.4% were hospitalized (Pang, Joensuu, and Vesikari 1999). Calicivirus is not considered an important source of severe gastroenteritis in children in either developed or developing countries (Caul 1996a; Kapikian, Estes, and Chanock 1996), and NLV infection is probably frequently asymptomatic in very young children (Talal et al. 2000). Mortality is generally rare unless preexisting conditions are present. Deaths have been associated with nursing home and hospital outbreaks (Cunney et al. 2000; Kaplan et al. 1982a, 1982b; Marx et al. 1999).

Sensitive Subpopulations

Norwalk virus has been found in the stools of HIV-positive people but it does not appear to be any more frequent or severe in this population compared with the immunocompetent. As mentioned earlier, outbreaks of NLV often occur among the elderly, but these outbreaks may result from group exposure rather than an increased biological susceptibility. In the case of SLV infection, the elderly may become more susceptible, if their immunity wanes, than other adults who still have immunity from childhood infection. Researchers studying the role of NLV outbreaks in Maryland nursing homes estimated that 80% of the gastroenteritis outbreaks during one winter season were caused by NLVs (Green et al. 2002), making them a serious health problem in this already frail population.

Evidence for genetic susceptibility exists based on volunteer studies showing that some people tend to keep developing illness after multiple challenges and some never develop illness after multiple challenges (Blacklow, Herrmann, and Cubitt 1987; Moe et al. 1999; Parrino et al. 1977). In work recently presented by Moe et al. (1999), volunteers with anti-Norwalk IgG were more likely to become infected and with lower doses than those without preexisting antibody. Because positive serum antibody levels seem to be unrelated to the risk of illness, it is possible that there is a genetically determined receptor that influences the ability of the virus to colonize the intestine or some other genetically determined protective factor.

Table 7-1 summarizes selected studies that looked at the prevalence of caliciviruses in populations around the world. Note that various assays specific to different strains are performed on both serum and stool samples; consequently, these studies are difficult to compare with each other.

Chronic Sequelae

No chronic sequelae are known to occur from human calicivirus infections.

Transmission

The mode of transmission for human caliciviruses is fecal–oral, either person-to-person or via ingestion of contaminated water (drinking and recreational) and food. Fomites and aerosolized vomit may also facilitate spread (CDC 2001). Until recently, contaminated drinking water had been identified as the source of exposure using descriptive epidemiology rather than laboratory data because of insufficient environmental detection methods (Green 1997). However, molecular detection techniques have linked caliciviruses to many common sources of outbreaks, such as food, water, and food handlers (Glass et al. 2000). Kapikian and colleagues (1996) estimated that approximately one quarter of the 96 waterborne outbreaks reported to the CDC between 1975 to 1981 were related to Norwalk virus. Other sources of human calicivirus outbreaks have included raw shellfish, lettuce, fruit salad, potato salad, cole slaw, melon, celery, cold meats, bakery products, cooked ham, sandwiches, and commercial ice (Caul 1996b; Green 1997). In New York, between 1980 and 1994, 339 seafood-related outbreaks were reported (Wallace et al. 1999). Of those with a confirmed etiology (n = 148), NLVs accounted for 42% of the outbreaks and 42% of the illnesses. Three multistate outbreaks of NLV were traced to oysters harvested in Louisiana (Berg et al. 2000). In an incident of secondary food contamination, more than 100 people in Wales got sick from custard that had been prepared with contaminated water (Brugha et al. 1999). Aerosol transmission from vomitus, especially in group settings, might be an additional route of exposure (Caul 1996a; Kapikian, Estes, and Chanock 1996; Marks et al. 2000).

Case reports suggest that certain animal calicivirus strains can cross species and infect humans, but data are limited (Smith et al. 1998). A recently published study found that through PCR detection methods, 33/75 veal calf fecal samples and 2/100 swine fecal samples were positive for NLV RNA, raising the suggestion of zoonotic transmission (van der Poel et al. 2000). Caliciviruses related to NLVs were found in calves from 45% of the dairy farms tested in The Netherlands (Koopmans et al. 2000). Zoonotic

Table 7-1: Calicivirus prevalence studies

Reference/Location	Sample	Results/Assay
Bon et al. 1999 (France)	Stool samples from 414 children with GE	61% of total virus-related = rotavirus, 14% = HuCv (Genotype 2), 6% = astrovirus, 3% = adenovirus 40/41 (ELISA, RT-PCR)
Caul 1996b (England/Wales)	90,405 stool samples identified with enteric viruses from 1990–1995	83% of total = rotavirus, 8.2% = adenovirus, 8.8% = *Caliciviridae* → of 8.8% (n = 7,947): SRSV = 64.8%, HuCV = 8.8%, astrovirus = 26.4%
Cubitt and Jiang 1996 (London)	206 stool samples of hospitalized children	0.5% (1/206) = MxV, 0/206 = NV (rEIA)
Cubitt, Green, and Payment 1998 (Canada)	566 sera samples; ages 9–79	65–100% = HaV (increased with age), 53–100% = NV
Dimitrov et al. 1997 (Kuwait)	433 serum samples; ages 0–100	98% overall = NV; highest prevalence in the 50–79 age group (rEIA); 96% = MxV
	151 serum samples from foreign workers	98% = NV; 95% = MxV

NOTE: Genotypes 1 and 2 now comprise the genera of Norwalk-like viruses (genogroups I and II); genotype 3 comprises Sapporo-like viruses.

KEY:

ELISA: Enzyme-linked immunosorbent assay
EM: Electron microscopy
GE: Gastroenteritis
HaV: Hawaii virus
HuCv: Human calicivirus (old terminology for SLV)
MxV: Mexico virus
NLV: Norwalk-like virus
NV: Norwalk virus
rEIA: Recombinant enzyme immunoassay
RT-PCR: Reverse-transcriptase polymerase chain reaction
SapV: Sapporo virus
SLV: Sapporo-like virus
SRSV: Small round structured virus (old terminology for NLV)

Table continued on next page.

Table 7-1: Calicivirus prevalence studies (continued)

Reference/Location	Sample	Results/Assay
Evans et al. 1998 (England/Wales)	1,568 outbreaks; 40,000+ people	43% of outbreaks associated with SRSV; 64% of all outbreaks associated with person-to-person transmission (mostly SRSVs occurring in residential homes/hospitals); 22% of all outbreaks reported as foodborne
Fankhauser et al. 1998 (USA)	90 outbreaks	96% of outbreaks associated with NLV (RT-PCR); 43% in nursing homes/hospitals; 26% restaurants or catered meals; 37% identified as foodborne
Gray et al. 1993 (England)	3,250 serum samples; all ages	73.3% = NV, increasing with age; 24.6% = 6–11 months; 89.7% = 60+ years
Hedlund, Rubilar-Abreu, and Svensson 2000 (Sweden)	3,700 stool samples from 676 outbreaks; all ages	89% of 455 viral outbreaks = NLV; 8 outbreaks = SLV (RT-PCR); 66 outbreaks identified as food or water etiology; 60% of patients = 70–90 years.
Honma et al. 1998 (Japan/Southeast Asia)	155 stool samples from sick children <10 (Japan)	1.3% (2/155) = MxV (ELISA)
	Sera prevalence in adults (Japan and Southeast Asia)	MxV prevalence low until it reaches 50% at school age; adolescence = 80%; adults = 82–88% (Assay of serum may detect greater range of strains than stool samples?)
Iritani et al. 2000 (Japan)	350 stool samples from 64 outbreaks; all ages	52% (182/350) samples from 73% (47/64) outbreaks were NLV (RT-PCR)
Jiang et al. 1995a (USA)	1,200 stool samples of children with diarrhea in day care over 1-year surveillance	2% (24/1,200) = MxV (rELISA, RT-PCR)
Jiang et al. 1995b (Mexico)	Serum samples from 200 children monitored from birth to 2 years	85% = NV by age 2 (rEIA)
Jing et al. 2000 (Beijing)	1,109 serum samples; all ages	Overall 89% = NV, 91% = MxV; lowest seroprevalence at 7–11 months; almost 100% prevalence by 8–9 years for both strains (EIA)

Table continued on next page.

Table 7-1: Calicivirus prevalence studies (continued)

Reference/Location	Sample	Results/Assay
Kirkwood and Bishop 2001 (Australia)	354 children admitted to hospital with acute GE	9% (32/354) = NLV; 0.6% (2/354) = SLV (RT-PCR of stool)
Lew et al. 1994 (Finland)	Serum samples of 159 children tested over 2-year surveillance	49% incidence over surveillance (rELISA); Probability of infection related to IgG titer—antibody to NV could be protective?
Maguire et al. 1999 (UK)	550 stool samples from 94 outbreaks	68% of outbreaks = SRSV; 98% = Genotype 2 (98% of those = Grimsby) (EM, RT-PCR)
Matson et al. 1989 (USA)	375 stool samples of children with diarrhea in day care	2.9% (11/375) = SapV (ELISA (screen), EM (confirm) Incidence rate = ½ of rotavirus; higher than *Campylobacter, Salmonella*, and *Shigella*
	86 stool samples of well children in day care	0 = SapV
Myrmel et al. 1996 (Norway)	1,017 serum samples of military recruits	29.5% = NV
Nakata et al. 1998 (Kenya)	1,431 stool samples from sick children <6 from a clinic	2.2% (32/1,431) = SapV, 0.1% (1/1,186) = NV, 0% (0/246) = MxV (rEIA)
	193 serum samples	Acquisition of all viruses occurred by 1–2 years of age; NV = 60%; MxV and SapV = 90% adults
Nakayama et al. 1996 (Japan)	209 stool samples of non-RV, nonbacterial GE from hospital out-patient clinics	12% = NLV (RT-PCR)
	378 stool samples of all GE patients from hospital outpatient clinics	7% = NLV
	17 stool samples from outbreaks	15/17 = NLV; Genotype 2 was most commonly identified (overall)

Table continued on next page.

Table 7-1: Calicivirus prevalence studies (continued)

Reference/Location	Sample	Results/Assay
Numata et al. 1994 (Japan)	159 stool samples of children <10 with acute GE from SRSV	0.6% (1/159) = NV (rELISA)
Pang et al. 2000 (Finland)	1,477 stool samples from children with GE; 2 months–2 years	20% = NLV; 9% = SLV (RT-PCR)
Parker, Cubitt, and Jiang 1995 (London)	338 serum samples of children 0–16	70% had MxV infection by 2 years, 12% had NV infection by 2 years (rEIA); children had maternal antibodies for first months of life
Payment, Franco, and Fout 1994 (Canada)	Serum samples from ages 9–60+ over 15-month surveillance	55% = NV at 9–19 years, 79% = NV at 20–39 years, 87% = NV at 40–49 years, 84% = NV at 50–59 years 100% = NV at 60+ years (EIA); incidence of 33% during study period
Pelosi et al. 1999 (Italy)	1,729 serum samples	91.2% = Genotype 2 (Lordsdale); 28.7% = Genotype 1 (Southampton) (ELISA, RT-PCR)
Pujol et al. 1998 (Venezuela)	1,120 fecal samples of children with diarrhea	0.4% (4/1,120) = NV (rEIA)
	Serum samples from urban, rural, and Amerindian populations	47–53% urban = NV, 83% rural = NV, 73–93% Amerindian = NV; 50% of children seropositive by age 5
Saito et al. 1998 (Japan)	119 stool samples with sporadic diarrhea	44% sporadic cases = Yuri (Genotype 2)
	46 stool samples from outbreaks	52% outbreak cases = Yuri (RT-PCR)

Table continued on next page.

Table 7-1: Calicivirus prevalence studies (continued)

Reference/Location	Sample	Results/Assay
Smit et al. 1997 (South Africa)	Family-based cohort and antenatal clinic cohort	96–99% = NV and MxV in both cohorts (rEIA)
	Infant and children cohort	100% of children had adult levels of NV and MxV antibodies
	276 stool samples of infants and children with GE	1.8% = NV, 4.3% = MxV (rEIA)
Taylor et al. 1996 (South Africa)	Serum samples of cohort of European and African descent; all ages	37% = NV at 7–11 months (rEIA); 62% = NV at 40 years; equal prevalence for European and African descent
Wolfaardt et al. 1997 (South Africa)	1,296 stool samples with sporadic GE	3.3% (43/1,296) = HuCv; of 3.3%, 81% = MxV-like, 8% = NVL; 11% = SapV-like (EM, rEIA, RT-PCR)
Wright et al. 1998 (Australia)	6,226 stool samples of ill patients of all ages	3.6% (223/6,226) = SRSV, 0.15% (9/6,226) = classical HuCv (EM; RT-PCR); SRSVs prevalent in all ages, frequently associated with nursing homes and hospitals; based on RT-PCR analysis, Genotypes 1 and 3 were rare; of Genotype 2, Mexico, Toronto, Lordsdale, and Camberwell were found; the latter was most common.

exposure is not likely to be a major route of transmission but worth further investigation.

Secondary spread has occurred among family members during community foodborne and waterborne outbreaks and among people in hospital, day-care, or nursing home settings and even during a football game (Becker et al. 2000; Benenson 1995; Cubitt 1994; Kapikian, Estes, and Chanock 1996). Seven of 38 outbreaks associated with Norwalk gastroenteritis were thought to have possible secondary cases (Kapikian, Estes, and Chanock 1996), but the source of the exposure was not specific in these outbreaks. In a nursing home outbreak, no infection point source was identified, but two major risk factors were exposure to ill residents or ill household members (Marx et al. 1999). Secondary attack rates in the literature have been 32% and 11% for Norwalk and Snow Mountain outbreaks, respectively (Kapikian, Estes, and Chanock 1996; Morens et al. 1979). Agents with low infectious dose and prolonged viral shedding increase the probability of secondary spread. Although NLV genogroups have been associated with relatively high secondary spread, secondary spread in SLVs appears to be negligible (Caul 1996b). This supports the hypothesis that people develop immunity to this genogroup after being infected in early childhood; whereas, infection with NLV produces very short-term or no immunity to later infection.

Dose Response

Because it is such a common cause of gastrointestinal outbreaks, researchers have assumed that NLVs have a low infective dose, but historically, difficulty in isolating the virus has made data difficult to obtain.

Infective Dose

The infective dose of Norwalk virus was derived from volunteer studies. Illness was shown to be induced orally with a $10^{-4.7}$ dilution of a Norwalk virus stool suspension (Kapikian, Estes, and Chanock 1996). Others estimate that only 10 to 100 viral units are necessary to infect a susceptible host (Caul 1996b; Cubitt 1994; Glass et al. 2000). This was recently proved in volunteers who became infected after ingesting 10 PCR-detectable units (Moe et al. 1999). Although the characteristic projectile vomiting might cause aerosol transmission, the oral infective dose failed to cause illness in adult volunteers challenged intranasally (Dolin et al. 1971, 1972).

Probability of Illness Based on Infection

A longitudinal study of Norwalk virus incidence in college students showed that 46% of those who became infected showed symptoms of illness (Johnson et al. 1990). In an experimental study of 50 volunteers challenged with Norwalk virus, 82% became infected and 68% of these infected people developed symptoms (Graham et al. 1994). In a recent volunteer dose study, 44 people were challenged with Norwalk virus, 20 became infected, and 17 (85%) developed symptoms (Moe et al. 1999). The presence of preexisting antibodies in any of the subjects did not protect them from acquiring infection; in fact, in Moe's report, people challenged with Norwalk virus who had preexisting antibody were more likely to become infected.

Data from all kinds of outbreaks have demonstrated attack rates ranging from 24 to 100% over different strains and different age groups (Cubitt 1994). Primary attack rates for drinking water associated outbreaks have ranged from 31 to 87% (Cannon et al. 1991; Goodman et al. 1982; Lawson et al. 1991; Taylor, Gary, and Greenberg 1981; Wilson et al. 1982).

Detection Methods

Caliciviruses cannot be isolated in cell culture or animal models, which has slowed the acquisition of data on incidence and prevalence. In recent years, RT-PCR, genogroup-specific enzyme-linked immunosorbent assays (ELISAs), and new recombinant enzyme immunoassays (rEIA) have contributed to the molecular-level knowledge of human caliciviruses. However, many classifications are still considered preliminary, and no doubt, many have yet to be identified (Green et al. 2000). Continued development of assays for type-specific antibody detection is important if researchers are to characterize the transmission and occurrence of calicivirus.

Detection of calicivirus in the environment is confined to RT-PCR and conventional electron microscopy (EM) or immune electron microscopy (IEM). Since EM requires 10^5 to 10^6 particles/gram for detection (Cubitt et al. 1994), RT-PCR, which can confirm a positive test based on a single virus particle, is the most efficient technique for environmental use. Molecular detection is really the only sensitive method available for environmental samples, but there is no standardized methodology available. Nevertheless, RT-PCR is costly, both financially and in sensitivity, but techniques to improve environmental detection are in development (Huang et al. 2000; Loisy et al. 2000; Myrmel, Rimstad, and Wasteson 2000). Also, PCR can only detect the genetic material of the virus, not whether it is viable or infective. The major obstacle in detection methods is the genetic diversity of the genogroups (Cubitt 1996; Dinulos and Matson 1994). Specific PCR primers have been developed to detect different genogroups, and the development of new primers continues to be the subject of laboratory research (Schwab et al. 2001). More than 100 sequences are available, and many more are under development (Hardy 1999).

Environmental Occurrence

Human calicivirus concentrations in US source waters are currently unknown. Nevertheless, because viruses do not multiply outside their host environment, their frequency distribution in source water is dependent on both

- the initial load discharged in the water and
- the ability of the virus to survive (De Zuane 1997).

Further, since the route of transmission is fecal–oral, Hunter (1997) suggested that the distribution of the initial load is comparable to the distribution of human fecal contamination in the source water. Groundwater and surface water survey research, however, has been unable to statistically identify surrogate indicators such as fecal or total coliform counts for any of the viruses in the *Caliciviridae.* Sewage in The Netherlands contained numerous Norwalk-like genetic clusters at levels up to 10^7 particles/L. RT-PCR confirmed that samples from area patients matched samples found in the sewage (Lodder et al. 1999). Norwalk viruses have also been found in canal waters throughout the Florida Keys, estuarine recreational water, and river water (Griffin et al. 1999; Wyn-Jones et al. 2000).

Many outbreaks of NLV have occurred in the winter (Mounts et al. 2000), but Norwalk-related outbreaks associated with contaminated drinking water have occurred in both the summer and winter seasons (Cannon et al. 1991; Kapikian, Estes, and Chanock 1996; Goodman et al. 1982; Lawson et al. 1991; Taylor, Gary, and Greenberg 1981; Wilson et al. 1982). Heavy rains were found to contaminate wells in the Georgia Norwalk outbreak in 1982 (Goodman et al. 1982) and the Philadelphia and Delaware Norwalk outbreaks in 1987 (Cannon et al. 1991). Investigators speculated that runoff from nearby septic tanks and a pig farm was the source of contamination in these outbreaks.

Water Treatment

Current filtration methods may not be adequate to remove viruses from the source water due to their small size (23 to 35 nm). However, indirect virus removal through the capture of host organisms or clusters of virus-colloidal matter using coagulation–flocculation and settling has been found to be 90% effective under controlled circumstances. This removal rate was exceeded (90 to 99%) when coagulation and sedimentation were followed by sand filtration (De Zuane 1997). Thus, virus removal may fail due to

- high turbidity levels,
- accumulation of virus clusters around the filter,
- lack of timely or controlled backwashing procedures, or
- high overflow rates in the settling basins since the floc is still infectious.

Two Norwalk-related outbreaks, one at an elementary school in Washington state (Taylor, Gary, and Greenberg 1981) and one at an Arizona resort (Lawson et al. 1991), occurred in conjunction with chlorination system failure and wastewater overflows.

At low turbidity levels, disinfection methods such as chlorination, bromination, iodination, and ozonation are considered the most effective method for virus inactivation from drinking water (De Zuane 1997). However, chemical concentration and contact time must be at optimal levels to inactivate the virus particles. The National Academy of Sciences found that virus removal by chlorine, for instance, was most effective with a free chlorine residual level of 0.4 mg/L and a 30-minute contact time (NRC 1977). Nevertheless, a prospective epidemiological study by Payment and colleagues (1994) found no statistical difference between the attack rates of Norwalk infections in a group of volunteers who drank tap water and a group of volunteers (ages 9 to 60+) who drank tap water filtered through a reverse-osmosis filtration unit. Data from outbreaks and a volunteer study indicated that Norwalk virus might be more resistant to chlorine than other viruses like poliovirus and rotavirus (Keswick et al. 1985). Keswick and colleagues found that 3.75 ppm of free chlorine did not inactivate Norwalk virus—evidenced by five of eight volunteers seroconverting after being challenged with treated virus. However, data from recent studies show that the human caliciviruses are less resistant than other enteric viruses to inactivation methods such as free chlorine and ozonolysis (Shin, Battigelli, and Sobsey 1999). Virus inactivation has also been noted during the water-softening process using lime and a constant pH level of 11 (De Zuane 1997).

Therefore, traditional drinking water treatment methods with either chlorination or ozonation may be adequate to kill human caliciviruses, but further research is necessary.

Membrane filtration technology, such as micro-, ultra-, and nanofiltration, as well as reverse osmosis, has the capacity to remove water contaminants down to the ion level—greater than conventional water treatment. Ultrafiltration, with a nominal pore size of 0.01 μm, can achieve 4-log virus removal or greater (Najm and Trussell 1999; Taylor and Wiesner 1999). This technology will continue to gain in popularity as costs become more competitive and regulatory approval is assured.

Little is known about the effect of residual chlorine or the presence of biofilm on viruses in the distribution system. Since viruses do not multiply outside their human hosts, amplification does not occur in the distribution system. Because detection methods for environmental testing are poor, it is difficult to assess the effect of posttreatment contamination during distribution.

Conclusions

Human caliciviruses are the most common cause of nonbacterial gastroenteritis outbreaks in the United States; however, though waterborne outbreaks of caliciviruses have been identified, they are probably more frequently transmitted through food and person-to-person means. Little is known about their occurrence in drinking water, but adequately treated water is probably a less important route of exposure.

Bibliography

Ando, T., J.S. Noel, and R.L. Fankhauser. 2000. Genetic classification of "Norwalk-like viruses." *J. Infect. Dis.*, 181(Suppl)2:S336–S348.

Becker, K.M., C.L. Moe, K.L. Southwick, and J.N. MacCormack. 2000. Transmission of Norwalk virus during football game. *N. Engl. J. Med.*, 343:1223–1227.

Benenson, A.S., ed. 1995. *Control of Communicable Diseases Manual.* 16th ed. Washington, DC: American Public Health Association.

Berg, D.E., M.A. Kohn, T.A. Farley, and L.M. McFarland. 2000. Multi-state outbreaks of acute gastroenteritis traced to fecal-contaminated oysters harvested in Louisiana. *J. Infect. Dis.*, 181(Suppl)2:S381–S386.

Blacklow, N.R., J.E. Herrmann, and W.D. Cubitt. 1987. Immunobiology of Norwalk virus. In *Novel Diarrhoea Viruses.* Chichester, England: John Wiley & Sons.

Bon, F., P. Fascia, M. Dauvergne, D. Tenenbaum, H. Planson, A.M. Petion, P. Pothier, and E. Kohli. 1999. Prevalence of group A rotavirus, human calicivirus, astrovirus, and adenovirus type 40 and 41 infections among children with acute gastroenteritis in Dijon, France. *J. Clin. Microbiol.*, 37:3055–3058.

Brugha, R., I.B. Vipond, M.R. Evans, Q.D. Sandifer, R.J. Roberts, R.L. Salmon, E.O. Caul, and A.K. Mukerjee. 1999. A community outbreak of food-borne small round-structured virus gastroenteritis caused by a contaminated water supply. *Epidemiol. Infect.*, 122:145–154.

Cannon, R.O., J.R. Poliner, R.B. Hirschhorn, D.C. Rodeheaver, P.R. Silverman, E.A. Brown, G.H. Talbot, S.E. Stine, S.S. Monroe, and D.T. Dennis. 1991. A multistate outbreak of Norwalk virus gastroenteritis associated with consumption of commercial ice. *J. Infect. Dis.*, 164:860–863.

Caul, E.O. 1996a. Viral gastroenteritis: small round structured viruses, caliciviruses and astroviruses. Part I. The clinical and diagnostic perspective. *J. Clin. Pathol.*, 49:874–880.

Caul, E.O. 1996b. Viral gastroenteritis: small round structured viruses, caliciviruses and astroviruses. Part II. The epidemiological perspective. *J. Clin. Pathol.*, 49:959–964.

Centers for Disease Control and Prevention (CDC). 2001. Norwalk-like viruses: Public health consequences and outbreak management. *MMWR*, 50:1–17.

Cubitt, W.D. 1994. Caliciviruses. In *Viral Infections of the Gastrointestinal Tract.* Edited by A.Z. Kapikian. New York: Marcel Dekker.

Cubitt, W.D. 1996. Historical background and classification of caliciviruses and astroviruses. *Arch. Virol.*, 12:225–235.

Cubitt, W.D., K.Y. Green, and P. Payment. 1998. Prevalence of antibodies to the Hawaii strain of human calicivirus as measured by a recombinant protein based immunoassay. *J. Med. Virol.*, 54:135–139.

Cubitt, W.D., and X. Jiang. 1996. Study on occurrence of human calicivirus (Mexico strain) as cause of sporadic cases and outbreaks of calicivirus-associated diarrhoea in the United Kingdom, 1983–1995. *J. Med. Virol.*, 48:273–277.

Cubitt, W.D., X.J. Jiang, J. Wang, and M.K. Estes. 1994. Sequence similarity of human caliciviruses and small round structured viruses. *J. Med. Virol.*, 43:252–258.

Cunney, R.J., P. Costigan, E.B. McNamara, B. Hayes, E. Creamer, M. LaFoy, N.A. Ansari, and N.E. Smyth. 2000. Investigation of an outbreak of gastroenteritis caused by Norwalk-like virus, using solid-phase immune electron microscopy. *J. Hosp. Infect.*, 44:113–118.

de Wit, M.A., M.P. Koopmans, L.M. Kortbeek, N.J. van Leeuwen, J. Vinje, and Y.T. Duynhoven. 2001. Etiology of gastroenteritis in sentinel general practices in the Netherlands. *Clin. Infect. Dis.*, 33:280–288.

De Zuane, J. 1997. Microbiological Parameters. In *Handbook of Drinking Water Quality*. New York: Van Nostrand Reinhold.

Dimitrov, D.H., S.A. Dashti, J.M. Ball, E. Bishbishi, K. Alsaeid, X. Jiang, and M.K. Estes. 1997. Prevalence of antibodies to human caliciviruses (HuCVs) in Kuwait established by ELISA using baculovirus-expressed capsid antigens representing two genogroups of HuCVs. *J. Med. Virol.*, 51:115–118.

Dinulos, M.B., and D.O. Matson. 1994. Recent developments with human caliciviruses. *Pediatr. Infect. Dis. J.*, 13:998–1003.

Dolin, R., N.R. Blacklow, H. DuPont, R.F. Buscho, R.G. Wyatt, J.A. Kasel, R. Hornick, and R.M. Chanock. 1972. Biological properties of Norwalk agent of acute infectious nonbacterial gastroenteritis. *Proc. Soc. Exp. Biol. Med.*, 140:578–583.

Dolin, R., N.R. Blacklow, H. DuPont, S. Formal, R.F. Buscho, J.A. Kasel, R.P. Chames, R. Hornick, and R.M. Chanock. 1971. Transmission of acute infectious nonbacterial gastroenteritis to volunteers by oral administration of stool filtrates. *J. Infect. Dis.*, 123:307–312.

Evans, H.S., P. Madden, C. Douglas, G.K. Adak, S.J. O'Brien, T. Djuretic, P.G. Wall, and R. Stanwell-Smith. 1998. General outbreaks of infectious intestinal disease in England and Wales: 1995 and 1996. *Commun. Dis. Public Health*, 1:165–171.

Fankhauser, R.L., J.S. Noel, S.S. Monroe, T. Ando, and R.I. Glass. 1998. Molecular epidemiology of "Norwalk-like viruses" in outbreaks of gastroenteritis in the United States. *J. Infect. Dis.*, 178:1571–1578.

Farkas, T., X. Jiang, M.L. Guerrero, W. Zhong, N. Wilton, T. Berke, D.O. Matson, L.K. Pickering, and G. Ruiz-Palacios. 2000. Prevalence and genetic diversity of human caliciviruses (HuCVs) in Mexican children. *J. Med. Virol.*, 62:217–223.

Glass, R.I., J. Noel, T. Ando, R. Fankhauser, G. Belliot, A. Mounts, U.D. Parashar, J.S. Bresee, and S.S. Monroe. 2000. The epidemiology of enteric caliciviruses from humans: a reassessment using new diagnostics. *J. Infect. Dis.*, 181(Suppl)2:S254–S261.

Goodman, R.A., J.W. Buehler, H.B. Greenberg, T.W. McKinley, and J.D. Smith. 1982. Norwalk gastroenteritis associated with a water system in a rural Georgia community. *Arch. Environ. Health*, 37:358–360.

Graham, D.Y., X. Jiang, T. Tanaka, A.R. Opekun, H.P. Madore, and M.K. Estes. 1994. Norwalk virus infection of volunteers: new insights based on improved assays. *J. Infect. Dis.*, 170:34–43.

Gray, J.J., X. Jiang, P. Morgan-Capner, U. Desselberger, and M.K. Estes. 1993. Prevalence of antibodies to Norwalk virus in England: detection by enzyme-linked immunosorbent assay using baculovirus-expressed Norwalk virus capsid antigen. *J. Clin. Microbiol.*, 31:1022–1025.

Green, K.Y. 1997. The role of human caliciviruses in epidemic gastroenteritis. *Arch. Virol. Suppl.*, 13:153–65.

Green, K.Y., T. Ando, M.S. Balayan, T. Berke, I.N. Clarke, M.K. Estes, D.O. Matson, S. Nakata, J.D. Neill, M.J. Studdert, and H.J. Thiel. 2000. Taxonomy of the caliciviruses. *J. Infect. Dis.*, 181(Suppl)2:S322–S330.

Green, K.Y., G. Belliot, J.L. Taylor, J. Valdesuso, J.F. Lew, A.Z. Kapikian, and F.-Y.C. Lin. 2002. A predominant role for Norwalk-like viruses as agents of epidemic gastroenteritis in Maryland nursing homes for the elderly. *J. Infect. Dis.*, 185:133–146.

Griffin, D.W., C.J. Gibson, E.K. Lipp, K. Riley, J.H. Paul, and J.B. Rose. 1999. Detection of viral pathogens by reverse transcriptase PCR and of microbial indicators by standard methods in the canals of the Florida Keys. *Appl. Environ. Microbiol.*, 65:4118–4125.

Häfliger, D., P. Hübner, and J. Lüthy. 2000. Outbreak of viral gastroenteritis due to sewage-contaminated drinking water. *Int. J. Food Microbiol.*, 54:123–126.

Han, D.P., H.W. Lee, J.H. Sohn, B.I. Yeh, J.W. Choi, and H.W. Kim. 2000. The new genotypic human calicivirus isolated in Seoul. *Exp. Mol. Med.*, 32:6–11.

Hardy, M.E. 1999. Norwalk and "Norwalk-like viruses" in epidemic gastroenteritis. *Clin. Lab. Med.*, 19:675–690.

Hedlund, K.O., E. Rubilar-Abreu, and L. Svensson. 2000. Epidemiology of calicivirus infections in Sweden, 1994–1998. *J. Infect. Dis.*, 181(Suppl)2:S275–S280.

Honma, S., S. Nakata, K. Numata, K. Kogawa, T. Yamashita, M. Oseto, X. Jiang, and S. Chiba. 1998. Epidemiological study of prevalence of genogroup II human calicivirus (Mexico virus) infections in Japan and Southeast Asia as determined by enzyme-linked immunosorbent assays. *J. Clin. Microbiol.*, 36:2481–2484.

Huang, P.W., D. Laborde, V.R. Land, D.O. Matson, A.W. Smith, and X. Jiang. 2000. Concentration and detection of caliciviruses in water samples by reverse transcription-PCR. *Appl. Environ. Microbiol.*, 66:4383–4388.

Hunter, P.R. 1997. Viral Gastroenteritis. In *Waterborne Disease: Epidemiology and Ecology*. Chichester, England: John Wiley & Sons.

Iritani, N., Y. Seto, K. Haruki, M. Kimura, M. Ayata, and H. Ogura. 2000. Major change in the predominant type of "Norwalk-like viruses" in outbreaks of acute nonbacterial gastroenteritis in Osaka City, Japan, between April 1996 and March 1999. *J. Clin. Microbiol.*, 38:2649–2654.

Jiang, X., D. Cubitt, J. Hu, X. Dai, J. Treanor, D.O. Matson, and L.K. Pickering. 1995a. Development of an ELISA to detect MX virus, a human calicivirus in the snow Mountain agent genogroup. *J. Gen. Virol.*, 76:2739–2747.

Jiang, X., D.O. Matson, W.D. Cubitt, and M.K. Estes. 1996. Genetic and antigenic diversity of human calicivirus (HuCVs) using RT-PCR and new EIAs. *Arch. Virol.*, 12:251–262.

Jiang, X., D.O. Matson, F.R. Velazquez, J.J. Calva, W.M. Zhong, J. Hu, G.M. Ruiz-Palacios, and L.K. Pickering. 1995b. Study of Norwalk-related viruses in Mexican children. *J. Med. Virol.*, 47:309–316.

Jing, Y., Y. Qian, Y. Huo, L.P. Wang, and X. Jiang. 2000. Seroprevalence against Norwalk-like human caliciviruses in Beijing, China. *J. Med. Virol.*, 60:97–101.

Johnson, P.C., J.J. Mathewson, H.L. DuPont, and H.B. Greenberg. 1990. Multiple-challenge study of host susceptibility to Norwalk gastroenteritis in US adults. *J. Infect. Dis.*, 161:18–21.

Kapikian, A.Z., M.K. Estes, and R.M. Chanock. 1996. Norwalk Group of Viruses. In *Fundamental Virology*. Edited by B.N. Fields, D.M. Knipe, and P.M. Howley. Philadelphia: Lippincott-Raven.

Kaplan, J.E., R. Feldman, D.S. Campbell, C. Lookabaugh, and G.W. Gary. 1982a. The frequency of a Norwalk-like pattern of illness in outbreaks of acute gastroenteritis. *Am. J. Public Health,* 72:1329–1332.

Kaplan, J.E., G.W. Gary, R.C. Baron, N. Singh, L.B. Schonberger, R. Feldman, and H.B. Greenberg. 1982b. Epidemiology of Norwalk gastroenteritis and the role of Norwalk virus in outbreaks of acute nonbacterial gastroenteritis. *Ann. Intern. Med.,* 96:756–761.

Keswick, B.H., T.K. Satterwhite, P.C. Johnson, H.L. DuPont, S.L. Secor, J.A. Bitsura, G.W. Gary, and J.C. Hoff. 1985. Inactivation of Norwalk virus in drinking water by chlorine. *Appl. Environ. Microbiol.,* 50:261–264.

Kirkwood, C.D., and R.F. Bishop. 2001. Molecular detection of human calicivirus in young children hospitalized with acute gastroenteritis in Melbourne, Australia, during 1999. *J. Clin. Microbiol.,* 39:2722–2724.

Koopmans, M., J. Vinjé, M. de Wit, I. Leenen, W. van der Poel, and Y. van Duynhoven. 2000. Molecular epidemiology of human enteric caliciviruses in The Netherlands. *J. Infect. Dis.,* 181(Suppl)2:S262–S269.

Kramer, M.H., B.L. Herwaldt, G.F. Craun, R.L. Calderon, and D.D. Juranek. 1996. Surveillance for waterborne-disease outbreaks—United States, 1993–1994. *MMWR,* 45:1–33.

Lawson, H.W., M.M. Braun, R.I. Glass, S.E. Stine, S.S. Monroe, H.K. Atrash, L.E. Lee, and S.J. Englender. 1991. Waterborne outbreak of Norwalk virus gastroenteritis at a southwest US resort: role of geological formations in contamination of well water. *Lancet,* 337:1200–1204.

Levy, D.A., M.S. Bens, G.F. Craun, R.L. Calderon, and B.L. Herwaldt. 1998. Surveillance for waterborne-disease outbreaks—United States, 1995–1996. *MMWR,* 47:1–34.

Lew, J.F., J. Valdesuso, T. Vesikari, A.Z. Kapikian, X. Jiang, M.K. Estes, and K.Y. Green. 1994. Detection of Norwalk virus or Norwalk-like virus infections in Finnish infants and young children. *J. Infect. Dis.,* 169:1364–1367.

Lodder, W.J., J. Vinjé, R. van De Heidi, A.M. de Roda Husman, E.J. Leenen, and M.P.G. Koopmans. 1999. Molecular detection of Norwalk-like caliciviruses in sewage. *Appl. Environ. Microbiol.,* 65:5624–5627.

Loisy, F., P. Le Cann, M. Pommepuy, and F. Le Guyader. 2000. An improved method for the detection of Norwalk-like caliciviruses in environmental samples. *Lett. Appl. Microbiol.,* 31:411–415.

Maguire, A.J., J. Green, D.W. Brown, U. Desselberger, and J.J. Gray. 1999. Molecular epidemiology of outbreaks of gastroenteritis associated with small round-structured viruses in East Anglia, United Kingdom, during the 1996–1997 season. *J. Clin. Microbiol.*, 37:81–89.

Marks, P.J., I.B. Vipond, D. Carlisle, D. Deakin, R.E. Fey, and E.O. Caul. 2000. Evidence for airborne transmission of Norwalk-like virus (NLV) in a hotel restaurant. *Epidemiol. Infect.*, 124:481–487.

Marx, A., D.K. Shay, J.S. Noel, C. Brage, J.S. Bresee, S. Lipsky, S.S. Monroe, T. Ando, C.D. Humphrey, E.R. Alexander, and R.I. Glass. 1999. An outbreak of acute gastroenteritis in a geriatric long-term-care facility: combined application epidemiological and molecular diagnostic methods. *Infect. Control Hosp. Epidemiol.*, 20:306–311.

Matson, D.O., M.K. Estes, R.I. Glass, A.V. Bartlett, M. Penaranda, E. Calomeni, T. Tanaka, S. Nakata, and S. Chiba. 1989. Human calicivirus-associated diarrhea in children attending day care centers. *J. Infect. Dis.*, 159:71–78.

Matsui, S.M., and H.B. Greenberg. 2000. Immunity to calicivirus infection. *J. Infect. Dis.*, 181(Suppl)2:S331–S335.

Moe, C., D. Rhodes, S. Pusek, F. Tseng, W. Heizer, C. Kapoor, B. Gilliam, P. Stewart, M. Harb, S. Miller, M. Sobsey, J. Herrmann, N. Blacklow, and R. Calderon. 1999. Determination of Norwalk virus dose–response in human volunteers. Presented at Health Effects Stakeholder Meeting for the Stage 2 DBPR and LT2ESWTR, February 12, 1999. Washington, DC.

Moore, A.C., B.L. Herwaldt, G.F. Craun, R.L. Calderon, A.K. Highsmith, and D.D. Juranek. 1993. Surveillance for waterborne disease outbreaks—United States, 1991–1992. *MMWR*, 42:1–22.

Morens, D.M., R.M. Zweighaft, T.M. Vernon, G.W. Gary, J.J. Eslien, B.T. Wood, R.C. Holman, and R. Dolin. 1979. A waterborne outbreak of gastroenteritis with secondary person-to-person spread. Association with a viral agent. *Lancet*, 1:964–966.

Mounts, A.W., T. Ando, M. Koopmans, J.S. Bresee, J. Noel, and R.I. Glass. 2000. Cold weather seasonality of gastroenteritis associated with Norwalk-like viruses. *J. Infect. Dis.*, 181(Suppl)2:S284–S287.

Myrmel, M., E. Rimstad, and Y. Wasteson. 2000. Immunomagnetic separation of a Norwalk-like virus (genogroup I) in artificially contaminated environmental water samples. *Int. J. Food Microbiol.*, 62:17–26.

Myrmel, M., E. Rimstad, M. Estes, E. Skjerve, and Y. Wasteson. 1996. Prevalence of serum antibodies to Norwalk virus among Norwegian military recruits. *Int. J. Food Microbiol.*, 29:233–240.

Najm, I., and R.R. Trussell. 1999. New and emerging drinking water treatment technologies. In *Identifying Future Drinking Water Contaminants.* Washington, DC: National Academy Press.

Nakata, S., S. Honma, K. Numata, K. Kogawa, S. Ukae, N. Adachi, X. Jiang, M.K. Estes, Z. Gatheru, P.M. Tukei, and S. Chiba. 1998. Prevalence of human calicivirus infections in Kenya as determined by enzyme immunoassays for three genogroups of the virus. *J. Clin. Microbiol.*, 36:3160–3163.

Nakata, S., K. Kogawa, K. Numata, S. Ukae, N. Adachi, D.O. Matson, M.K. Estes, and S. Chiba. 1996. The epidemiology of human calicivirus/ Sapporo/82/Japan. *Arch. Virol.*, 12(Suppl):263–270.

Nakayama, M., Y. Ueda, H. Kawamoto, Y. Han-jun, K. Saito, O. Nishio, and H. Ushijima. 1996. Detection and sequencing of Norwalk-like viruses from stool samples in Japan using reverse transcription-polymerase chain reaction amplification. *Microbiol. Immunol.*, 40:317–320.

National Research Council (NRC). 1977. *Drinking Water and Health,* Vol. 1. Washington, DC: National Academy of Sciences.

Numata, K., S. Nakata, X. Jiang, M.K. Estes, and S. Chiba. 1994. Epidemiological study of Norwalk virus infections in Japan and Southeast Asia by enzyme-linked immunosorbent assays with Norwalk virus capsid protein produced by the baculovirus expression system. *J. Clin. Microbiol.*, 32:121–126.

Pang, X.L., S. Honma, S. Nakata, and T. Vesikari. 2000. Human caliciviruses in acute gastroenteritis of young children in the community. *J. Infect. Dis.*, 181(Suppl)2:S288–S294.

Pang, X.L., J. Joensuu, and T. Vesikari. 1999. Human calicivirus-associated sporadic gastroenteritis in Finnish children less than two years of age followed prospectively during a rotavirus vaccine trial. *Pediatr. Infect. Dis. J.*, 18:420–426.

Pang, X.L., S.Q. Zeng, S. Honma, S. Nakata, and T. Vesikari. 2001. Effect of rotavirus vaccine on Sapporo virus gastroenteritis in Finnish infants. *Pediatr. Infect. Dis. J.*, 20:295–300.

Parker, S.P., W.D. Cubitt, and X. Jiang. 1995. Enzyme immunoassay using baculovirus-expressed human calicivirus (Mexico) for the measurement of IgG responses and determining its seroprevalence in London, UK. *J. Med. Virol.*, 46:194–200.

Parrino, T.A., D.S. Schreiber, J.S. Trier, A.Z. Kapikian, and N.R. Blacklow. 1977. Clinical immunity in acute gastroenteritis caused by Norwalk agent. *N. Engl. J. Med.*, 297:86–89.

Payment, P., E. Franco, and G.S. Fout. 1994. Incidence of Norwalk virus infections during a prospective epidemiological study of drinking water-related gastrointestinal illness. *Can. J. Microbiol.*, 40:805–809.

Pelosi, E., P.R. Lambden, E.O. Caul, B. Liu, K. Dingle, Y. Deng, and I.N. Clarke. 1999. The seroepidemiology of genogroup 1 and genogroup 2 Norwalk-like viruses in Italy. *J. Med. Virol.*, 58:93–99.

Pujol, F.H., G. Vasquez, A.M. Rojas, M.E. Fuenmayor, C.L. Loureiro, I. Perez-Schael, M.K. Estes, and F. Liprandi. 1998. Norwalk virus infection in Venezuela. *Ann. Trop. Med. Parasitol.*, 92:205–211.

Saito, H., S. Saito, K. Kamada, S. Harata, H. Sato, M. Morita, and Y. Miyajima. 1998. Application of RT-PCR designed from the sequence of the local SRSV strain to the screening in viral gastroenteritis outbreaks. *Microbiol. Immunol.*, 42:439–446.

Schvoerer, E., F. Bonnet, V. Dubois, A.M. Rogues, J.P. Gachie, M.E. Lafon, and H.J. Fleury. 1999. A hospital outbreak of gastroenteritis possibly related to the contamination of tap water by a small round structured virus. *J. Hosp. Infect.*, 43:149–154.

Schwab, K.J., F.H. Neill, F. Le Guyader, M.K. Estes, and R.L. Atmar. 2001. Development of a reverse transcription-PCR-DNA enzyme immunoassay for detection of "Norwalk-like" viruses and hepatitis A virus in stool and shellfish. *Appl. Environ. Microbiol.*, 67:742–749.

Shin, G., D. Battigelli, and M.D. Sobsey. 1999. Inactivation of Norwalk virus by free chlorine disinfection of water. Poster Q-131 presented at ASM 99th General Meeting, Chicago, IL. May 30–June 3, 1999.

Smit, T.K., A.D. Steele, I. Peenze, X. Jiang, and M.K. Estes. 1997. Study of Norwalk virus and Mexico virus infections at Ga-Rankuwa Hospital, Ga-Rankuwa, South Africa. *J. Clin. Microbiol.*, 35:2381–2385.

Smith, A.W., D.E. Skilling, N. Cherry, J.H. Mead, and D.O. Matson. 1998. Calicivirus emergence from ocean reservoirs: zoonotic and interspecies movements. *Emerg. Infect. Dis.*, 4:13–20.

Talal, A.H., C.L. Moe, A.A. Lima, K.A. Weigle, L. Barrett, S.I. Bangdiwala, M.K. Estes, and R.L. Guerrant. 2000. Seroprevalence and seroincidence of Norwalk-like virus infection among Brazilian infants and children. *J. Med. Virol.*, 61:117–124.

Taylor, J.W., G.W. Gary Jr., and H.B. Greenberg. 1981. Norwalk-related viral gastroenteritis due to contaminated drinking water. *Am. J. Epidemiol.*, 114:584–592.

Taylor, M.B., S. Parker, W.O. Grabow, and W.D. Cubitt. 1996. An epidemiological investigation of Norwalk virus infection in South Africa. *Epidemiol. Infect.*, 116:203–206.

Taylor, J.S., and M. Wiesner. 1999. Membranes. In *Water Quality and Treatment: A Handbook of Community Water Supplies.* 5th ed. Edited by American Water Works Association. New York: McGraw-Hill.

van der Poel, W.H.M., J. Vinjé, R. van der Heide, M. Herrera, A. Vivo, and M.P.G. Koopmans. 2000. Norwalk-like calicivirus genes in farm animals. *Emerg. Infect., Dis.*, 6:36–41.

Wallace, B.J., J.J. Guzewich, M. Cambridge, S.F. Altekruse, and D.L. Morse. 1999. Seafood-associated disease outbreaks in New York, 1980–1994. *Am. J. Prev. Med.*, 17:48–54.

Wilson, R., L.J. Anderson, R.C. Holman, G.W. Gary, and H.B. Greenberg. 1982. Waterborne gastroenteritis due to the Norwalk agent: clinical and epidemiologic investigation. *Am. J. Public Health*, 72:72–74.

Wolfaardt, M., M.B. Taylor, H.F. Booysen, L. Engelbrecht, W.O. Grabow, and X. Jiang. 1997. Incidence of human calicivirus and rotavirus infection in patients with gastroenteritis in South Africa. *J. Med. Virol.*, 51:290–296.

Wright, P.J., I.C. Gunesekere, J.C. Doultree, and J.A. Marshall. 1998. Small round-structured (Norwalk-like) viruses and classical human caliciviruses in southeastern Australia, 1980–1996. *J. Med. Virol.*, 55:312–320.

Wyn-Jones, A.P., R. Pallin, C. Dedoussis, J. Shore, and J. Sellwood. 2000. The detection of small round-structured viruses in water and environmental materials. *J. Virol. Methods*, 87:99–107.

CHAPTER 8

Coxsackievirus in Drinking Water

By Martha A. Embrey

Executive Summary

Problem Formulation

Occurrence of Infection

- CDC surveillance reports showed that of 3,209 positive nonpolio enterovirus samples reported between 1993 and 1996, the predominant coxsackievirus species were B5 (11.5%), B1 (6.6%), and B2 (6.2%). Of the 18,000 enteroviruses reported in the United States between 1970 and 1979, 24% were coxsackieviruses B1–B5. WHO's surveillance between 1975 and 1983 found that 10% of nonpolio enterovirus infections were coxsackievirus group A and 25% were group B.

Role of Waterborne Exposure

- Coxsackieviruses are found in all types of water: lakes, rivers, marine environments, wastewater, and chlorinated drinking water. Theoretically, coxsackievirus can have a waterborne transmission; however, little evidence supports this. Most outbreaks appear to result from person-to-person spread.

Degree of Morbidity and Mortality

- Coxsackievirus primarily causes meningitis, unspecified febrile illness, hand-foot-and-mouth disease, and conjunctivitis. Most disease resolves on its own within a week of infection.

- Susceptibles such as newborns can have a fatality rate of up to 10%, but otherwise, fatality is uncommon.

Detection Methods in Water/Clinical Specimens

- Isolation of coxsackieviruses in cell culture, followed by serotyping by neutralization is still the "gold standard" of detection. Most coxsackievirus group A species are difficult to culture and are traditionally detected using a newborn mouse model. Molecular methods like PCR are being used increasingly in clinical and environmental research and although it is more sensitive, it is not standardly used in either type of research. PCR is appealing for clinical applications because of its rapid results; however, for environmental samples (e.g., drinking water), culturing is the only way to show infectivity.

Mechanisms of Water Contamination

- Coxsackieviruses are shed in human fecal matter, which is the only route of water contamination.

Concentrations at Intake

- Coxsackieviruses are found in all types of water sources—from sewage to drinking water. Enterovirus concentrations reported in sewage have varied, up to 13,000 pfu/L and in drinking water up to 0.0006 most probable number of infective units/L.

Efficacy of Water Treatment

- Standard water treatment with chlorine disinfectant decreases concentrations significantly but not necessarily by 100%. Free chlorine of 0.3–0.5 ppm reportedly results in rapid inactivation.
- Water temperature, contact time, and level of organic material in the source water are all related to disinfection efficacy.
- Membrane filtration technology has the potential of removing virtually all viral pathogens.

Survival/Amplification in Distribution

- Not applicable in viruses, though survival within biofilms is an untested possibility.

Routes of Exposure

- Fecal–oral: primarily person-to-person directly or via fomites.
- Theoretically through water or food ingestion; aerosolization.

Dose Response

Infectious Dose

- The best study available on infectious doses of enteroviruses was a volunteer study performed with a species of echovirus. The study of 149 volunteers ingesting echovirus 12 concluded that 919 pfu are necessary to infect 50% of an exposed population, and estimated 17 pfu are necessary to infect 1% of an exposed population. The infection rate was dose related up to 10,000 pfu, but only for infection—not time to or length of viral shedding or the development of illness.

Probability of Illness Based on Infection

- The probability of illness based on infection varies widely by serotype and the host's age and antibody status. Overall, a large percentage of those infected are asymptomatic. In a prospective study, 21% of neonates who were infected with nonpolio enterovirus in their first month became ill; serotype had a large influence on the incidence of illness. An outbreak of coxsackie-induced pleurodynia in teenagers resulted in 20% of those exposed reporting illness.

Efficacy of Medical Treatment

- No antiviral drugs effectively prevent or treat coxsackievirus infections. Gammaglobulin therapy and exchange transfusions have been used with limited success in the critically ill. Traditional drug therapies are used to treat the effects of enterovirus-related heart disease but not the infection itself.

Secondary Spread

- Secondary spread of coxsackievirus infections is common. One source estimated the attack rate in family members of an index case to be 76% for coxsackievirus. Viral shedding generally lasts from a week to a month. This relatively long period increases the probability of secondary spread. Frequently, secondary infection is asymptomatic.

Chronic Sequelae

- Outcomes from coxsackievirus infections are mostly acute. Data is equivocal on whether severe central nervous system effects (e.g., seizure) in young children can cause cognitive, behavioral, or neurological sequelae later in life. The largest and most well designed study to date showed no effect. A large cohort study recently associated childhood enteroviral central nervous system infections (especially with coxsackievirus B5) with adult-onset schizophrenia. One study suggests that placental infection can lead to serious problems in newborns including severe mental defects and seizures.

- Coxsackievirus infection in children has been a suggested precursor to insulin-dependent diabetes.
- Coxsackievirus B5 is a major cause of viral myocarditis, which is primarily acute; however, myocarditis can become chronic and infection is also associated with chronic dilated cardiomyopathy.

Introduction

The *Picornaviridae* family comprises five genera including enterovirus, hepatovirus, rhinovirus, cardiovirus, and aphthovirus. Along with polioviruses 1–3, echoviruses 1–9, 11–27, and 29–34, and enteroviruses 68–71, coxsackieviruses A1–22, A24, and B1–6 comprise the major enterovirus phylogenetic groups. The first three groups were divided subjectively, and since 1970, new species have simply been allocated an enterovirus number, starting with 68. With the advent of cell culturing in the late 1940s, enteroviruses were isolated from patients suffering a wide variety of syndromes, yet it was unclear whether the association was causal. Coxsackieviruses were originally isolated from patients with suspected poliomyelitis in Coxsackie, New York.

Much of our knowledge of enteroviruses comes from the extensive study of poliovirus, and indeed, much of the information in the literature is presented as representative of the enterovirus genus as a whole and not specifically to the individual virus groups, such as echovirus or coxsackievirus. Multiple viruses can cause many of the same syndromes, though certain diseases are primarily caused by one species or group. Even though this chapter focuses on coxsackievirus, it presents a significant amount of data on enteroviruses as a whole, which encompass all the phylogenetic groups, including coxsackievirus.

The enteric viruses replicate in the alimentary tract (hence the name); however, they generally do not cause significant gastrointestinal symptoms. Infection is mostly asymptomatic (Minor 1998; Modlin 1986), but disease outcomes can range from fatal central nervous system complications to mild, unspecified febrile illness. Specific disease syndromes include paralytic poliomyelitis, hand-foot-and-mouth disease, encephalitis, aseptic meningitis, myocarditis, conjunctivitis, and respiratory illness, among others. Generally, certain serotypes are associated with particular symptoms or syndromes, although this is not exclusive. For example, coxsackievirus is most often associated with myocarditis, but other enteroviruses can occasionally cause myocarditis.

Epidemiology

Enterovirus infections occur both sporadically and epidemically within a population, and certain strains tend to follow one or the other pattern (CDC 2000). Some tend to show up regularly in populations (e.g., coxsackievirus B5 and echovirus 9), while others, like B1, surface for a season or two then disappear (CDC 1997; Modlin 1986; Moore et al. 1984). Babies have a high incidence rate of enterovirus serotypes that are circulating at the time of their birth. On the other hand, older children (5–10 years old) can be the most affected age group in a population when an uncommon serotype emerges because of their lack of exposure and immunity during infancy (Gondo et al. 1995; Modlin 1986; Rotbart 1995b; Yamashita et al. 1994).

Incidence and Prevalence

A large, prospective study looking at enterovirus incidence in newborns was published in 1984 (Jenista, Powell, and Menegus 1984). The authors tested 666 newborns and 629 mothers who were not excreting nonpolio enteroviruses within the first day after delivery. After hospital discharge, the babies were tested one to four times a week for a month using culture specimens. The incidence of enterovirus infection within that first month of life was 12.8% with 79% of the infants asymptomatic. The overall infection prevalence at the end of the testing period was 5.3%. Twenty-one percent of the babies who tested positive had to return to the hospital, compared with 4% of the infants overall. Risk factors associated with infection in this cohort were low socioeconomic status and lack of breastfeeding (both $p < 0.0001$). The only factors that influenced whether or not infection produced symptoms were the virus serotype (e.g., 0% for echovirus 22 and 54% for echovirus 5) and the presence of maternal antibodies.

A seroepidemiology study of 1,944 children <15 years of age showed that 92% had antibodies against at least one of the six strains of coxsackievirus B virus tested (Leogrande 1991). The positive response rates fluctuated with the age of the children. They were high at birth and in the first few months of life (97%), decreased to 82% between 6 months and 1 year, and steadily rose again to 98% at 10–11 years. Following that, there was a large decrease to 82% from 11–13 years. The most common antibodies in the 1,944 samples were against B2 (54%), B5 (53%), and B1 (51%). B6 had the lowest prevalence at 31%.

Enterovirus Surveillance

Surveillance systems, by nature, skew reports of disease toward including cases with more serious complications. Therefore, surveillance of enteroviral disease overestimates the true incidence of serious outcomes, especially related to the central nervous system (Morens and Pallansch 1995). Also, the detection of enterovirus in a stool sample does not necessarily correspond with disease, because fecal shedding occurs for a long time after infection.

The World Heath Organization (WHO) has had a viral infection surveillance program since; 1963 (Minor 1998). The data are subject to the foibles of passive reporting; however, some useful information on the prevalence of various clinical outcomes emerges. WHO reported almost 60,000 nonpolio enterovirus infections between 1975 and 1983. Of the total, 65% (n = 38,191) were attributed to echovirus, 10% to coxsackievirus A (n = 5,787), 25% to coxsackievirus B (n = 14,934), and 0.4% to enteroviruses 68–71 (Minor 1998). Overall, coxsackievirus B4 was associated with the widest range of syndromes; the most commonly reported was A9, which is also the easiest to detect in cell culture (Minor 1998). The predominating coxsackievirus serotypes were A9, A16, and B4, which are commonly associated with skin and central nervous system infections. Almost 80% of the reported coxsackievirus A and B occurred in the three youngest age groups: <1 year, 1–4 years, and 5–14 years, concentrating most in the 1–4 year group (Minor 1998). The United Kingdom's enteroviral prevalence picture was almost identical. Between 1975 and 1994, 61% of reported nonpolio enterovirus infections were attributed to echovirus, 29% to coxsackievirus B, and 10% to coxsackievirus A (Maguire et al. 1999). Isolation was highest in infants between 1 and 2 months old.

The US Centers for Disease Control and Prevention (CDC) has maintained a surveillance database for enteroviruses since 1951, and it estimates that every year in the United States, 30 million nonpolio enterovirus infections cause an unknown number of cases of aseptic meningitis, hand-foot-and-mouth disease, and upper-respiratory disease (CDC 1997). In their most recent report of nonpolio enterovirus infections (1997–1998), states reported a total of 1,700 nonpolio isolates (CDC 2000). Of those isolates, the predominant coxsackievirus serotypes were B1–3 and A9. Only 25.3% of the reports included a clinical diagnosis: aseptic meningitis (37.6%), encephalitis (4.1%), pneumonia or other respiratory disease (9.3%), and paralysis (0.2%).

Canada has a broader surveillance program that tracks viral and selected nonviral infections in Canada (Wilson and Weber 1995). In 1993, 38 laboratories contributed data from 66,447 virus-positive samples (from 1.44 million total viral and select antiviral samples). Coxsackievirus B1 and B4 showed dramatic decreases in 1993 reporting compared with previous years: 101 to 8 and 103 to 6 from 1992 and 1991, respectively.

In the 20-year period between 1975 and 1994, laboratories in England and Wales reported more than 40,000 culture-confirmed samples of enteroviruses (Maguire et al. 1999). Sixty-one percent were echoviruses, 29% were coxsackievirus B, and 10% were coxsackievirus A. Supporting the seasonal nature of enteroviral infections, 71% of the reports were from July through mid-December.

Outbreaks

Many outbreaks associated with enterovirus and specifically coxsackievirus have been reported—usually in the form of aseptic meningitis, acute hemorrhagic conjunctivitis, and hand-foot-and-mouth disease. Following is a summary of several of these reports:

- In 1991, a high school football team suffered an outbreak of pleurodynia that was attributed to coxsackievirus B1 (Ikeda et al. 1993). Twenty percent of the players ($n = 17$) reported illness. The risk factors for illness centered around the use of common water containers, including eating ice cubes (RR 9.2) and drinking water from the shared cooler (RR 6.3). It is possible that one infected player contaminated the common water supply and spread the infection to other players.
- The island of St. Croix in the US Virgin Islands experienced an outbreak of acute hemorrhagic conjunctivitis in September and October of 1998 (CDC 1998). By the end of the outbreak period, 1,051 cases had been identified. The mean age of the patients was 13.5 years (range: 3.5 months–81 years). Thirty-eight percent of the cases were between 6 and 17 years; 40% were 18 years or older; the rest were between 0 and 5 years. The preliminary testing indicated that coxsackievirus A24 variant was responsible for the outbreak. The A24 variant has also been responsible for pandemics of acute hemorrhagic conjunctivitis in other areas around the world, mainly Southeast Asia and India. However, by the mid-1980s, epidemics had spread widely to other areas (Kosrirukvongs et al. 1996; Morens and Pallansch 1995; Nayak et al. 1996).
- Coxsackievirus A16 caused an outbreak of hand-foot-and-mouth disease in England and Wales in 1994 (Bendig and Fleming 1996). Most of the 952 identified cases were between 1 and 4 years. The predominant symptoms included rashes primarily on hands but also on feet and in mouths. Most cases were mild, and secondary spread among family members was rare.
- Taiwan had a huge enteroviral epidemic in 1998. Several enteroviruses circulated during the epidemic, but enterovirus 71 caused most of the serious sequelae and almost 100 deaths

(Ho et al. 1999). When 64 patients with coxsackievirus A16 were compared with 120 patients with enterovirus 71 in a children's hospital, 32% of the enterovirus 71 patients had complications compared with 6% of the other patients (Chang et al. 1999). Fourteen patients died and 5% had continuing sequelae—all infected with enterovirus 71.

Clinical

Coxsackie and other enteroviruses are usually acquired through the fecal–oral route, where they infect the gastrointestinal tract after ingestion. The consequent minor viremia can seed many different organ systems, resulting in a variety of clinical manifestations affecting the central nervous system, liver, lungs, or heart.

Aseptic Meningitis

Aseptic meningitis refers to meningitis of a nonbacterial nature and is most commonly caused by enterovirus infections. Meningitis is more typically seen in children <1 year old (Rorabaugh et al. 1993) and is associated with the summer months (Coyle 1999). Echoviruses and coxsackievirus group B are more frequently associated with meningitis than the other enteroviruses (Rorabaugh et al. 1993; Rotbart 1995b). The symptoms are usually acute and consist of fever, headache, stiff neck, and irritability. The self-limited illness typically lasts less than a week with complete recovery in a matter of weeks. Adult patients will occasionally continue reporting symptoms for several weeks (Rotbart 1995a). In a cohort study of 52 young children (<2 years) with meningitis, 9.0% developed more advanced neurological complications such as seizures and coma, the rest recovered without incident. The complications were more commonly seen in children >3 months old and were equally attributable to echovirus and coxsackievirus group B (Rorabaugh et al. 1993). The neonate who develops meningitis symptoms in the first few days of life is at extreme risk for serious morbidity and death (possibly up to a 10% mortality rate) (Dagan 1996; Rotbart 1995a). Newborns who die as a result of these early enterovirus infections usually succumb not to central nervous system complications but to liver failure in the case of echovirus or myocarditis from coxsackievirus (Modlin 2000; Rotbart 1995b). Group B coxsackievirus serotypes 2–5 are frequently associated with overwhelming systemic neonatal infections (Modlin 2000). Enterovirus meningitis that occurs in patients who are beyond the neonatal stage rarely results in these types of severe outcomes (Rotbart 1995a). Studies have been conducted to test the long-term sequelae on children who had meningitis infection early in life (see section on Chronic Sequelae).

Febrile Illness

The most common clinical manifestation of an enterovirus infection is unspecified febrile illness with or without rashes (Rotbart and Hayden 2000). Other symptoms that may appear are vomiting, rash, and respiratory illness. All enterovirus types can cause this type of febrile illness, but the prevalence varies among the different types and serotypes (Dagan 1996). The incubation period ranges from <2 days to 2 weeks, and symptoms in young children last from <24 hours to longer than a week, though the usual course is 3–4 days (Dagan 1996; Modlin 2000). Usually more than one enterovirus serotype is circulating in the population during the season, so patients can suffer from more than one febrile episode caused by different serotypes (Dagan and Menegus 1995). Because it is rarely reported, little is known about the presentation of nonspecific febrile illness in older children and adults.

Myocarditis

Coxsackievirus group B, especially B3, is one of the most commonly known viral infectants of heart muscle. Infections frequently go undiagnosed, so the true incidence of the disease is unknown. However, it is estimated that one third to one half of myocarditis infections and around one third of dilated cardiomyopathy cases are caused by enteroviruses (Martino et al. 1995) (Table 8-1). Of 12,747 routine autopsies in Sweden, 1% reportedly had evidence of myocarditis (Friman et al. 1995). Dilated cardiomyopathy has an estimated incidence of between 1 and 10 cases per year per 100,000 people (Martino et al. 1995). It is estimated that up to 5% of all symptomatic coxsackievirus (primarily group B) infections induce heart infections and disease (Martino et al. 1995; Minor 1998). A significant proportion of infections are chronic but subclinical, which can lead to dilated cardiomyopathy, the number one cause of heart transplantation worldwide. Over a person's lifetime, repeated exposures to enteroviruses increase the chance of acquiring an infection with cardiac symptoms. It has been suggested that 70% of the overall population has been exposed to cardiotropic viruses—half of which resulted in primarily subclinical myocarditis (O'Connell in Martino et al. 1995). The virus initially can cause an acute reaction, but an overactive immune response can lead to chronic disease that is autoimmune in nature (Caforio and McKenna 1996; Suddaby 1996). In children, it can appear as acute with heart failure. Adults appear to be more prone to develop chronic dilated cardiomyopathy (Martino et al. 1995). Those with persistent ventricular dysfunction face a 20% one-year mortality rate (Liu et al. 1996).

Table 8-1: Risk factors for enterovirus-associated heart disease

Factor	Myocarditis	Dilated Cardiomyopathy
Age	Yes 20–39 years; infants <6 months	Yes 40+ years
Gender	Yes Male	Yes Male
Race	No	No
Geography	No	Possibly increased in underdeveloped and tropical countries

Carthy et al. (1997) suggested that coxsackievirus B3 alters the immune response, which delays viral clearance after an acute infection. Host and genetic factors can influence susceptibility of disease persistence and progression. Treatment is best if it is early and sustained. It is generally supportive in nature, though immunosuppressive and antiviral therapies are sometimes used (Liu et al. 1996).

The causal link between enteroviral myocarditis and dilated cardiomyopathy has been elucidated by the results of recent breakthroughs in molecular detection and analysis (Bowles and Towbin 1998; Kawai 1999). Recently, reverse transcriptase-polymerase chain reaction (RT-PCR) has been used to study enteroviral genetic sequences in patients with infectious myocarditis complications. In 35 patients with suspected heart muscle disease, 42.9% of cardiac biopsy samples were coxsackievirus B3 positive compared with 7.1% in patients with other cardiac conditions (OR = 9.75) (Archard et al. 1998). Using RT-PCR methodology, tissue from 19 patients with dilated cardiomyopathy and 14 with chronic coronary disease were tested for enterovirus. Eleven of 19 (58%) in the first group and 8 of 14 (57%) in the latter group were positive compared with no positives in either the 35 heart-healthy control group or in the 33 controls with myocardial infarction (Andreoletti et al. 1996).

Baboonian and Treasure (1997) performed a meta-analysis of data that used molecular techniques to analyze the relationship between acute myocarditis and dilated cardiomyopathy with enterovirus infection. Eleven of 12 studies found an infection relationship with acute myocarditis (OR = 4.4) and 11 of 17 were positive for dilated cardiomyopathy (OR = 3.8). The latter association was weaker than the first but still positive overall (Baboonian and Treasure 1997). Fujioka and colleagues (2000) tested specimens from 26 patients with dilated cardiomyopathy for a variety of viruses. Only coxsackievirus B were found in nine samples, and seven of those were determined to be active infections versus latent persistence. Six of those seven patients died within 6 months. (A summary of molecular detection studies can be found in Martino et al. 1995.)

Pleurodynia

Pleurodynia, or Bornholm disease, is a type of muscle inflammation that is characterized by stabbing pain in the muscles of the chest and abdomen. It is caused principally by coxsackievirus group B. It is seen more often in older children and young adults and is occasionally seen in epidemic form (White and Fenner 1994). Though the associated pain can be severe, pleurodynia resolves completely without sequelae.

Respiratory Disease

Worldwide, enteroviruses account for approximately 2 to 15% of viral respiratory disease (Chonmaitree and Mann 1995). After unspecified febrile illness, respiratory illness is the second most common syndrome associated with the enteroviruses (Chonmaitree and Mann 1995). Most enteroviruses can cause a respiratory illness (the "common cold") that clinically cannot be distinguished from rhinovirus, respiratory syncytial virus, or adenovirus infection (Minor 1998; Modlin 2000). One identifying feature is that respiratory disease from enterovirus occurs primarily during the summer months, whereas the others are predominantly winter illnesses (Chonmaitree and Mann 1995; Minor 1998). Most summertime colds in children are caused by the enteroviruses (Modlin 2000). Types of coxsackievirus associated with upper respiratory tract infection include A9 and B1–B4. A21 has been used as the prototype enterovirus in adult volunteer studies of respiratory infection (Chonmaitree and Mann 1995; Modlin 2000).

Other clinical presentations include ear infection, tonsillitis and pharyngitis, herpangina, croup, bronchitis, and pneumonia. The successful transmission of enteroviral respiratory infection depends on the route of transmission, type of person-to-person contact, and the virus serotype (Chonmaitree and Mann 1995). Aerosolization appears to be more efficient than ingestion in inducing respiratory infection, and though contagiousness is high within households, infection after casual contact is relatively low (Chonmaitree and Mann 1995). Respiratory diseases caused by enteroviruses are mild and require no treatment; about one half to two thirds of infected people do not develop symptoms; however, the percentage of asymptomatic respiratory infection may be higher in echovirus than in coxsackievirus (Chonmaitree and Mann 1995). Respiratory illness may be less common in infected infants than in older children and adults (Dagan and Menegus 1995).

Encephalitis

An estimated 20,000 cases of viral encephalitis occur in the United States annually, with herpes simplex being the most commonly identified source, only accounting for about 10% (Rotbart 1995b). Arboviruses and enteroviruses are probably the next most common causes; however, the etiology usually remains unknown (Modlin 2000; Rotbart 1995b). The primary cause of enteroviral encephalitis is enterovirus 71; however, other serotypes, including coxsackievirus A9, B2, B3, B5, and B6, have also been implicated (Minor 1998; Modlin 2000). The true incidence of enterovirus-associated encephalitis is difficult to ascertain, because the virus is rarely cultured from patients with the disorder. Brain biopsy has been one of the only reliable ways to determine enterovirus as a source of encephalitis, but the advent of PCR testing will undoubtedly improve the ability to gather information (Rotbart 1995a). Enterovirus encephalitis is considered a more serious illness than enterovirus meningitis (Rotbart 1995a), though patients beyond the neonatal period usually make a full recovery (Modlin 2000).

Exanthems

Exanthems that are usually rubelliform or maculopapular in nature occur during enterovirus infections—usually coxsackieviruses A9, A16, and B5 and echoviruses 4, 9, 16, and 25 (Hill 1996; Minor 1998). Rashes are often transient and are accompanied by other symptoms such as fever, malaise, and aseptic meningitis (Minor 1998). In a review of 82 infants <3 months old, 14% developed a rash as part of their coxsackievirus group B infections (Dagan and Menegus 1995). Hand-foot-and-mouth disease is a very common syndrome related primarily to coxsackievirus A16 but it is related to other coxsackieviruses as well. It is highly contagious and seen mostly in children in the summer months. In addition to the formation of vesicles on the hands, feet, and mouth, the disease usually presents with a low-grade fever.

Hepatitis

Hepatitis results from infection with coxsackievirus B and echoviruses mainly in neonates but occasionally in older children or adults (Minor 1998; Modlin 2000). Hepatitis syndrome in infants causes significant morbidity and mortality. In a review of 61 cases of perinatal echovirus infections, babies who developed hepatitis as sequelae had a fatality rate of 83% (Modlin 1986).

Paralytic Disease

Poliovirus remains the leading cause of viral paralysis in the world (Minor 1998). Sporadic cases have been reportedly linked with coxsackievirus A7 and B2–B6 and echovirus types 3, 4, 6, 11, and 19 (Minor 1998).

Treatment

Until recently, no antiviral drugs effectively treated or prevented nonpolio enterovirus infections. Pleconaril is a broad-spectrum, antipicornaviral agent that is close to Food and Drug Administration approval for use with aseptic meningitis and respiratory illness. Adult patients were challenged with coxsackievirus A21 after receiving pleconaril or placebo (Schiff and Sherwood 2000). In those who received the pleconaril, symptoms and viral shedding were significantly decreased.

Isolating ill patients to contain the spread of infection is not effective because so many asymptomatic people shed the virus and the spread from an index patient to secondary contacts is rapid. Obviously, controlling a specific environment, such as a neonatal unit in a hospital, can reduce an outbreak situation.

In very serious neonatal enterovirus infections (especially in group B coxsackievirus infections but also echoviruses), γ-globulin, maternal plasma, and exchange transfusions have been used intravenously with limited success (Dagan 1996; Minor 1998; Rotbart 1995a). Pleconaril has also been used successfully with critically ill neonates (Aradottir, Alonso, and Shulman 2001). Agammaglobulinemic patients have also experienced some improvement in their conditions through γ-globulin therapy (Rotbart 1995a). In healthy patients, enteroviral meningitis can usually be treated with palliative care until recovery is complete (Dagan 1996; Rotbart 1995b).

The diagnosis of myocarditis is difficult until a myopathic process occurs. Myocarditis-induced heart failure is treated with conventional therapy like diuretics, beta-blockers, and angiotensin-converting enzyme inhibitors (Anandasabapathy and Frishman 1998; Caforio and McKenna 1996). Treatments for the infection itself are still under investigation, but therapies under consideration include nonsteroidal anti-inflammatory agents, immunoglobulins, and immunosuppressants.

Susceptible Subpopulations

Immunocompromised

Generally, enteroviral infections do not discriminate between healthy and immune-compromised hosts. Though severe disease can occur in the immunocompromised, enterovirus is not a major infection in people with AIDS, cancer, etc. (Morens and Pallansch 1995). There is a notable exception, however. The enteroviruses are cleared by the host by antibody-mediated mechanisms, in contrast with other viruses, which work through cellular immune mechanisms. Because of this difference, people without effective humoral immunity (e.g., x-linked agammaglobulinemia) are sensitive to infection and present with a much different clinical expression than immunocompetent people. In many of these patients, the infection is fatal (Modlin 2000). Patients with common variable immunodeficiency or agammaglobulinemia, which are both primary immunodeficiency diseases, can develop chronic meningoencephalitis from enterovirus infection that lasts for years and can result in death (Rotbart 1995b; Sicherer and Winkelstein 1998). Twelve of 16 patients with chronic meningoencephalitis responded positively after treatment with pleconaril (Rotbart and Webster 2001). Intravenous treatment with antibodies helps some patients; however, the virus may persist despite this type of therapy (Rotbart 1995a). Expectedly, patients undergoing allogeneic bone marrow transplants have also developed severe complications from enterovirus infections, including the report of an outbreak of coxsackievirus A1 that resulted in increased mortality in marrow transplant recipients (Schwarer et al. 1997; Townsend et al. 1982).

Children

Enterovirus infections in newborns are common; data suggest that enteroviruses (especially echoviruses and group B coxsackieviruses) may be the most prevalent cause of infection in neonates (Abzug 1995). Many enterovirus outbreaks in newborns result in asymptomatic infections (Abzug 1995; Dagan 1996). Generally, serious complications from enteroviral infections are less common in young children than in older children and adults (Abzug 1995; Morens and Pallansch 1995), though the reasons are unclear. One exception to this is when fetuses are infected transplacentally (or neonates are infected in the first day or two after birth); this route of transmission puts them at higher risk for serious illness (Modlin 2000; Abzug 1995). Though the humoral immune system of neonates can respond to enteroviral infections, their macrophage function cannot stop viral replication until a few weeks after birth. It also appears that the timing of maternal infection plays a role in neonatal infection outcomes, as the development of maternal antibodies passively acquired by the neonate has a critical influence (Modlin 2000). One community-based study demonstrated a 13% enterovirus incidence in neonates <1 month old, with 21% of the infected children becoming ill (Jenista, Powell, and Menegus 1984). Risk factors that are associated with serious clinical outcomes in neonates include infection with particular serotypes (especially coxsackievirus B2–B5), male gender, prematurity and low birth weight, and prepartum illness in the mother (Abzug 1995; Modlin 1986). The most serious complications associated with enterovirus infections in neonates are hepatitis and myocarditis. In a review of the literature, Modlin (1986) reported that 67% of children classified as having severe echovirus infection developed hepatitis, and 83% of them died. In a recent study of placental tissue, 6 placentas infected with coxsackievirus yielded infants with severe respiratory problems, mental defects, developmental delays, and seizures compared with 10 controls and 6 placentas with known viral infection that tested negative for coxsackievirus (Euscher et al. 2001).

Myocarditis can appear alone or in combination with central nervous system infections or hepatic disease. Sequelae include arrhythmia, cardiomegaly, congestive heart failure, and pericarditis. Neonatal myocarditis is primarily acute and rarely evolves into chronic calcific myocarditis with chronic heart failure. Myocarditis in the neonate results in mortality less than 50% of the time (Modlin 2000).

Other

Selenium deficiency in the host affects the inflammatory response in cardiac tissue during infection with coxsackievirus (Beck and Matthews 2000). The immune response, not the virus itself, causes heart damage during infection, so this nutritional deficiency may make a host more susceptible to cardiac damage. Exercise has also been associated with an increase in infection rates and disease severity in enteroviral infections (Minor 1998; Morens and Pallansch 1995).

Chronic Sequelae

Coxsackievirus infection causes a variety of clinical syndromes, most commonly aseptic meningitis and conjunctivitis but also herpangina, encephalitis, and myopericarditis. Occasionally, the clinical outcome of a central nervous system-related illness will be serious and involve complex seizures, intracranial pressure, or coma (Rorabaugh et al. 1993). Several studies have investigated the risk of long-term neurological and cognitive sequelae from these acute manifestations—the evidence appears to be equivocal. An older study reported that children who had been hospitalized with enterovirus infection in their first year scored significantly lower in I.Q., language and speech, and head circumference (Sells, Carpenter, and Ray 1975). However, the study only included 19 children with matched controls. A similar study (Wilfert et al. 1981) found that nine children who had enteroviral meningitis during the first 3 months of life were significantly underdeveloped in language function but showed no difference in intellectual ability, hearing function, or head circumference. Bergman and colleagues (1987) compared 42 children who had enteroviral meningitis during their first year of life with their siblings for neurological, psychological, language, and academic indicators. They found no differences in any functioning scores between the case children and their control siblings. A large prospective cohort study is reported by Rorabaugh and coworkers (1993). They assembled a cohort of 277 children <2 years old with aseptic meningitis and tracked their acute illnesses and neurologic complications. Of the 277, neurologic signs were noted in 52 patients; 25 patients (about 10% of the total cohort) had significant neurologic complications such as seizures, coma, or intracranial pressure. These complications generally afflicted children >3 months. The children were followed up four times over 42 months and given neurologic and developmental tests. The children with central nervous system complications did not score any differently than the children without complications. Of the 169 cases that had confirmed etiologies, 55% were caused by coxsackievirus group B and 38% by the echoviruses. Fewer than 5% were caused by other enteroviruses or adenoviruses. The complications occurred equally in the coxsackievirus B- and the echovirus-caused cases, and no single serotype dominated.

An interesting study examined the connection between childhood central nervous system infections and adult-onset schizophrenia or other psychoses in a population-based cohort of 11,017 in Finland (Rantakallio et al. 1997). In this cohort study, followed for almost 30 years, 102 had a viral central nervous system infection and 129 had schizophrenia or another psychosis. After adjusting for risk factors, the odds ratio for childhood viral infection and schizophrenia was 4.8 (95% CI: 1.6–14.0). Coxsackievirus B5 was emphasized as a particularly important type.

If a neonate experiences hepatitis resulting from enterovirus infection and if it survives the acute attack, the child may continue to have hepatic disease throughout infancy. However, liver function usually improves in time, and the child will continue to develop normally without long-term sequelae (Abzug 1995). Also, though neonatal myocarditis can evolve into chronic calcific myocarditis and heart failure, most survivors of neonatal myocarditis do not appear to have long-term cardiac problems (Martino et al. 1995).

Acute myopericarditis caused by enteroviral infection can lead to chronic cardiac conditions. Electrocardiographic abnormalities (10–20%), cardiomegaly (5–10%), and chronic congestive heart failure are examples of permanent heart injury occurring in about one third of patients (Modlin 2000). These conditions can lead to chronic dilated cardiomyopathy, which is the second leading cause of congestive heart failure. Persistent, active infection with coxsackievirus B has been associated with poor clinical outcome in patients with dilated cardiomyopathy (Fujioka et al. 2000).

Various chronic neuromuscular conditions have been reported as enteroviral sequelae. Most of these associations have centered on certain serotypes of group A and all group B coxsackieviruses and, of course, polioviruses. Rare case reports give accounts of paralysis, transverse myelitis, and chronic fatigue syndrome occurring after coxsackievirus infection. Though a few cases are reported as permanent conditions, realistically, little is known about the length of time these conditions may have persisted in affected patients (Dalakas 1995; Figueroa et al. 1989; Gear 1984; Takahashi et al. 1995; Yamashita et al. 1994). Several studies suggest that coxsackievirus group B may be a precipitating factor in chronic fatigue syndrome; however, there is some question as to their detection methodology. Nairn and colleagues (1995) compared sera of 100 patients with chronic fatigue syndrome and 100 healthy people. Using PCR, 42% of the ill patients were positive for coxsackievirus B compared to only 9% of the healthy group. Apart from PCR, the comparisons were made with an older neutralization assay, and the researchers found that the neutralization assay was not able to differentiate between the two groups, making its use questionable. These emerging molecular detection techniques will shine additional light on these possible associations.

Many studies have been conducted looking at the association between enteroviruses and insulin-dependent diabetes mellitus (IDDM); data supporting both sides of an enteroviral link with IDDM have been published. The main focus has been the coxsackievirus–IDDM connection; however, echoviruses have also appeared as potential problems. Study designs have included cross-sectional, case-control, prospective cohort, retrospective cohort, and animal studies. Much of the mechanistic research has focused on beta-cell autoimmunity or destruction. Most epidemiological studies have centered on measuring serum antibody levels in children with IDDM, their mothers, and their siblings. The weight of evidence indicates that enteroviral infection is a risk factor for IDDM, and though echoviruses have been occasionally implicated, coxsackieviruses are implicated most frequently (Hyoty, Hiltunen, and Lonnrot 1998; Roivainen et al. 1998).

Transmission

Humans are the only known reservoir for the (human) enteroviruses, which are shed primarily in the feces but can also be shed from the throat—typically for 1–2 weeks (Rotbart 1995a). Unlike other gastrointestinal viruses, enterovirus colonization does not change the ecology of the gut; therefore, the fecal shedding periods for the enteroviruses are relatively long compared with other gastrointestinal viruses, which frequently last only as long as the associated symptoms (Morens and Pallansch 1995). Children infected with poliovirus and coxsackievirus can shed for weeks or months after initial infection; for echovirus, the shedding period in children is shorter—seldom longer than 1 month (Minor 1998; Rotbart 1995a). Adults shed enteroviruses for less time than children did overall—1 to 2 weeks. The viruses are frequently shed by asymptomatic people, and asymptomatic infection has been reported at 76% for coxsackievirus (Minor 1998).

Transmission occurs directly through the fecal–oral route or indirectly through food or water (this is theoretical because little evidence of this type of transmission exists) (Modlin 2000; Morens and Pallansch 1995). The viruses are occasionally spread through aerosolized respiratory droplets or blood (Modlin 2000; Morens and Pallansch 1995). Though enteroviruses are often found in sewage, water, and shellfish, transmission from these sources has never been documented (Modlin 2000). Transmission can also occur vertically, from mother to child *in utero*, either transplacentally, from the amniotic fluid, or in the birth canal (Abzug 1995; Modlin 1986; Morens and Pallansch 1995; Rotbart and Hayden 2000). Vertical transmission from mother to child can have serious consequences for the infected newborn, because babies infected this way are at high risk for serious illness (Dagan 1996; Modlin 1986). In addition to the transplacental route, neonates can also become infected by family members or hospital staff. Transmission of enteroviruses between mother and child are estimated to be between 30 and 50% (Modlin 1988).

Secondary Spread

Secondary spread is common in all enterovirus infections; however, coxsackievirus and poliovirus have a higher rate than echovirus. The secondary attack rates of infection in susceptible family members of as high as 92% for polioviruses, 76% for coxsackieviruses, and 43% for echoviruses have been reported (Minor 1998; Modlin 2000). Many household members infected through secondary spread do not become sick. Coxsackieviruses and polioviruses are shed for longer periods of time than echovirus, which may influence their attack rates. The following factors have been found to affect virus spread to household members: (1) duration of virus excretion, (2) household size, (3) number of siblings, (4) socioeconomic status, (5) type of virus, and (6) immunity status of household members (Morens and Pallansch 1995). In a review of nursery outbreaks, Modlin (1986) found that infants who acquired nosocomial echovirus infections had less severe illness than the index cases, but this may not be generalized to coxsackievirus infections. Much of the hospital-based secondary spread appears to be transmitted through health care workers (Modlin 1986). Attack rates in these nursery outbreaks have been as high as 50% (Modlin 1998).

Dose Response

Infectious Dose

Schiff et al. (1984) conducted a controlled study looking for infectious dose information on enteroviruses—using echovirus 12 as a prototype. In the absence of other volunteer studies with different serotypes, we can only assume this infectious dose data represent other enteroviruses, including coxsackie. One hundred forty-nine volunteers were given 0–330,000 plaque forming units (pfu) of echovirus 12 in nonchlorinated water. A dose response was seen in the rate of infection in doses up to 10,000 pfu. The dose, however, did not affect the time to viral shedding (the first week after inoculation in all infected volunteers) or duration of shedding (up to 26 days). None of the volunteers developed significant illness. The investigators developed a probit analysis and generated a dose response based on their results. Based on the curve, they estimated a dose of 919 pfu as necessary to infect 50% of volunteers. From that figure, they extrapolated an estimated dose necessary to infect 1% of volunteers as 17 pfu (Table 8-2).

Table 8-2: Results of volunteer feeding studies of echovirus 12

		Number of Volunteers		
Dose (pfu)	Total	Intestinal Shedding	Seroconversion	Infection (% of group exposed)
0	34	0	0	0
330	50	14	7	30
1,000	20	8	3	45
3,300	26	18	11	73
10,000	12	11	3	100
33,000	4	2	1	50
330,000	3	2	1	67

Adapted from Schiff et al. 1984.

Volunteer studies from up to 40 years ago and using poliovirus estimated smaller infective doses (e.g., 4, 72, and 1 pfu), but those studies used attenuated virus in a study population of children, which may account for the discrepancy. The second part of the Schiff et al. study looked at the effect of previous infection on reinfection rate. Of 18 previously infected volunteers, 72% became reinfected with echovirus after a dose of 1,500 pfu. These results suggest that antibody status does not necessarily affect reinfection rate or number of shedding days. The paper gave no indication of the period between primary infection and rechallenge. Other studies have found a link between increased antibody levels and decreased infection rates (Chonmaitree and Mann 1995; Minor 1998).

In outbreaks, young children are more frequently infected by enteroviruses. A case series of 14 neonates showed a mild nonspecific disease in 8 and disseminated, overwhelming disease in 6, 3 of whom died (Isacsohn et al. 1994). Half of the children were infected vertically through their mothers at birth; the other half were assumed to be nosocomial. Coxsackieviruses from groups A and B have been associated with nursery outbreaks, which have had attack rates of up to 50% reported (Modlin 1988). The tendency of children to have higher attack rates than adults probably results from their increased exposure opportunities and susceptible immune status. In a cohort study of 75 babies infected with nonpolio enteroviruses, the incidence of symptomatic infection was 0% in those with antibodies and 38% in those without antibodies (Jenista, Powell, and Menegus 1984). Except for coxsackievirus A16, which is more often severe in young people, enteroviral illness is frequently milder in children than it is in adults. (Neonates, who were discussed earlier in this chapter, are the exception.) (Helfand et al. 1994; Hill 1996; Morens and Pallansch 1995).

Virulence

Generally, enteric viruses invade the intestinal epithelium, where they cause destruction of enterocytes, leading to illness. Viruses (and bacteria) use different mechanisms to evade a host's gastrointestinal defenses. Among virulence factors common to enteroviruses are a resistance to acid and bile, which helps them get through the stomach and the gut, and resistance to proteolytic enzymes, which also increases their survival and passage into the large intestine (Duncan and Edberg 1995).

Infectivity and virulence vary by genotype and phenotype. A study investigating the effect of in vivo conditions (e.g., gastric acidity) on enteroviruses found differences among the serotypes; echovirus 22 appeared more sensitive to inactivation in body fluids, though the infectivity of coxsackieviruses A9 and B4 were also decreased (Piirainen, Hovi, and Roivainen

1998). All the viruses remained infective after a 2-hour exposure to pH 2 at 37° C, but after 24 hours, they were all inactivated.

Recent research in a murine model showed that a selenium deficiency increased the virulence of a coxsackievirus B3 strain and allowed the conversion of a nonvirulent coxsackievirus B3 strain to virulence (Levander and Beck 1999). The authors believed it was the first report of a nutritional factor in the host influencing the genome of the invading pathogen. Other work suggests a similar result from vitamin E deficiency (Beck 2000).

Environmental Occurrence

Enteroviruses are ubiquitous in the environment. They are found in sewage, fresh water, marine water, shrimp and shellfish, and even chlorinated water. Enteroviruses are far more prevalent in the summer and early fall seasons (in temperate climates). In tropical climates, they circulate more evenly throughout the year or are sometimes more prevalent during the rainy season (Minor 1998; Modlin 2000; Morens and Pallansch 1995).

Survivability

Enteroviruses are hardy. They survive for a long time on hands, fomites, and in water (Morens and Pallansch 1995). Yates and colleagues (1985) seeded groundwater samples with three viruses, including echovirus 1 and poliovirus 1, and measured the factors affecting survival. Measurements included pH, nitrates, turbidity, and hardness. They found that temperature was the only variable significantly correlated with viral decay rates and that none of the viruses tested remained viable significantly longer than the others.

The long-term survival of three strains of enterovirus (coxsackievirus B3, echovirus 7, and poliovirus 1) was tested in freshwater from five sites with different physical characteristics (Hurst, Benton, and McClellan 1989). The five sites included an artificial lake, a pond, large- and medium-sized rivers, and a small creek. The average inactivation rates were 6.5–7.0 $\log_{10}$ units over 8 weeks at 22° C, 4.0–5.0 $\log_{10}$ units over 12 weeks at 1° C, and 0.4–0.8 $\log_{10}$ units over 12 weeks at –20° C. The effect of temperature was significant ($p < 0.00001$), as were the serotypes and the water source at each of the three temperatures. Other operative factors included hardness and conductivity (which were detrimental to survival), turbidity and suspended solids (which were beneficial to survival), and, at 22° C, the number of generations of bacterial growth supported by the water sample (which was detrimental to survival).

Occurrence in Water Sources

The enteroviruses have been isolated from all types of water sources: rivers, raw sewage, wastewater, oceans, and groundwater. Table 8-3 details selected studies on enteroviral, and specifically, coxsackievirus occurrence in the environment.

Descriptions of enterovirus concentrations have usually not been serotype specific. Recent ability to more easily detect serotypes has resulted in more detailed analyses. Keswick and colleagues (1984) did an extensive study of the occurrence of viruses in conventionally treated drinking water coming from poor-quality source water. Seven percent of the 37 finished water samples tested positive for enterovirus (see Table 8-3). Payment and colleagues (1988) reported enterovirus averages of between 1 and 145 mpniu/L (most probable number of infectious units per liter) from river waters in Quebec. In a river in Japan, concentrations of enteroviruses detected over a 63-month period averaged 41 pfu/L, with a range of 1 to 190 pfu/L (Tani et al. 1995). A total of five coxsackievirus serotypes were isolated, with B1–B5 being found most often. A 1981 study reported on the occurrence of enterovirus species, including coxsackievirus B3 and B4, in community swimming pools (Keswick, Gerba, and Goyal 1981). Ten of 14 samples contained virus; 3 samples were positive in the presence of residual chlorine; 2 of those residual levels exceeded the 0.4-ppm standard. Seven of seven of the wading pool samples were positive, which is not surprising considering the water is usually warm and the bathers are usually toddlers.

Detection Methods

Clinical

A problem with diagnosing enterovirus infection is that it mimics other common bacterial or viral infections. The symptoms of most enteroviral diseases are indistinguishable from symptoms of other types of infection (e.g., encephalitis/herpes simplex, respiratory illness/rhinovirus). Also, the systemic enteroviral infection developed by some neonates is clinically indistinguishable from bacterial infections and sepsis (Dagan 1996; Rotbart 1995a). Antibiotic treatment and continued in-hospital observation are standard for suspected meningitis infections until a bacterial etiology is ruled out (Dagan 1996; Rotbart and Romero 1995). It is important for the clinician to establish etiology—or at least to rule certain things out—so appropriate treatments can be given when available. Additionally, the quick diagnosis of viral disease can avert unnecessary treatment. Sometimes the

Table 8-3: Occurrence of entero- (coxsackie-) viruses in the environment

Reference/Location	Environmental Source	Serotype	Occurrence (%)*	Notes
Hovi, Stenvik, and Rosenlew 1996 Finland	Raw sewage	Cox A9	0.6	1,161 samples taken between 1971 and 1992; peak occurrence in early fall; cell culture
		Cox B1	3.5	
		Cox B2	11.1	
		Cox B3	7.8	
		Cox B4	18.7	
		Cox B5	17.3	
Hurst 1991 Varied	Natural freshwater	Varied	47.2 (median value of summary)	Summary of 16 studies of occurrence of enteric viruses (mainly enteroviruses) in natural freshwater
	Finished drinking water		0 (median value of summary)	Summary of nine studies of occurrence of enteroviruses in drinking water treated with coagulation, sedimentation, filtration, and disinfection; four studies did find samples that were positive
Keswick et al. 1984 Unknown	Raw source water	Varied (entero)	56.0	18 samples; dry season; average turbidity = 6.4; cell culture
	Clarified source water		35.0	23 samples; dry season; average turbidity = 5.6; cell culture
	Finished drinking water		41.0	39 samples; dry season; average turbidity = 1.2; cell culture
	Raw source water		7.0	14 samples; rainy season; average turbidity = 26; cell culture
	Clarified source water		29.0	7 samples; rainy season; average turbidity = 10; cell culture

*Percentage of samples testing positive.

Table continued on next page.

Table 8-3: Occurrence of entero- (coxsackie-) viruses in the environment (continued)

Reference/Location	Environmental Source	Serotype	Occurrence (%)*	Notes
	Filtered source water		15.0	15 samples; rainy season; average turbidity = 6.8; cell culture
	Finished drinking water		7.0	37 samples; rainy season; average turbidity = 9.6; cell culture
Krikelis et al. 1985 Greece	Raw sewage	Cox B1	5.2	191 samples taken over 15 months; peak occurrence in September; average concentration of enteroviruses = 291 CPU/L; highest = 900 CPU/L; cell culture
		Cox B2	5.2	
		Cox B4	10.0	
		Cox B5	9.0	
Payment, Trudel, and Plante 1985 Canada	Raw source water	Varied (mainly entero)	79.0	152 samples over 1 year; average concentration 3.36 mpncu/L; cell culture.
	Prechlorinated source water		65.0	17 samples over 1 year; average concentration 0.072 mpncu/L; cell culture.
	Sedimented source water		20.0	119 samples over 1 year; average concentration 0.016 mpncu/L; cell culture.
	Filtered source water		14.0	119 samples over 1 year; average concentration 0.001 mpncu/L; cell culture.
	Ozonated source water		9.0	45 samples over 1 year; average concentration 0.0003 mpncu/L; cell culture.
	Finished drinking water		9.0	138 samples over 1 year; average concentration 0.0006 mpncu/L; cell culture; included Cox B5

*Percentage of samples testing positive.

Table continued on next page.

Table 8-3: Occurrence of entero- (coxsackie-) viruses in the environment (continued)

Reference/Location	Environmental Source	Serotype	Occurrence (%)*	Notes
Payment, Fortin, and Trudel 1986 Canada	Raw sewage	Cox B1	5.3	19 samples; average concentration 0.3 mpniu/L; cell culture
		Cox B2	5.3	19 samples; average concentration <0.1 mpniu/L; cell culture
		Cox B3	10.5	19 samples; average concentration 1.6 mpniu/L; cell culture
		Cox B4	57.9	19 samples; average concentration 25.3 mpniu/L; cell culture
		Cox B5	15.8	19 samples; average concentration 0.4 mpniu/L; cell culture
	Primary sewage effluent	Cox B1	10.5	19 samples; average concentration 0.8 mpniu/L; cell culture
		Cox B2	5.3	19 samples; average concentration <0.1 mpniu/L; cell culture
		Cox B3	10.5	19 samples; average concentration 0.1 mpniu/L; cell culture
		Cox B4	52.6	19 samples; average concentration 4.6 mpniu/L; cell culture
		Cox B5	15.8	19 samples; average concentration 1.9 mpniu/L; cell culture
	Secondary sewage effluent	Cox B1	5.3	19 samples; average concentration <0.1 mpniu/L; cell culture
		Cox B2	5.3	19 samples; average concentration <0.1 mpniu/L; cell culture
		Cox B3	5.3	19 samples; average concentration <0.1 mpniu/L; cell culture
		Cox B4	42.1	19 samples; average concentration 0.9 mpniu/L; cell culture
		Cox B5	0	19 samples; cell culture

*Percentage of samples testing positive.

Table continued on next page.

Table 8-3: Occurrence of entero- (coxsackie-) viruses in the environment (continued)

Reference/Location	Environmental Source	Serotype	Occurrence (%)*	Notes
Reynolds et al. 1998 Hawaii	Primary sewage effluent	Varied	100	12 samples collected; average concentration = 3,300 pfu/L; highest = 13,000 pfu/L; cell culture
	Marine water at sewage outfall discharge site		28.6	7 samples collected; average concentration = 0.04 pfu/L; highest = 0.044 pfu/L; cell culture
	Canal freshwater		17.0	12 samples collected; average concentration = 0.035 pfu/L; highest = .045 pfu/L; cell culture
	Swimming beach water		8.0	50 samples collected; average concentration = .055 pfu/L; highest = 0.1 pfu/L; cell culture
	Stream water		0	4 samples; cell culture
van Olphen et al. 1984	Raw source water	Mainly entero	67.0	45 samples; cell culture.
	Partially treated drinking water		20.0	55 samples; cell culture
	Finished drinking water		0	100 samples of 500 L

*Percentage of samples testing positive.

season of the year may offer hints to etiology, because enteroviral infections occur much more frequently in the summer and early fall months.

Isolation of the enterovirus in cell culture followed by serotype identification with neutralization is still the "gold standard" for diagnosis (Rigonan, Mann, and Chonmaitree 1998; Rotbart 1995a), but culturing presents a lot of problems for the diagnostician. No particular cell line is appropriate for all enterovirus serotypes; however, most laboratories use a combination of monkey kidney cell lines and a human diploid fibroblast line (Dagan 1996; Modlin 2000; Rotbart 1995a). Most of the coxsackievirus group A strains are difficult to culture and have traditionally been detected using newborn mice (Gjoen and Bruu 1997). Both culturing and live animal methods are technically challenging and labor/time intensive. Also, one fourth to two thirds of clinical samples cultured from patients with characteristic enterovirus infection are negative but are positive using RT-PCR (Rotbart 1995a; Stellrecht et al. 2000).

Because the enteroviruses do not share an antigen, immunoassays have been slow to develop as clinical detection methods (Rotbart 1995a); however, Rigonan and colleagues (1998) reported 93% sensitivity and 90% specificity compared with standard neutralization using immunofluorescent assay with commercially available antibody blends. Work with coxsackievirus group B is promising because the group shares an antigen (Rotbart and Romero 1995). The ability to use easily available reagents is an appealing feature of that assay, and perhaps more validation will be forthcoming. Electron microscopy has no application in detecting enteroviruses (Rotbart and Romero 1995). PCR has been used broadly—to detect many or all serotypes—or in a more limited way by detecting a few or one single serotype (Rotbart and Romero 1995). RT-PCR is used mainly to study the molecular epidemiology of enteroviruses, such as strain prominence and genetic shifts. RT-PCR is definitely a promising alternative to culturing, especially for meningitis cases where quick diagnosis is critical. Tanel et al. (1996) reported on a comparison of RT-PCR and viral culture in the cerebrospinal fluid of meningitis cases. They found the RT-PCR assay to be 77.8% sensitive and 100% specific compared with a sensitivity of 66.7% for culture alone. Other researchers report RT-PCR sensitivity as 85% and culture sensitivity as 24% (Gorgievski-Hrisoho et al. 1998). Another benefit of RT-PCR is the reduced time it takes to get results. Cell culture usually takes 3 to 15 days compared with PCR, which can be finished in the same or next day (Furione et al. 1998; Gorgievski-Hrisoho et al. 1998). In a cost–benefit analysis of using PCR as a clinical tool, Schlesinger and colleagues (1994) also found PCR to be more sensitive and specific than viral culture. To analyze cost, they went on to estimate from retrospective records that if PCR had been used instead of culture for diagnosis, the time saved would have resulted in briefer hospitalizations—an average of 1.2 days fewer per child—representing a significant cost savings.

There is little need to distinguish among the specific enterovirus serotypes in a clinical diagnosis, because the diseases are often not serotype specific. One important factor to consider in clinical diagnosis is the sample site. The nasopharynx and the gastrointestinal tract can be colonized for weeks or months, thereby giving a false-positive test result. The central nervous system, blood, and the genitourinary tract are infection specific—a positive result from one of these sites is likely to be associated with illness (Rotbart and Romero 1995).

Environmental

The same advantages and disadvantages of detection methods described for clinical usage also apply for environmental usage. Cell culturing is the standard method and is the only current method available that evaluates viral infectivity. However, as detailed above, it is slow and not nearly as specific or sensitive as newer molecular methods. PCR is especially vulnerable to the presence of inhibitory substances typically found in environmental samples; this can significantly decrease its effectiveness. From their work, Reynolds et al. (1998) concluded that RT-PCR is more useful only when the samples are either relatively free of inhibitory compounds or the viral load is so great that the inhibitors are diluted. Therefore, PCR works well for sewage but not necessarily for environmental source water. In another study, Reynolds et al. (2001) combined cell culture and RT-PCR to look at sewage, marine water, and surface drinking water sources. They found 11 positive samples when the methods were combined and only three when RT-PCR was used alone. Finally, when a sample has numerous inhibitors, PCR can only test low equivalent volumes compared with cell culture; therefore, samples with low virus concentrations but high inhibitor levels may not test positive because of the small amount of testable sample. However, when conditions are equivalent, PCR is generally a more accurate measure than conventional cell culturing.

Researchers are busy trying to simplify and improve molecular detection techniques because of their recognized potential to detect virus in water. Puig et al. (1994) concluded that nested PCR is even more sensitive than one-step PCR methods in detecting enteroviruses from river water and sewage samples. However, they noted that nested PCR is still susceptible to false positives from contamination with amplified DNA. Overall, they concluded that nested PCR can detect fewer than 10 particles of virus, which is 100 to 1,000 times greater than cell culture. The ability to detect small amounts of virus is useful when evaluating the quality of different water sources, including drinking water.

With any detection method, the viruses must first be separated and concentrated from the environmental samples—sometimes from hundreds or thousands of gallons at a time. Various researchers have described particular techniques to concentrate enteroviruses using glass wool and glass powder to adsorb the viral particles (Gantzer et al. 1997; Hugues et al. 1993; Senouci, Maul, and Schwartzbrod 1996) and different types of filters and filter processes (Gilgen et al. 1995; Guttman-Bass and Nasser 1984; Li et al. 1998; Shieh, Baric, and Sobsey 1997). The results vary depending on water source, variation in chemical usage, enterovirus serotype, etc. In their study to detect low levels of enteroviruses in sewage, Shieh and colleagues (1997) point out that concentration and recovery techniques that were designed for use in cell culture are inadequate for use with PCR and increase the natural substances in environmental samples that make PCR difficult to perform. Several researchers are working to overcome the problem of inhibition in samples; some are combining cell culture with PCR in an effort to capture the desirable qualities of each method (Reynolds et al. 1998; Straub, Pepper, and Gerba 1995).

RT-PCR has also been used to detect enteroviruses in shellfish (Casas and Sunen 2001; LeGuyader et al. 2000). Of oyster and mussel samples tested, enteroviruses were found in 19 and 45% of samples, respectively.

Water Treatment

Enteroviruses can stay viable for years at –20° C and for weeks at –4° C, but they gradually lose their infectivity at room temperature and quickly lose infectivity as temperatures rise. Infectious coxsackievirus B3 was undetectable after 15 minutes at 55°C (Gantzer, Levi, and Schwartzbrod 1996). Enteroviruses are easily inactivated by ultraviolet light (if there is no organic matter) and through desiccation (Minor 1998). Free residual chlorine of 0.3 to 0.5 ppm reportedly results in rapid loss of viability as well, though the water's condition (e.g., high pH, low temperature, and increased organic matter) can affect the efficacy of chlorine disinfection (Minor 1998). Water utilities maintain about 1.1 mg/L as a residual (Rice, Clark, and Johnson 1999). Hurst (1991) summarized four studies that reported on the efficiency of the entire conventional drinking water process in removing enteric viruses (generically stated) and coliphage. He defined conventional treatment as coagulation, sedimentation, filtration, and disinfection. The median removal efficiency after the full treatment process was ≥95.1%. The study reported on plants in Canada, Mexico, and the United States, so it is likely that treatment processes, treatment quality, and raw water quality varied. The enterovirus removal figures after full treatment in the two different US studies were >93.0% and >95.4%. Payment and colleagues (1985) greatly increased removal efficiency by prechlorinating the raw water, which resulted in an immediate 95.0% removal and increased to 99.97% after completion of full-scale treatment. The prechlorination reduction was achieved after 5 minutes at a residual chlorine level of 0.5 to 1.0 mg/L. However, the drinking water process reduced the average viral concentrations from 125 to 0.003 mpncu/L, or 4–5 $\log_{10}$, at the maximum efficiency level. The overall virus reduction was about 4 $\log_{10}$. The treatment eliminated other indicators such as total coliforms more effectively (6 $\log_{10}$ reduction). In 138 finished water samples of 1,000 L/sample from six different plants, 12 samples contained virus (8.7%), including coxsackievirus B3–B5. The average virus concentration was 0.0006 mpncu/L, and the highest concentration reported in finished water was 0.02 mpncu/L.

Researchers are busy testing ways to improve virus removal from drinking water. Efforts include filtering finished water through oxidized coal (Cloete, Da Silva, and Nel 1998) and using copper in domestic plumbing (Colquhoun et al. 1995). The latter study tested the virucidal effect of plumbing material on sample enteroviruses, including coxsackievirus B4. In a comparison with glass, polybutylene, polyethylene, and silicone, copper was up to two orders of magnitude better at reducing viruses than the other materials, which could be significant when trying to control biofilm colonization.

Membrane filtration technology, such as micro-, ultra-, and nanofiltration, as well as reverse osmosis, has the capacity to remove water contaminants down to the ion level—greater than conventional water treatment. Ultrafiltration, with a nominal pore size of 0.01 μm, can achieve 4-log virus removal or greater (Najm and Trussell 1999; Taylor and Wiesner 1999). This technology will continue to gain in popularity as costs become more competitive and regulatory approval is assured.

Wastewater treatment has direct bearing on drinking water treatment because wastewater is usually discharged into drinking water sources and the amount of viral load in the source water affects the ability to treat drinking water. Irving and Smith (1981) tested wastewater for adenoviruses, reoviruses, and enteroviruses over 1 year. The average concentration of enteroviruses in raw sewage was 1,400 IU/L. Effluent after primary treatment averaged 1,250 IU/L, an 11% decrease, and secondary effluent averaged 100 IU/L, a 93% decrease. The process removed enteroviruses more efficiently than adenoviruses or reoviruses. The coxsackieviruses most often isolated in the sewage were, in descending order, B3, B2, B5, and B4.

Payment and colleagues (1986) conducted a similar study in Quebec. Their raw sewage averaged 95.1 mpniu/L of enteric viruses. After a primary settling stage, viral reduction was 75%, and the secondary effluent was reduced by 98%. Group B coxsackieviruses, polioviruses, and reoviruses were most frequently isolated, while the echoviruses tested were rarely encountered. As with drinking water treatment, the quality of sewage treatment facilities, procedures, and location will substantially affect a plant's ability to effectively treat its water, making it difficult to generalize the reduction efficiency reported in studies to all wastewater.

Concerns about viral retention in groundwater have been sparked by the large number of waterborne disease outbreaks traced to groundwater (Moore et al. 1993). Sobsey and colleagues (1995) evaluated soil's ability to reduce the viral load in groundwater. They used four types of soil ranging from clay loam to sand and they evaluated four viruses: echovirus 1, poliovirus 1, hepatitis A, and an indicator virus. Generally, soil filtration reduced poliovirus the most and echovirus the least. Viruses were also removed more efficiently at a higher incubation temperature. Overall, clay was the most efficient at virus reduction; it completely (>99.9 to >99.999%) removed all viruses from groundwater and primary sewage effluent at both 5° C and 25° C. Course sand, on the other hand, reduced viruses no more than 80% from primary effluent and no more than 98% from groundwater.

Conclusions

Nonpolio enteroviruses, which include coxsackieviruses, cause common infections in children and adults. Notably, coxsackievirus group B is the most common cause of viral myocarditis, though it is also one of the most common causes of aseptic meningitis. Coxsackievirus A manifests itself most commonly in conjunctivitis and hand-foot-and-mouth disease, especially in children. In all but neonates and those with agammaglobulinemia, enterovirus infections are self-limited and last less than 1 week. However, coxsackieviruses' link with IDDM and chronic dilated cardiomyopathy are noteworthy and should be considered significant chronic outcomes of infection. Because it is excreted in human fecal matter, the possibility of foodborne or waterborne transmission exists; however, this route seems to be of little or no importance in the spread of infection. Numerous outbreaks of coxsackievirus illness, such as acute hemorrhagic conjunctivitis and hand-foot-and-mouth disease, occur worldwide and are almost always caused by person-to-person transmission. Coxsackieviruses are found in all kinds of water, from raw sewage to chlorinated drinking water. Generally, conventional water treatment with chlorine disinfection inactivates coxsackievirus, though that cannot be assumed to be 100%.

Overall, though coxsackieviruses are associated with significant health outcomes, they appear to be a minor threat as a waterborne disease.

Bibliography

Abzug, M.J. 1995. Perinatal Enterovirus Infections. In *Human Enterovirus Infections*. Edited by H.A. Rotbart. Washington, DC: ASM Press.

Anandasabapathy, S., and W.H. Frishman. 1998. Innovative drug treatments for viral and autoimmune myocarditis. *J. Clin. Pharmacol.*, 38:295–308.

Andreoletti, L., D. Hober, C. Decoene, M.C. Copin, P.E. Lobert, A. Dewilde, C. Stankowiac, and P. Wattre. 1996. Detection of enteroviral RNA by polymerase chain reaction in endomyocardial tissue of patients with chronic cardiac diseases. *J. Med. Virol.*, 48:53–59.

Aradottir, E., E.M. Alonso, and S.T. Shulman. 2001. Severe neonatal enteroviral hepatitis treated with pleconaril. *Pediatr. Infect. Dis. J.*, 20:457–459.

Archard, L.C., M.A. Khan, B.A. Soteriou, H. Zhang, H.J. Why, N.M. Robinson, and P.J. Richardson. 1998. Characterization of Coxsackie B virus RNA in myocardium from patients with dilated cardiomyopathy by nucleotide sequencing of reverse transcription-nested polymerase chain reaction products. *Hum. Pathol.*, 29:578–584.

Baboonian, C., and T. Treasure. 1997. Meta-analysis of the association of enteroviruses with human heart disease. *Heart*, 78:539–543.

Beck, M.A. 2000. Nutritionally induced oxidative stress: effect on viral disease. *Am. J. Clin. Nutr.*, 71(6 Suppl):1676S-81S.

Beck, M.A., and C.C. Matthews. 2000. Micronutrients and host resistance to viral infection. *Proc. Nutr. Soc.*, 59:581–585.

Bendig, J.W., and D.M. Fleming. 1996. Epidemiological, virological, and clinical features of an epidemic of hand, foot, and mouth disease in England and Wales. *Commun. Dis. Rep. CDR Rev.*, 6:R81–R86.

Bergman, I., M.J. Painter, E.R. Wald, D. Chiponis, A.L. Holland, and H.G. Taylor. 1987. Outcome in children with enteroviral meningitis during the first year of life. *J. Pediatr.*, 110:705–709.

Bowles, N.E., and J.A. Towbin. 1998. Molecular aspects of myocarditis. *Curr. Opin. Cardiol.*, 13:179–184.

Caforio, A.L., and W.J. McKenna. 1996. Recognition and optimum management of myocarditis. *Drugs*, 52:515–525.

Carthy, C.M., D. Yang, D.R. Anderson, J.E. Wilson, and B.M. McManus. 1997. Myocarditis as systemic disease: new perspectives on pathogenesis. *Clin. Exp. Pharmacol. Physiol.*, 24:997–1003.

Casas, N., and E. Sunen. 2001. Detection of enterovirus and hepatitis A virus RNA in mussels (*Mytilus* spp.) by reverse transcriptase-polymerase chain reaction. *J. Appl. Microbiol.*, 90:89–95.

Centers for Disease Control and Prevention (CDC). 1997. Nonpolio enterovirus surveillance—United States, 1993–1996. *MMWR*, 46:748–750.

Centers for Disease Control and Prevention (CDC). 1998. Acute hemorrhagic conjunctivitis—St. Croix, U.S. Virgin Islands, September–October 1998. *MMWR*, 47:899–901.

Centers for Disease Control and Prevention (CDC). 2000. Enterovirus surveillance—United States, 1997–1999. *MMWR*, 49:913–916.

Chang, L.Y., T.Y. Lin, Y.C. Huang, K.C. Tsao, S.R. Shih, M.L. Kuo, H.C. Ning, P.W. Chung, and C.M. Kang. 1999. Comparison of enterovirus 71 and coxsackie-virus A16 clinical illnesses during the Taiwan enterovirus epidemic, 1998. *Pediatr. Infect. Dis. J.*, 18:1092–1096.

Chonmaitree, T., and L. Mann. 1995. Respiratory Infections. In *Human Enterovirus Infections*. Edited by H.A. Rotbart. Washington, DC: ASM Press.

Cloete, T.E., E. Da Silva, and L.H. Nel. 1998. Removal of waterborne human enteric viruses and coliphages with oxidized coal. *Curr. Microbiol.*, 37:23–27.

Colquhoun, K.O., S. Timms, J. Slade, and C.R. Fricker. 1995. Domestic plumbing materials and their effect on viruses. *Water Sci. Technol.*, 32:I–III

Coyle, P.K. 1999. Overview of acute and chronic meningitis. *Neurol. Clin.*, 17:691–710.

Dagan, R. 1996. Nonpolio enteroviruses and the febrile young infant: epidemiologic, clinical and diagnostic aspects. *Pediatr. Infect. Dis. J.*, 15:67–71.

Dagan, R., and M.A. Menegus. 1995. Nonpolio Enteroviruses and the Febrile Infant. In *Human Enterovirus Infections*. Edited by H.A. Rotbart. Washington, DC: ASM Press.

Dalakas, M.C. 1995. Enteroviruses and Human Neuromuscular Diseases. In *Human Enterovirus Infections*. Edited by H.A. Rotbart. Washington, DC: ASM Press.

Duncan, E.H., and S.C. Edberg. 1995. Host-microbe interaction in the gastrointestinal tract. *Crit. Rev. Microbiol.*, 21:85–100.

Euscher, E., J. Davis, I. Holzman, and G.J. Nuovo. 2001. Coxsackie virus infection of the placenta associated with neurodevelopmental delays in the newborn. *Obstet. Gynecol.*, 98:1019–26

Figueroa, J.P., D. Ashley, D. King, and B. Hull. 1989. An outbreak of acute flaccid paralysis in Jamaica associated with echovirus type 22. *J. Med. Virol.*, 29:315–319.

Friman, G., L. Wesslen, J. Fohlman, J. Karjalainen, and C. Rolf. 1995. The epidemiology of infectious myocarditis, lymphocytic myocarditis and dilated cardiomyopathy. *Eur. Heart J.*, 16(Suppl)O:36–41.

Fujioka, S., Y. Kitaura, A. Ukimura, H. Deguchi, K. Kawamura, T. Isomura, H. Suma, and A. Shimizu. 2000. Evaluation of viral infection in the myocardium of patients with idiopathic dilated cardiomyopathy. *J. Am. Coll. Cardiol.*, 36:1920–1926.

Furione, M., M. Zavattoni, M. Gatti, E. Percivalle, N. Fioroni, and G. Gerna. 1998. Rapid detection of enteroviral RNA in cerebrospinal fluid (CSF) from patients with aseptic meningitis by reverse transcription-nested polymerase chain reaction. *New Microbiol.*, 21:343–351.

Gantzer, C., Y. Levi, and L. Schwartzbrod. 1996. Effect of heat on the survival of infectious coxsackievirus B3 and its genome in water. *Zentralbl. Hyg. Umweltmed.*, 199:76–83.

Gantzer, C., S. Senouci, A. Maul, Y. Levi, and L. Schwartzbrod. 1997. Enterovirus genomes in wastewater: concentration on glass wool and glass powder and detection by RT-PCR. *J. Virol. Methods*, 65:265–271.

Gear, J.H. 1984. Nonpolio causes of polio-like paralytic syndromes. *Rev. Infect. Dis.*, 6(Suppl)2:S379–S384.

Gilgen, M., B. Wegmuller, P. Burkhalter, H.P. Buhler, U. Muller, J. Luthy, and U. Candrian. 1995. Reverse transcription PCR to detect enteroviruses in surface water. *Appl. Environ. Microbiol.*, 61:1226–1231.

Gjoen, K.V., and A.L. Bruu. 1997. Specific detection of Coxsackie viruses A by the polymerase chain reaction. *Clin. Diagn. Virol.*, 8:183–188.

Gondo, K., K. Kusuhara, H. Take, and K. Heda. 1995. Echovirus type 9 epidemic in Kagoshima, Southern Japan: seroepidemiology and clinical observation of aseptic meningitis. *Pediatr. Infect. Dis. J.*, 14:787–791.

Gorgievski-Hrisoho, M., J.D. Schumacher, N. Vilimonovic, D. Germann, and L. Matter. 1998. Detection by PCR of enteroviruses in cerebrospinal fluid during a summer outbreak of aseptic meningitis in Switzerland. *J. Clin. Microbiol.*, 36:2408–2412.

Guttman-Bass, N., and A. Nasser. 1984. Simultaneous concentration of four enteroviruses from tap, waste, and natural waters. *Appl. Environ. Microbiol.*, 47:1311–1315.

Helfand, R.F., A.S. Khan, M.A. Pallansch, J.P. Alexander, H.B. Meyers, R.A. DeSantis, L.B. Schonberger, and L.J. Anderson. 1994. Echovirus 30 infection and aseptic meningitis in parents of children attending a child care center. *J. Infect. Dis.*, 169:1133–1137.

Hill, W.M.J. 1996. Are echoviruses still orphans? *Br. J. Biomed. Sci.*, 53:221–226.

Ho, M., E.R. Chen, K.H. Hsu, S.J. Twu, K.T. Chen, S.F. Tsai, J.R. Wang, and S.R. Shih. 1999. An epidemic of enterovirus 71 infection in Taiwan. Taiwan Enterovirus Epidemic Working Group. *N. Engl. J. Med.*, 341:929–935.

Hovi, T., M. Stenvik, and M. Rosenlew. 1996. Relative abundance of enterovirus serotypes in sewage differs from that in patients: clinical and epidemiological implications. *Epidemiol. Infect.*, 116:91–97.

Hugues, B., M. Andre, J.L. Plantat, and H. Champsaur. 1993. Comparison of glass wool and glass powder methods for concentration of viruses from treated waste waters. *Zentralbl. Hyg. Umweltmed.*, 193:440–449.

Hurst, C.J. 1991. Presence of enteric viruses in freshwater and their removal by the conventional drinking water treatment process. *Bull. World Health Org.*, 69:113–119.

Hurst, C.J., W.H. Benton, and K.A. McClellan. 1989. Thermal and water source effects upon stability of enteroviruses in surface freshwaters. *Can. J. Microbiol.*, 35:474–480.

Hyoty, H, M. Hiltunen, and M. Lonnrot. 1998. Enterovirus infections and insulin dependent diabetes mellitus—evidence for causality. *Clin. Diagn. Virol.*, 9:77-84.

Ikeda, R.M., S.F. Kondracki, P.D. Drabkin, G.S. Birkhead, and D.L. Morse. 1993. Pleurodynia among football players at a high school. An outbreak associated with coxsackievirus B1. *JAMA*, 270:2205–2206.

Irving, L.G., and F.A. Smith. 1981. One-year survey of enteroviruses, adenoviruses, and reoviruses isolated from effluent at an activated-sludge purification plant. *Appl. Environ. Microbiol.*, 41:51–59.

Isacsohn, M., A.I. Eidelman, M. Kaplan, A. Goren, B. Rudensky, R. Handsher, and Y. Barak. 1994. Neonatal coxsackievirus group B infections: experience of a single department of neonatology. *Isr. J. Med. Sci.*, 30:371–374.

Jenista, J.A., K.R. Powell, and M.A. Menegus. 1984. Epidemiology of neonatal enterovirus infection. *J. Pediatr.*, 104:685–690.

Kawai, C. 1999. From myocarditis to cardiomyopathy: mechanisms of inflammation and cell death: learning from the past for the future. *Circulation,* 99:1091–1100.

Keswick, B.H., C.P. Gerba, H.L. DuPont, and J.B. Rose. 1984. Detection of enteric viruses in treated drinking water. *Appl. Environ. Microbiol.,* 47:1290–1294.

Keswick, B.H., C.P. Gerba, and S.M. Goyal. 1981. Occurrence of enteroviruses in community swimming pools. *Am. J. Public Health,* 71:1026–1030.

Kosrirukvongs, P., R. Kanyok, S. Sitritantikorn, and C. Wasi. 1996. Acute hemorrhagic conjunctivitis outbreak in Thailand, 1992. *Southeast Asian J. Trop. Med. Public Health,* 27:244–249.

Krikelis, V., N. Spyrou, P. Markoulatos, and C. Serie. 1985. Seasonal distribution of enteroviruses and adenoviruses in domestic sewage. *Can. J. Microbiol.,* 31:24–25.

Le Guyader, F., L. Haugarreau, L. Miossec, E. Dubois, and M. Pommepuy. 2000. Three-year study to assess human enteric viruses in shellfish. *Appl. Environ. Microbiol.,* 66:3241–3248.

Leogrande, G. 1991. Studies on the epidemiology of child infections in the Bari area. I. Current state of immunity against Coxsackie B viruses. *Ric. Clin. Lab.,* 21:95–103.

Levander, O.A., and M.A. Beck. 1999. Selenium and viral virulence. *Br. Med. Bull.,* 55:528–533.

Li, J.W., X.W. Wang, Q.Y. Rui, N. Song, F.G. Zhang, Y.C. Ou, and F.H. Chao. 1998. A new and simple method for concentration of enteric viruses from water. *J. Virol. Methods,* 74:99–108.

Liu, P., T. Martino, M.A. Opavsky, and J. Penninger. 1996. Viral myocarditis: balance between viral infection and immune response. *Can. J. Cardiol.,* 12:935–943.

Maguire, H.C., P. Atkinson, M. Sharland, and J. Bendig. 1999. Enterovirus infections in England and Wales: laboratory surveillance data: 1975 to 1994. *Commun. Dis. Public Health,* 2:122–125.

Martino, T.A., P. Liu, M. Petric, and M.J. Sole. 1995. Enteromyocarditis and dilated cardiomyopathy: A review of clinical and experimental studies. In *Human Enterovirus Infections.* Edited by H.A. Rotbart. Washington, DC: ASM Press.

Minor, P. 1998. Picornaviruses. In *Virology.* Edited by B.W.J. Mahy and L. Collier. London: Arnold.

Modlin, J.F. 1986. Perinatal echovirus infection: insights from a literature review of 61 cases of serious infection and 16 outbreaks in nurseries. *Rev. Infect. Dis.*, 8:918–926.

Modlin, J.F. 1988. Perinatal echovirus and group B coxsackievirus infections. *Clin. Perinatol.*, 15:233–246.

Modlin, J.F. 2000. Coxsackieviruses, echoviruses, and newer enteroviruses. In *Principles and Practice of Infectious Diseases.* Edited by G.L. Mandell, J.E. Bennett, and R. Dolin. Philadelphia: Churchill Livingstone, Inc.

Moore, A.C., B.L. Herwaldt, G.F. Craun, R.L. Calderon, A.K. Highsmith, and D.D. Juranek. 1993. Surveillance for waterborne disease outbreaks—United States, 1991–1992. *MMWR*, 42:1–22.

Moore, M., M.H. Kaplan, J. McPhee, D.J. Bregman, and S.W. Klein. 1984. Epidemiologic, clinical, and laboratory features of Coxsackie B1–B5 infections in the United States, 1970–79. *Public Health Rep.*, 99:515–522.

Morens, D.M., and M.A. Pallansch. 1995. Epidemiology. In *Human Enterovirus Infections.* Edited by H.A. Rotbart. Washington, DC: ASM Press.

Nairn, C., D.N. Galbraith, and G.B. Clements. 1995. Comparison of coxsackie B neutralisation and enteroviral PCR in chronic fatigue patients. *J. Med. Virol.*, 46:310–313.

Najm, I., and R.R. Trussell. 1999. New and emerging drinking water treatment technologies. In *Identifying Future Drinking Water Contaminants.* Washington, DC: National Academy Press.

Nayak, N., S.K. Gupta, G.V. Murthy, G. Satpathy, and S. Mohanty. 1996. Community-based investigation of an outbreak of acute viral conjunctivitis in urban slums. *Trop. Med. Int. Health*, 1:667–671.

Payment, P., F. Affoyon, and M. Trudel. 1988. Detection of animal and human enteric viruses in water from the Assomption River and its tributaries. *Can. J. Microbiol.*, 34:967–973.

Payment, P., S. Fortin, and M. Trudel. 1986. Elimination of human enteric viruses during conventional waste water treatment by activated sludge. *Can. J. Microbiol.*, 32:922–925.

Payment, P., M. Trudel, and R. Plante. 1985. Elimination of viruses and indicator bacteria at each step of treatment during preparation of drinking water at seven water treatment plants. *Appl. Environ. Microbiol.*, 49:1418–1428.

Piirainen, L., T. Hovi, and M. Roivainen. 1998. Variability in the integrity of human enteroviruses exposed to various simulated in vivo environments. *Microb. Pathog.*, 25:131–137.

Puig, M., J. Jofre, F. Lucena, A. Allard, G. Wadell, and R. Girones. 1994. Detection of adenoviruses and enteroviruses in polluted waters by nested PCR amplification. *Appl. Environ. Microbiol.*, 60:2963–2970.

Rantakallio, P., P. Jones, J. Moring, and L. Von Wendt. 1997. Association between central nervous system infections during childhood and adult onset schizophrenia and other psychoses: a 28-year follow-up. *Int. J. Epidemiol.*, 26:837–843.

Reynolds, K.A., C.P. Gerba, M. Abbaszadegan, and L.L. Pepper. 2001. ICC/PCR detection of enteroviruses and hepatitis A virus in environmental samples. *Can. J. Microbiol.*, 47:153–157.

Reynolds, K.A., K. Roll, R.S. Fujioka, C.P. Gerba, and I.L. Pepper. 1998. Incidence of enteroviruses in Mamala Bay, Hawaii using cell culture and direct polymerase chain reaction methodologies. *Can. J. Microbiol.*, 44:598–604.

Rice, E.W., R.M. Clark, and C.H. Johnson. 1999. Chlorine inactivation of *Escherichia coli* O157:H7. *Emerg. Infect. Dis.*, 5:461–463.

Rigonan, A.L., L. Mann, and T. Chonmaitree. 1998. Use of monoclonal antibodies to identify serotypes of enterovirus isolates. *J. Clin. Microbiol.*, 36:1877–1881.

Roivainen, M., M. Knip, H. Hyoty, P. Kulmala, M. Hiltunen, P. Vahasalo, T. Hovi, and H.K. Akerblom. 1998. Several different enterovirus serotypes can be associated with prediabetic autoimmune episodes and onset of overt IDDM. Childhood Diabetes in Finland (DiMe) Study Group. *J. Med. Virol.*, 56:74–78.

Rorabaugh, M.L., L.E. Berlin, F. Heldrich, K. Roberts, L.A. Rosenberg, T. Doran, and J.F. Modlin. 1993. Aseptic meningitis in infants younger than 2 years of age: acute illness and neurologic complications. *Pediatrics*, 92:206–211.

Rotbart, H.A. 1995a. Enteroviral infections of the central nervous system. *Clin. Infect. Dis.*, 20:971–981.

Rotbart, H.A. 1995b. Meningitis and Encephalitis. In *Human Enterovirus Infections*. Edited by H.A. Rotbart. Washington, DC: ASM Press.

Rotbart, H.A., and F.G. Hayden. 2000. Picornavirus infections: a primer for the practitioner. *Arch. Fam. Med.*, 9:913–920.

Rotbart, H.A., and J.P. Romero. 1995. Laboratory Diagnosis of Enteroviral Infections. In *Human Enterovirus Infections*. Edited by H.A. Rotbart. Washington, DC: ASM Press.

Rotbart, H.A., and A.D. Webster. 2001. Treatment of potentially life-threatening enterovirus infections with pleconaril. *Clin. Infect. Dis.*, 32:228–235.

Schiff, G.M., and J.R. Sherwood. 2000. Clinical activity of pleconaril in an experimentally induced coxsackievirus A21 respiratory infection. *J. Infect. Dis.*, 181:20–26.

Schiff, G.M., G.M. Stefanovic, E.C. Young, D.S. Sander, J.K. Pennekamp, and R.L. Ward. 1984. Studies of echovirus-12 in volunteers: determination of minimal infectious dose and the effect of previous infection on infectious dose. *J. Infect. Dis.*, 150:858–866.

Schlesinger, Y., M.H. Sawyer, and G.A. Storch. 1994. Enteroviral meningitis in infancy: potential role for polymerase chain reaction in patient management. *Pediatrics,* 94:157–162.

Schwarer, A.P., S.S. Opat, A.M. Watson, D. Spelman, F. Firkin, and N. Lee. 1997. Disseminated echovirus infection after allogeneic bone marrow transplantation. *Pathology,* 29:424–425.

Sells, C.J., R.L. Carpenter, and C.G. Ray. 1975. Sequelae of central-nervous-system enterovirus infections. *N. Engl. J. Med.*, 293:1–4.

Senouci, S., A. Maul, and L. Schwartzbrod. 1996. Comparison study on three protocols used to concentrate poliovirus type 1 from drinking water. *Zentralbl. Hyg. Umweltmed.*, 198:307–317.

Shieh, Y.S., R.S. Baric, and M.D. Sobsey. 1997. Detection of low levels of enteric viruses in metropolitan and airplane sewage. *Appl. Environ. Microbiol.*, 63:4401–4407.

Sicherer, S.H., and J.A. Winkelstein. 1998. Primary immunodeficiency diseases in adults. *JAMA,* 279:58–61.

Sobsey, M.D., R.M. Hall, and R.L. Hazard. 1995. Comparative reductions of hepatitis A virus, enteroviruses and coliphage MS2 in miniature soil columns. *Water Sci. Technol.*, 1:203–209.

Stellrecht, K.A., I. Harding, F.M. Hussain, N.G. Mishrik, R.T. Czap, M.L. Lepow, and R.A. Venezia. 2000. A one-step RT-PCR assay using an enzyme-linked detection system for the diagnosis of enterovirus meningitis. *J. Clin. Virol.*, 17:143–149.

Straub, T.M., I.L. Pepper, and C.P. Gerba. 1995. Comparison of PCR and cell culture for detection of enteroviruses in sludge-amended field soils and determination of their transport. *Appl. Environ. Microbiol.*, 61:2066–2068.

Suddaby, E.C. 1996. Viral myocarditis in children. *Crit. Care Nurse,* 16:73–82.

Takahashi, S., A. Miyamoto, J. Oki, H. Azuma, and A. Okuno. 1995. Acute transverse myelitis caused by ECHO virus type 18 infection. *Eur. J. Pediatr.*, 154:378–380.

Tanel, R.E., S.Y. Kao, T.M. Niemiec, M.J. Loeffelholz, D.T. Holland, L.A. Shoaf, E.R. Stucky, and J.C. Burns. 1996. Prospective comparison of culture vs genome detection for diagnosis of enteroviral meningitis in childhood. *Arch. Pediatr. Adolesc. Med.*, 150:919–924.

Tani, N., Y. Dohi, N. Kurumatani, and K. Yonemasu. 1995. Seasonal distribution of adenoviruses, enteroviruses and reoviruses in urban river water. *Microbiol. Immunol.*, 39:577–580.

Taylor, J.S., and M. Wiesner. 1999. Membranes. In *Water Quality and Treatment: A Handbook of Community Water Supplies.* 5th ed. Edited by American Water Works Association. New York: McGraw-Hill.

Townsend, T.R., E.A. Bolyard, R.H. Yolken, W.E. Beschorner, C.A. Bishop, W.H. Burns, G.W. Santos, and R. Saral. 1982. Outbreak of Coxsackie A1 gastroenteritis: a complication of bone-marrow transplantation. *Lancet*, 1:820–823.

van Olphen, M., J.G. Kapsenberg, E. van de Baan, and W.A. Kroon. 1984. Removal of enteric viruses from surface water at eight waterworks in The Netherlands. *Appl. Environ. Microbiol.*, 47:927–932.

White, D.O., F.J. Fenner. 1994. Picornaviridae. In *Medical Virology*, 4th ed. Edited by D.O. White and F.J. Fenner. San Diego: Academic Press.

Wilfert, C.M., R.J. Thompson Jr., T.R. Sunder, A. O'Quinn, J. Zeller, and J. Blacharsh. 1981. Longitudinal assessment of children with enteroviral meningitis during the first three months of life. *Pediatrics*, 67:811–815.

Wilson, G.A., and J.M. Weber. 1995. Laboratory reports of human viral and selected nonviral agents in Canada—1993. *Can. Med. Assoc. J.*, 153:51–53.

Yamashita, K., K. Miyamura, S. Yamadera, N. Kato, M. Akatsuka, M. Hashido, S. Inouye, and S. Yamazaki. 1994. Epidemics of aseptic meningitis due to echovirus 30 in Japan. A report of the National Epidemiological Surveillance of Infectious Agents in Japan. *Jpn. J. Med. Sci. Biol.*, 47:221–239.

Yates, M.V., C.P. Gerba, and L.M. Kelley. 1985. Virus persistence in groundwater. *Appl. Environ. Microbiol.*, 49:778–781.

Chapter 9

Cyanobacteria in Drinking Water

by Martha A. Embrey

Executive Summary

Problem Formulation

Occurrence of Intoxication

- Worldwide, the number of humans acutely affected by cyanobacterial toxins is low compared with other waterborne contaminants. However, because of decreasing water quality, the potential for an increase in incidents is high.

Role of Waterborne Exposure

- Cyanobacteria are found in all types of water: lakes, rivers, marine environments, and drinking water reservoirs. Most acute exposures result from recreational water use; low levels in drinking water are associated with an increase in hepatocellular cancer in certain exposed populations.

Degree of Morbidity and Mortality

- Cyanobacteria act primarily as hepatotoxins and neurotoxins. They are extremely potent toxins and, therefore, have the potential to be fatal to humans. However, acute oral or dermal exposures have not resulted in any known human deaths. Reported illnesses in humans exposed to cyanobacterial toxins range from dermatitis and diarrhea to subclinical and frank hepatitis. Illness resolves on its own. Chronic exposures—as through drinking water—are linked with liver cancer in humans and other kinds of cancers in animals.

- Cyanobacterial toxins have caused deaths when contaminated water was injected directly into the bloodstreams of kidney dialysis patients.

Detection Methods in Water/Clinical Specimens

- The most obvious way to detect cyanobacteria in water is visually—seeing evidence of algal blooms in surface water. Standard microscopy can be used to speciate environmental samples. Quick and easy ways to detect toxicity are still under development. Mouse bioassays have traditionally been used for this purpose, but this method is undesirable for many reasons. Newer methods that are more frequently seen now include the use of liquid chromatography and enzyme-linked immunosorbent assays. Polymerase chain reaction is also under development.

Mechanisms of Water Contamination

- Water conditions such as nutrient levels and temperature contribute to the formation of algal blooms, which have the potential to release cyanobacterial toxins.

Concentrations at Intake

- Cyanobacterial toxins are ubiquitous, though their occurrence is dependent on the conditions that contribute to algal bloom formation. Concentrations vary widely depending on the species of bloom and the stage of the bloom's formation and deterioration. Toxin concentrations have been reported as ranging from 0.2 µg/L to 8.5 mg/L.

Efficacy of Water Treatment

- There is some question as to the efficacy of standard drinking water treatment (e.g., coagulation, sedimentation, disinfection, and filtration) for removing all but large concentrations of cyanobacterial toxins, though current methods are effective enough to prevent any acute lethal effects.
- Evidence on the efficacy of chlorine on the microcystins is equivocal; chlorine is ineffective on anatoxin-a. Activated carbon treatment appears to be the best removal method for treated water.
- Preventing the formation of blooms in the source water is the best way to assure cyanobacteria-free drinking water.
- Membrane filtration technology has the potential to remove virtually any cyanobacteria or their toxins from drinking water.

Survival/Amplification in Distribution

- Unknown.

Routes of Exposure

- Primarily ingestion through drinking water or recreational water contact, also, dermal exposure and possibly aerosolization.

Dose Response

Toxic Dose

- Unknown for humans. The no observed adverse effect level (NOAEL) for mice dosed orally with microcystin-LR has been reported to be 40 µg/kg/day for 13 weeks. The NOAEL reported for mice dosed orally with anatoxin-a has been reported to be 0.1 mg/kg/day. Intraperitoneal and intranasal exposure is more potent than oral ingestion for both toxins.

Probability of Illness Based on Exposure

- Unknown for humans for oral ingestion.

Efficacy of Medical Treatment

- Unknown in an acute exposure; however, antihistamines and steroids may be helpful for allergic reactions. If given in a timely manner, activated charcoal or an emetic could have a positive effect on the toxic response. Rifampin has been tested in laboratory animals with limited success, but their possible use in humans is unknown.

Secondary Spread

- Not applicable.

Chronic Sequelae

- Outcomes from cyanobacterial toxins are mostly acute; however, the microcystins and other toxins have been shown to be tumor promoters, and epidemiological evidence suggests that low levels of microcystins in drinking water are associated with an increase in hepatocellular cancer.

Introduction

Though cyanobacteria, or blue-green algae, are listed on the US Environmental Protection Agency's (USEPA) contaminant candidate list (CCL) as microbial agents of concern in drinking water, their health effects and water ecology are more closely related to those of a chemical contaminant than an infectious microorganism. The dangers of cyanobacteria in water have been documented for more than 100 years. Common sense has dictated that because of aesthetic considerations (i.e., taste and odor), humans would not choose to drink or recreate in water with algal (or cyanobacterial) blooms, thereby protecting themselves from cyanobacteria's acute toxic effects. Animals apparently are not deterred by the foul taste and, consequently, are often affected by contaminated water. In fact, in an experimental study, mice chose to consume toxic cyanobacteria over clear water—swallowing the toxic material until they died (Lopez and Costas 1999). However, as questions have been raised about the possible risk of exposure to low-level concentrations of toxic cyanobacteria, these commonsense views have shifted. Also, changing ecology and an abundance of nutrients in the environment have increased the amount of eutrophication in lakes and reservoirs, leading to an increased occurrence of blooms in drinking water supplies. An increase in toxic blooms, caused mainly by climatic conditions in the last decade or so, have led to the water research communities in countries with problems, such as Australia and the United Kingdom, to pay great attention to this issue. Consequently, much of the research on the occurrence, detection, and effects of toxic cyanobacteria is being performed by these and other countries facing this emerging problem. At a recent conference on toxic cyanobacteria, cochair Wayne Carmichael, a cyanobacterial researcher stated, "Decreasing water quality throughout the world means that these blooms are present more often and for a longer duration, and people are using marginal water supplies more than they did in the past. They are forced to bathe in a bloom or to use water from a bloom.... We are now starting to see acute poisonings, contact irritations, accidental ingestion, and low-dose exposure [that may result in] liver cancer" (EHP 1999).

Cyanobacteria are classed separately and, although they were known commonly as "blue-green algae," they are not true algae because, unlike eukaryotic algae, they are prokaryotes, like bacteria. They are ubiquitous in the environment, occurring worldwide in freshwater rivers, lakes, streams, ponds, and reservoirs and in marine waters. Cyanobacteria differ in size and shape, ranging from a small unicellular structure to a filamentous form. Their structure usually depends on the type of mucilaginous material they produce—clumping together of individual cells or cells forming a long filamentous sheath.

Cyanobacterial species number in the thousands, and though a recent estimate named 46 toxin-producing species, the true number of species that produce toxins is unknown (WHO 1998). It is estimated that 50–70% of all algal blooms are toxic (Yoo et al. 1995), and more than 60 different toxins have been identified and described (Codd, Ward, and Bell 1997). The genera that are most frequently cited in relation to adverse health conditions are *Anabaena, Microcystis, Oscillatoria, Aphanizomenon,* and *Nodularia*. Different species exhibit different toxin-producing characteristics, but not all toxin-producing species or toxic water blooms will be toxic at all times. The toxins are divided into four functional categories: (1) neurotoxins, such as anatoxin-a and anatoxin–a(s); (2) hepatotoxins, such as microcystins; (3) general cytotoxins, such as cylindromspermopsin; and (4) lipopolysaccharide endotoxins. The human health effects can be acute and range from gastroenteritis and dermatitis to hepatitis. However, possible chronic health risks related to the tumor-promoting effects of the microcystins have been identified in the past 10 years or so, and these low-level chronic exposures are of the most concern.

An interesting side to considering cyanobacteria as a toxic contaminant is the beneficial aspect of cyanobacteria in human health. Extracts of blue-green algae have been found to have antitumor, antibiotic, and antifungal properties—clinical trials are ongoing looking at cyanobacterial extracts as a cancer treatment (EHP 1999). Blue-green algae has also been used for some time as a natural health promoter; more than 1 million people in the United States and Canada are estimated to use blue-green algae as a dietary supplement for this reason, though toxic contaminants should be a concern (Gilroy et al. 2000).

Epidemiology

Though the health effects of cyanobacterial toxins have been known for more than 100 years, investigations into possible cyanobacterial outbreaks or other epidemiological studies in humans have generally been incomplete or nonexistent (Elder, Hunter, and Codd 1993; Yoo et al. 1995). However, there is enough evidence from human incidents, animal studies, and knowledge about the toxins' molecular structure to prove that cyanobacteria can endanger human health.

A number of gastrointestinal outbreaks ultimately have been associated with cyanobacterial toxins in the water; these include events in Washington, D.C., in 1930; Charleston, West Virginia, the same year, involving 8,000–10,000 people; and New Jersey in 1940 (Bourke and Hawes 1983). In addition to ingestion through drinking water, many cases of acute gastrointestinal effects or allergic dermatitis in people who swam in or fell into contaminated water have been reported. Except in extreme cases (as are noted below), patients' symptoms dissipate without any sort of therapy. In addition to the circumstantial evidence of exposure, cyanobacterial gastroenteritis differs from microbial gastroenteritis because of its rapid onset—sometimes only hours after exposure (Billings 1981). Following is a series of reported outbreaks related to cyanobacteria.

- A large outbreak of hepatoenteritis affected 149 people (mainly children) in Queensland, Australia, in 1979 (Bourke and Hawes 1983). The outbreak occurred after an algal bloom had been treated with copper sulfate, which probably lysed the cyanobacterial cells, releasing toxins. Later tests showed the presence of *Cylindrospermopsis raciborskii* (a species endemic in the tropics) in the community's drinking water reservoir. More than 140 children required hospitalization for fluid replacement; all recovered completely within 4–26 days. Recent analysis tried to find the etiology of a prevalent ailment in the outback of Australia that had been termed "Barcoo fever," "Barcoo sickness," or "the Barcoo" (Hayman 1992). A comparison of the available records on this illness with known gastrointestinal diseases resulted in a hypothesis that cyanobacterial toxins, specifically *C. raciborskii*, were the cause of this once-common outback malady. The author pointed out that these toxic cyanobacteria have probably not disappeared, just lessened in incidence and severity.
- Blooms of *Nodularia* and *Anabaena* contaminating the Murray River in South Australia were associated with illnesses to nearly 200 people who used the river water in their households and/or for recreation (el Saadi et al. 1995; Soong et al. 1992). Symptoms for those with skin contact included rash, itching, mouth blistering, and eye irritation. Oral ingestion resulted in symptoms such as diarrhea, vomiting, sore throat, and headache. Those patients with exposures from both routes reported a mixture of the two types of symptoms. Most people in the area did not use the river water as their drinking water source but rather rainwater collected in cisterns. Subjects who drank chlorinated river water rather than rainwater were significantly more likely to develop gastrointestinal symptoms ($p = 0.008$). Those using untreated river

water for domestic purposes other than drinking were significantly more likely to report both gastrointestinal and dermal symptoms.

- In August 1979, a series of outbreaks related to recreational lakes occurred in Pennsylvania (Billings 1981). Fifty people or more suffered diarrhea, vomiting, headache, sore throat, eye irritation, and other hay fever-type symptoms. Circumstantial evidence led the investigators to believe the outbreaks were caused by an *Anabaena* bloom in one lake and an unidentified cyanobacterial toxin in the other lake (by the time of the investigation, no bloom was apparent). One of the lakes was closed to recreation until the bloom dissipated.
- Eight hundred fifty-two Australians were interviewed about their exposure to cyanobacteria during summer recreational activities (Pilotto et al. 1997). Comparing those exposed to the water (n = 777) with those who were not (n = 75), researchers found an association between increased symptoms and both the duration of water contact and cyanobacterial cell density. People who were exposed to more than 5,000 cells/mL for more than an hour had a significantly higher incidence of illness. However, symptoms were not related to the presence of hepatotoxins in the water.
- Soldiers training in water and canoes in England in 1989 were exposed to *Microcystis* blooms that produced gastrointestinal illness as well as dermal irritation, such as blistering sores (Yoo et al. 1995). A British national survey that year estimated that 60–70% of the widespread algal blooms in the area were toxin producers.

Clinical

The health effects of exposure to cyanobacteria can range from acute hepatotoxicity or neurotoxicity from ingestion to an allergic reaction from dermal contact. In recent years, concern over the connection between chronic, low-level ingestion and cancer promotion has been prominent.

Acute

Most of the data available on the acute response to cyanobacterial poisoning is in animals, particularly livestock, because no human deaths have been attributed to oral ingestion of cyanobacterial toxins (WHO 1998). Cyanobacterial poisons generally work through one of two mechanisms—either as neurotoxins or hepatotoxins. Of the four neurotoxins that have been studied most extensively, anatoxin-a and anatoxin-a(s) appear to be unique to cyanobacteria. The other two, saxitoxin and neosaxitoxin, are also produced primarily by species of marine algae but have also been found in freshwater (Lagos et al. 1999; Pereira et al. 2000). Anatoxin-a is produced by various species of *Anabaena* and *Oscillatoria*; anatoxin-a(s) is produced by *Oscillatoria*. Though chemically distinct, these two neurotoxins are nicotinic agonists that block neuromuscular transmission and cause death by respiratory arrest (Fawell et al. 1999b). Animals that are poisoned in this way usually suffer muscle twitching, cramping, and paralysis and, if the respiratory muscles are affected, convulsions and death. Saxitoxin and neosaxitoxin also disrupt the acetylcholine pathway between the nerves and muscles. Though they occur in some species of *Anabaena* and *Aphanizomenon*, they are more commonly associated with marine dinoflagellates (i.e., red tides), which cause paralytic seafood poisoning. The neurotoxins are potent but they do not remain in the body for long and exert no lasting effects.

Cyanobacterial hepatotoxins are more common and more geographically ubiquitous than the neurotoxins (Carmichael 1994). The two main types of cyanobacterial hepatotoxins are the microcystins and the nodularins from the *Mycrocystis* and *Nodularia* genera. The microcystins occur far more commonly. The hepatotoxins work by causing the hepatocytes to shrink, allowing blood to collect in the liver tissue, leading to local damage and often shock. The hepatotoxins disrupt enzymes that affect the development of proteins. Animals poisoned by hepatotoxins may suffer weakness, anorexia, and sometimes behavioral derangement. Coma and tremors precede death from hemorrhage or liver failure. Necropsy shows poisoned animals to have livers that are two to three times the normal size (Yoo et al. 1995).

Symptoms from acute exposure in humans due to contact with recreational water include gastroenteritis, headache, skin and eye irritation, hay fever symptoms or asthma, and fatigue. Because diagnosis and detection for cyanobacterial poisoning have been lacking, we do not know if these symptoms are related more to neurotoxin or hepatotoxin exposure (or both) (Yoo et al. 1995). Apparently, allergic reactions to cyanobacteria, resulting in dermal irritation and hay fever symptoms especially, are not uncommon, though they have only been reported in the literature as case reports (WHO 1998; Yoo et al. 1995). This includes a report of pneumonia associated with the inhalation of microcystins while canoeing (Turner et al. 1990). Lipopolysaccharide endotoxins have also been implicated in waterborne outbreaks in humans, usually with dermal irritation (Carmichael 1981).

Chronic

Acute dose–response curves for the microcystins are steep; therefore, acute liver damage will probably not be apparent until the dose is high enough to cause severe toxic effects (WHO 1998). People who are not aware of the risk factors could consume toxic levels over an extended period of time, because they will not experience symptoms; however, cumulative effects are possible, including progressive liver damage and the potential for tumor promotion (WHO 1998). The microcystins and nodularins block the protein phosphatases 1 and 2a, which act as molecular switches in cells. This inhibition results in a mechanism for tumor promotion that has been demonstrated in laboratory animals (Codd, Ward, and Bell 1997; Falconer and Buckley 1989; WHO 1998; Yoo et al. 1995) and is likely a risk to humans (FWR, Nov. 1994c; WHO 1998). In addition to evidence of liver toxicity, researchers have shown that microcystins stimulated colon tumor growth in experimental mice (Humpage et al. 2000).

Results of in vitro genotoxicity tests of microcystins extracted from a Chinese water source were recently published (Ding et al. 1999). The researchers tested the extract using three assays: (1) *Salmonella typhimurium* (Ames test), (2) single-cell gel electrophoresis, and (3) mouse micronucleus. The extracted sample (which was primarily microcystin-LR) was found strongly mutagenic by the Ames test and tested positive for DNA damage in both the single-cell gel and micronucleus assays. These results suggest that cyanobacterial toxins are highly genotoxic, which makes them potential carcinogens. Rao and Bhattacharya (1996) also reported the genotoxicity of microcystin-LR in mouse livers using a different assay than the Chinese researchers—fluorimetric analysis of DNA unwinding. Their results showed a dose–response effect that was also time dependent. A Japanese study in rats found that nodularin, like microcystin-LR, is not only a tumor promoter but can also act as an initiator, giving it the properties of a complete carcinogen (Ohta et al. 1994). Other researchers, however, have not found genotoxic effects in in vitro assays (Standridge, Karner, and Barnum 1999). Preliminary in vivo data suggest that the alkaloid toxin, cylindrospermopsin, which is produced by *C. raciborskii*, is a tumor initiator (Falconer and Humpage 2001).

Other studies have linked water mutagenicity with liver and stomach cancer in men living along the Huangpu River, China (Tao et al. 1991a, 1991b). The mortality rates for male stomach and liver cancer increased from the upper part of the river, to the middle, and to the lower part (67.7, 86.2, and 146.0/100,000 person years for stomach and 56.9, 67.7, and 81.3/100,000 person years for liver). These figures are consistent with the mutagenicity levels (by Ames test) of samples up and down the river, which were 0%, 70%, and 100% from upper end to lower end. No hypothesis was offered on what might be making the water mutagenic, though microcystins are a real possibility. These data provide plausible evidence to explain and validate epidemiological studies linking microcystin-contaminated water with high levels of liver cancer in certain Chinese populations.

An epidemiological study in New South Wales, Australia, investigated the effects of microcystin in the drinking water supply (Bourke et al. in Yoo et al. 1995). The bloom that was present in the water supply was being tracked as part of an ongoing survey of cyanobacteria, so measurements of its toxicity and dates of intentional chemical lysing were known. Researchers measured liver function enzymes in a group of people exposed to the drinking water (and controls who were not) before the bloom, during the bloom, and after the bloom. They found that only the exposed population suffered a statistically significant increase in serum gamma glutamyl transferase, which indicates hepatotoxicity (subclinical hepatitis). The increase occurred only at the time of the bloom.

Important epidemiological research has been going on in Chinese provinces (mainly Jian-Su, Qidong, and Guangxi) where the levels of liver cancer are some of the highest in the world. Identified risk factors include hyperendemic hepatitis B, hepatitis C, and the consumption of aflatoxin through contaminated corn and soy products (Yu 1995). Another risk factor that has been identified through six different epidemiological studies is drinking water source: People who drink from pond or ditch water experienced higher rates of hepatocellular cancer than those who take their water from deep wells. Additional studies have shown that pond and ditch water have typically been contaminated with cyanobacterial toxins, specifically, microcystin-LR, a potent tumor promoter (Harada et al. 1996; Ueno et al. 1996). Concentration levels have been reported as ranging from 60 pg/mL to 100 ng/mL, with samples testing negative from deep wells (Harada et al. 1996; Ueno et al. 1996). Yu (1995) reported that education on the risk factors and prevention of liver cancer in affected provinces has resulted in the government requiring hepatitis B vaccination for babies to lower the endemic rate, dietary changes to reduce aflatoxin exposure, and, in the Qidong province, more than 80% of people changing their source water from ponds or ditches to deep wells to reduce cyanobacterial exposure. Undoubtedly, the combination of these risk factors has worked synergistically to result in these areas' extremely high rate of liver cancer. Future years will show if these prevention strategies produce positive results.

Some outbreaks of chronic diarrheal disease have been possibly linked with cyanobacteria. An organism was found in the stools of 55 visitors to Nepal at a Kathmandu clinic (Shlim et al. 1991). Their illness consisted of prolonged, watery diarrhea; anorexia; fatigue; and weight loss; and the mean duration of illness was 43 days. Antimicrobial treatment did not alter the course of the illness. Based on a microscopic analysis, the isolated organism appeared to share some characteristics of cyanobacteria; however, no specific tests were done to confirm that diagnosis. Similar organisms were found in the stools of eight people with watery diarrhea who had traveled to tropical countries (four were HIV+) (Long et al. 1990). Again, based on microscopy, it appeared that the organisms could be some sort of cyanobacteria, but confirmatory data were unavailable. Based on the evidence, the outbreaks could also have resulted from a novel protozoal infection (Elder, Hunter, and Codd 1993).

Treatment

Treatment for hepatotoxic poisoning has focused on animals and livestock, and even then, little is known about efficacious treatment options. Powdered charcoal and cholestyramine have been suggested, and some drugs, including cyclosporin-A, rifampin, and silymarin, have been tested in laboratory animals with limited success (Dawson 1998; Yoo et al. 1995). Supportive treatment might include administering whole blood or glucose. We can only estimate how useful such treatments might be in the case of human toxicity. For humans who exhibit apparent allergic reactions after exposure to cyanobacteria, antihistamines or steroids may be helpful (Elder, Hunter, and Codd 1993).

No known medical therapy exists for anatoxin-a, which is the most common neurotoxin produced by cyanobacteria, though respiratory support may give the body enough time to detoxify itself. The effects of the neurotoxins are very fast, which may make diagnosis, without obvious evidence of consumption, difficult.

Susceptible Subpopulations

Little research has been conducted that looks at the effects of cyanobacterial exposure in all humans; consequently, data on possible sensitive subpopulations is sparse.

Developing Offspring

Animal toxicological research has been equivocal in showing that exposure to algal toxins during pregnancy may affect birth outcomes. In rats, increased fetal mortality, small fetuses, neurological effects, and congenital malformations were associated with exposure to *M. aeruginosa* toxin during pregnancy (Collins et al. 1981; Kirpenko, Sirenko, and Kirpenko 1981); microcystin-LR given to pregnant mice resulted in decreased fetal weight (attributed to maternal toxicity) but no other fetal effects. Another study in mice associated chronic oral administration of microcystins to both males and females (from pre-mating to 5 days postpartum) with neurological effects in their offspring (Falconer et al. 1988). More recent studies in mice showed that high doses of microcystin-LR in pregnancy—enough to cause maternal toxicity—did not cause any fetal effects, and the no observed adverse effects level (NOAEL) for any developmental toxicity was 600 µg/kg/day during days 6–15 of gestation (Fawell et al. 1999a). Likewise, 2.46 mg/kg/day anatoxin-a given to pregnant mice during days 6–15 did not cause any apparent fetal effects (FWR, Nov. 1994c).

A study was recently published investigating the possible association between exposure to cyanobacterial toxins and human birth outcomes in southeast Australia, where algal contamination of the drinking water supply has been a problem (Pilotto et al. 1999). Records of 32,700 births in 156 communities were compared with cyanobacterial occurrence in the drinking water during the gestational period. Though there were statistically significant associations between cyanobacterial levels in the first trimester and abnormally low birth weight, prematurity, and congenital malformations, the authors did not observe any dose–response effects and concluded that the data were not strong enough to provide evidence for a causal link.

Kidney Dialysis Patients

In the worst outbreak of cyanobacteria-related illness, 60 people who had undergone kidney dialysis in a Brazilian dialysis center died of liver failure after exposure to microcystins through their treatment (Jochimsen et al. 1998; Pouria et al. 1998). Of 130 exposed patients, 101 met the case definition for acute liver failure resulting from the exposure to the hepatotoxins (attack rate = 78%), and over half of them died. The affected dialysis center received water from the same reservoir as an unaffected dialysis center. The difference was that water supplied to the second center was filtered and chlorinated; whereas, the water supplied to the affected center remained untreated. Microcystins were found in the reservoir, the center's water, and in tissues of affected patients. In 1974 near Washington, D.C., 23 of 70 patients treated at a dialysis center developed chills, fever, and hypotension (Hindman et al. 1975). Investigation turned up high levels of endotoxin in the dialysis fluid and in the affected patients. Tap water contaminated with cyanobacteria was implicated as the cause. Previous case reports had also been published reporting illness associated with cyanobacteria contamination in dialysis water (Codd, Ward, and Bell 1997), but the Brazilian incident was by far the most severe.

Children

Children may be somewhat more sensitive to the effects of cyanobacteria for two reasons: (1) as with any toxin, children are likely to be affected more readily and at smaller doses because of their lower weight, and (2) their behavior may increase the possibility of exposure (e.g., children may be more likely to swim in a contaminated lake or pond and child swimmers are more likely to swallow more water more often).

Other

Additional populations sensitive to cyanobacterial hepatotoxins would be those who already have liver damage or risk factors associated with liver damage (e.g., hepatitis B and hepatitis C infection or alcoholism). However, we do not have enough information on cyanobacterial exposure in these people to calculate their precise health effects risk from exposure.

Transmission

There are three main ways for humans to be exposed to cyanobacterial toxins: (1) ingestion or dermal contact with recreational water, (2) ingestion of drinking water, or (3) ingestion through food products, such as shellfish and commercial dietary supplements. Exposure from recreational water is the most likely route of exposure, though health effects from chronic low levels of hepatotoxins in drinking water are now under more scrutiny (Codd, Ward, and Bell 1997; Elder, Hunter, and Codd 1993). The primary exposure risk factor would be environmental conditions that allow for the formation of algal blooms. Blooms appear most readily when winds are calm, water is warm (15–30°C), water pH is neutral to alkaline (6–0), and nutrients such as phosphorus and nitrogen are abundant (Carmichael 1994). In temperate climates, blooms occur most often during the summer and fall but can occur year-round in tropical areas (Codd, Ward, and Bell 1997). Because cyanobacteria often float to the surface to obtain light, they tend to clump together and collect along the shoreline, where the water is shallow, warm, and more protected from winds that can disperse the blooms. This makes contact more likely for animals that come to the edge of the water to drink and humans who swim or wade close to shore. In the presence of blooms, other risk factors include the amount of dermal contact and/or the amount of water ingested. Accordingly, an Australian government group delineates swimming, diving, sailboarding, water skiing, and wading as high-risk recreational activities and canoeing, sailing, and rowing as moderate risk (Yoo et al. 1995). Because no risk factors that differentiate toxic from nontoxic cyanobacteria are known, the presence of any algal bloom should be considered dangerous. However, even toxic cyanobacterial blooms are not always hazardous to animals (or presumably humans) for several possible reasons: (1) low concentrations of toxins within the bloom, (2) low biomass concentration of the bloom, and (3) variation in animal species' response or animal characteristics, including age and gender.

In marine environments, neurotoxins accumulate in seafood (especially shellfish) and cause paralytic shellfish poisoning. The same types of neurotoxins are present in freshwater (though not as commonly) and have the potential to contaminate mussels and fish (Negri and Jones 1995; Pereira et al. 2000). Additionally, some data suggest that microcystin-LR can be accumulated in edible fish (Magalhaes, Soares, and Azevedo 2001; Vasconcelos 1999) and mussels (Watanable et al. 1997). The amount of exposure from this route of transmission is unknown.

Since it is unlikely that conventionally treated drinking water would have high enough concentrations to cause an acute reaction, long-term exposure to small concentrations of the hepatotoxins is the primary exposure of concern for this route of transmission.

Dose Response

Cyanobacteria is different from microbial pathogens in that it does not infect the person who is exposed. It produces toxins, and its dose–response parameters are more closely related to chemicals than microbials.

Toxic Dose and Toxicity Studies

The most accurate data available on toxic doses of cyanobacteria comes from in vivo work in the laboratory. This work is described in the following paragraphs.

Microcystin-LR

- Microcystin-LR is 30–200 times more toxic by intraperitoneal injection than oral ingestion in mice (Fawell et al. 1999a; Yoshida et al. 1997). Intranasal aerosol exposure is as high as intraperitoneal exposure (Codd, Ward, and Bell 1997).

- The toxic effects of microcystin-LR administered orally and intraperitoneally in mice were similar: hepatocellular injuries with characteristics of hemorrhage and necrosis, characterized by hepatic apoptosis (Yoshida et al. 1997).
- In chronic (13 weeks in mice) liver toxicity studies, the NOAEL for microcystin-LR was 40 µg/kg/day in mice dosed orally. Extrapolated to humans, this NOAEL would allow a safety factor of 250 for an infant drinking 0.75 L/day with a concentration of 1 µg/L microcystin-LR (Fawell et al. 1999a).
- The LD_{50} (the dose required to cause death in 50% of those exposed) and minimum lethal doses for mice dosed intraperitoneally were reported at 100 and 50 µg/kg, respectively (Codd, Ward, and Bell 1997; Yoo et al. 1995).
- Yoshida et al. (1997) reported an LD_{50} of 10.9 mg/kg oral dose and 65.4 µg/kg intraperitoneally in BALB/c mice.
- The LD_{50} and minimum lethal doses for rainbow trout dosed intraperitoneally were reported at 840 and 10 µg/kg, respectively (Codd, Ward, and Bell 1997).

Anatoxin-a

- Intraperitoneal and intranasal exposure is more potent than oral ingestion in mice (Codd, Ward, and Bell 1997).
- Administration of an oral sublethal dose to mice over 28 days did not produce apparent toxicity (Fawell et al. 1999b).
- The NOAEL for mice was 0.1 mg/kg/day by oral gavage (Fawell et al. 1999b).
- The minimum lethal dose reported for mice was 186 µg/kg/day (Codd, Ward, and Bell 1997).
- Recovery from a single sublethal dose producing symptoms in mice was rapid and complete (FWR, Nov. 1994b).
- The LD_{50} and minimum lethal dose reported for rainbow trout was 1,400 and 12 µg/kg/day, respectively (Codd, Ward, and Bell 1997).

- The LD_{50} of anatoxin-a(s) delivered to mice intraperitoneally is 20 μg/kg (Yoo et al. 1995).

The comparison of mice and trout show that interspecies distinctions are important; mice appear to be more sensitive to hepatotoxins than fish, and fish are more sensitive to neurotoxins than mice.

Based on some of these results, the World Health Organization (WHO) has provisionally recommended a limit of 1 μg/L/day for microcystin in drinking water. This level is based on chronic animal studies plus a 1,000×-uncertainty factor for a 60-kg adult drinking 2 L/day for his or her life (Algal Toxins National Forum 1997; WHO 1998). The proposed level for anatoxin-a is similar, based on the NOAEL for mice plus a 1,000×-uncertainty factor for a 10-kg child drinking 1L/day. Three individual countries are working on maximum acceptable level guidelines for microcystins in drinking water: Australia, Canada, and Great Britain. Because the neurotoxins are considered less common and are unlikely to cause chronic health effects, they are not being addressed (Yoo et al. 1995). Australia has proposed a maximum acceptable level of 1.0 μg/L of microcystins or nodularins; Canada has proposed a similar level for total microcystins, specifically, 0.5 μg/L for microcystin-LR (Yoo et al. 1995). For recreational water, Australian researchers (including a government-sponsored task force on cyanobacteria) have recommended 15,000–20,000 cells/mL as a threshold for exposure (in Yoo et al. 1995). At these levels, water would appear discolored, indicating levels in excess of what is safe for human exposure (Yoo et al. 1995).

Detection Methods

Because it is easy to see algal blooms in water bodies, the biggest challenge in environmental detection is measuring toxicity. From a clinical standpoint, associating health effects with exposure to harmful cyanobacteria would most likely come from a history from the patient.

Clinical

Clinical detection of cyanobacterial exposure has been limited to the descriptive association between symptoms and contact with known algal blooms. Because cyanobacteria form a gelatinous sheath that protects them somewhat from external elements, they can sometimes be seen in gut contents and feces (Elder, Hunter, and Codd 1993; Yoo et al. 1995), which can then be speciated microscopically (Skulberg in Yoo et al.

1995). The identification of toxins in serum and liver tissue has also been reported (Jochimsen et al. 1998).

Environmental

The most obvious environmental detection method is the observation of algal blooms in drinking water source waters. Water contaminated with algae will also have an unpleasant odor and taste. However, it is important to detect water blooms as early as possible to ensure the best mitigation. Cyanobacteria float within the water column and may rise and accumulate on the water's surface. Cyanobacteria differ visually from green algae (which also accumulate on the surface of freshwater) by their paint-like look. Green algae (which are nontoxic) form stringy bunches that can be pulled out as strands. The colors can vary from light to dark green (*Anabaena, Aphanizomenon, Microcystis, Nodularia, Nostoc, Oscillatoria*) or reddish brown (*Oscillatoria*). The cells form a slimy covering that allows them to form colonies and adhere to each other. The smell of a newly formed bloom has been described as grassy, but as the bloom ages, it smells more like rotting garbage (Yoo et al. 1995).

A variety of methods similar to what is being used to detect microbial pathogens is under development for cyanobacteria. The most common detection method is microscopic analysis, which can identify the major toxigenic genera. The mouse bioassay has been the traditional method for measuring the toxicity of cyanobacterial samples. However, the amount of sample necessary for detection in this assay is 10^2–10^5 greater than some of the newer enzyme-linked immunosorbent assays (ELISAs), polymerase chain reaction (PCR), protein phosphatase inhibition assay (Carmichael and An 1999), and chromatographic methods; therefore, the mouse bioassay will significantly underestimate the occurrence of cyanobacterial toxins (Codd, Ward, and Bell 1997). Another disadvantage is the number of laboratory animals that would be needed for routine monitoring. To overcome some of these problems, researchers are looking at toxicity assays using other organisms such as brine shrimp larvae (Lahti et al. 1995) and locusts (Hiripi et al. 1998). However, none of these alternatives has been adopted for routine monitoring (Yoo et al. 1995).

Rudi and colleagues (1998a, 1998b) used magnetic beads to concentrate algal cells and purify DNA for PCR amplification using oligonucleotide gene probes. In their laboratory test, the detection limit (of *Microcystis* spp.) using this method was 100 cells/mL water sample; they found the same results when they tested environmental water samples. They noted that the method could be adapted to detect and quantify multiple target organisms using the single assay. Others are working to develop probes that will differentiate toxic and nontoxic samples (Litvaitis 1999).

Rivasseau and colleagues (1998) were able to detect microcystins at trace levels (30 ng/L in drinking water and 100–200 ng/L in surface water) using a combination of solid-phase extraction on octadecylsilica followed by high-performance liquid chromatography (HPLC). HPLC has also been used to detect anatoxin-a, traced to *Anabaena* and *Oscillatoria* species in Irish lake water (James, Sherlock, and Stack 1997). The authors reported a detection limit of 0.02 ng/mL.

Wisconsin researchers developed an ELISA to detect microcystin-LR and microcystin-RR at levels potentially as low as 95 pg/mL (McDermott, Feola, and Plude 1995). Using this method, they were able to detect measurable amounts of microcystin in Wisconsin waters ranging from 0.2–200 ng/mL. This ELISA test for algal hepatotoxins has now been commercialized (Karner et al. 2001). A simpler protein phosphatase inhibition assay detecting toxins directly from water is available with commercially available materials (Heresztyn and Nicholson 2001).

A group of experts convened by USEPA (USEPA 2001) concluded that analytical standards need to be developed for all algal toxins, except saxitoxins, which are available through the US Food and Drug Administration. They also stated the need for more commercially available analytical detection assays.

Environmental Occurrence

Microcystins are some of the most common toxins isolated from blooms in different areas of the world, such as Japan (Park et al. 1993), Portugal (Vasconcelos 1999), and the United Kingdom (FWR, March 1992b). In a test conducted specifically for microcystin-LR in 23 reservoir water samples, few were positive; those that were positive, were at concentrations close to 0.2 µg/L, which is the detection limit for the analytical detection method used (HPLC with ultraviolet [UV] detection) (FWR, Dec. 1992). However, while one sample with visible algal cells had a very high level (8.5 mg/L) for microcystin-LR, another similar sample from a different location at the site did not have detectable levels of that toxin. This supports the evidence that the ability of algal blooms to produce toxins is highly variable.

It appears that the occurrence of any cyanobacterial cells in source water used for drinking is variable as well. Treated water samples in the United Kingdom were tested for microcystin-LR and anatoxin-a, while their raw water supplies were analyzed for general algal cell numbers

(FWR, Nov. 1994a). Water samples were taken five times from six different treatment sites. No cells were detected in any of the drinking water samples; the level of cells in the raw water ranged from zero (two sites) to 1.8×10^4. Interestingly, the sites were chosen because of their potentially high numbers of cells and the fact that two sites had no cells detected at all was unexpected.

Researchers in Wisconsin closely examined 60 sets of seven samples from source and treated water representing six different water systems (Standridge, Karner, and Barnum 1999). They measured the amount of microcystin-LR in each sample using an ELISA. One hundred percent of the raw water samples contained measurable amounts of the microcystin (range: 11.3–4,135 ng/L). Though the finished water also tested positive, the concentrations represented a two to three log reduction after treatment (range: 0.0–11.1 ng/L). WHO's standard for drinking water is 1.0 µg/L or 1,000 ng/L—two to three orders of magnitude more than the concentrations in these samples (WHO 1998). Each plant used a pretreatment of oxidation and sedimentation followed by coagulation and filtration with activated carbon. The researchers determined that 61% of the microcystins were removed after pretreatment chemicals such as potassium permanganate were added (Karner et al. 2001). The subsequent steps of coagulation and sedimentation increased the reduction to 96%.

Finally, microcystin-LR and anatoxin-a were measured in reservoir and finished water samples at 10 sites in England, Northern Ireland, and Scotland. Microcystin-LR was not detectable in either type of sample. Anatoxin-a was found at low levels (up to 1.3 µg/L) in two treated water samples that came from source water that contained anatoxin-a.

Environmental Fate

Toxins are probably released into water bodies by cyanobacteria as part of the normal algal life cycle (Table 9-1); however, significant amounts may be released when the cells lyse following an algal bloom or with chemical treatment (e.g., addition of copper sulfate). Little is known about the fate of these toxins in their natural environment (e.g., whether they degrade or adsorb onto sediments). A 1992 study (FWR, March 1992b) described an investigation of the degradation of microcystin-LR. The toxin was measured in waters of different pH (4, 7, and 9) and lighting scenarios. After 39 days, the researchers reported that little of the microcystin-LR had degraded.

Some data suggest that some cyanobacterial toxins can remain in the environment even after the algal blooms have dispersed. Researchers in Finland tracked concentrations of microcystin-LR dissolved in lake water

Table 9-1: Cyanobacterial blooms testing positive for toxicity

Country	Number of Samples/Sites	% Positive Samples/Sites
Australia	231	42
Baltic Sea	25	72
Belgium	17	59
China*	989	17 (pond/ditch) 32 (river) 4 (shallow well)
Denmark†	392	68
Germany	86	88
Hungary	35	82
Netherlands	39	82
Norway, Sweden, and Finland	342	54
Slovenia‡	18	83
United Kingdom	136	63
Wisconsin, USA (1986)	308	27
Wisconsin, USA (1998)§	420	100 (raw water) <70 (treated water)

Adapted from Codd, Ward, and Bell 1997 and WHO 1998 unless noted.

*Ueno et al. 1996.
†Henriksen and Moestrup 1997.
‡Sedmak and Kosi 1997.
§Standridge, Karner, and Barnum 1999.

and in the particulate matter using HPLC (Lahti, Rapala, and Spink 1997). Not surprisingly, microcystin-LR was detected in lake water and in particulate matter during the decomposition of a bloom. However, the toxin was detectable at low concentrations weeks after the bloom had dissipated. Their data showed that the microcystin was more persistent dissolved in water than present in the particulate matter.

A UK report described a study on the persistence of anatoxin-a in reservoir water (Smith and Sutton 1994). The authors spiked reservoir water at different pH and light conditions and measured the presence of anatoxin-a using HPLC and UV, which has a detection level of close to 10 ng/L (FWR, March 1993a). Using this method, they found that anatoxin-a was stable for at least 21 days at a pH of 4. At pH levels from 8–10, the original sample had degraded by 95% after 14 days. When the samples were tested with and without reservoir sediments, the presence of sediment reduced the half-life to 5 days.

Water Treatment

There has been some question as to the efficacy of standard drinking water treatment (i.e., coagulation, sedimentation, disinfection, and filtration) in removing all but large concentrations of cyanobacterial toxins (Donati, Drikas, and Hayes 1994; FWR, Sept. 1991). In fact, some data suggest that the process of coagulation lyses cells, resulting in higher toxin levels than would otherwise be found (FWR, June 1991). These conventional treatments should effectively remove enough toxin from drinking water to prevent any acute lethal effects (Yoo et al. 1995). Karner and colleagues (2001) found that source water with high levels of microcystins was effectively treated (1–3 log removals) with combinations of pretreatment chemicals, coagulation, sedimentation, and filtration so that finished water easily met the WHO guidelines. Older studies found that chlorine at lower levels or with shorter contact times was ineffective at removing microcystin toxins from drinking water (Yoo et al. 1995). However, more recent data suggest that free chlorine treatment at robust doses does inactivate microcystins and cylindrospermopsin (Nicholson, and Rositano, and Burch 1994; Tsuji et al. 1997) but only at low pH or long contact times (FWR, April 1994; March 1993a; Senogles et al. 2000). Chloramination apparently is ineffective at removing hepatotoxins (Nicholson, Rositano, and Burch 1994), and chlorination does not affect anatoxin-a (FWR, April 1994, March 1993a; Yoo et al. 1995). Little information exists on whether or not algal toxins combined with chlorine disinfection result in hazardous disinfection by-products. Some data suggest that the by-products from chlorine and the isolated toxins are nontoxic, though administering chlorine to the entire cell itself causes it to lyse and produce trihalomethanes (Tsuji et al. 1997). The use of lime treatment in appropriate circumstances has been suggested as an alternative to chlorine (Lam, Prepas, and Spink 1995; Yoo et al. 1995). Ozone is effective at inactivating microcystin LR, LA, and anatoxin-a, but its effectiveness was dependent on the water quality; saxitoxins were ozone resistant under all conditions (Rositano et al. 2001). The pH of the water and the presence of naturally occurring organic matter strongly affects ozone's efficiency (Hitzfeld et al. 2000). Also, the concern about toxic disinfection by-products would remain with ozonation as with chlorination. Though one report did not find ozonation by-products (with microcystin-LR) to be toxic, more data would be needed to draw any conclusions.

Other treatment options are being tested, especially in areas where algal blooms have been a concern, such as Australia and the United Kingdom. Donati and colleagues (1994) looked at the absorption of microcystin-LR by eight different powdered activated carbons (PACs). They found that wood-based carbons were the most effective, followed by coal-based carbons. Coconut and peat moss carbons were the least effective. They associated this difference to the volume of mesopores in the carbon, which is a function of its source material. However, when this treatment was used with river water, the organics present in the water competed for absorption to the carbon and thereby reduced its ability to reduce microcystin levels.

Membrane filtration technology, such as micro-, ultra-, and nanofiltration, as well as reverse osmosis, has the capacity to remove water contaminants down to the ion level. Microfiltration, with a nominal pore size of 0.2 μm, can remove cyanobacteria better than conventional water treatment, and ultrafiltration, with a nominal pore size of 0.01 μm, can achieve up to a 6-log pathogen removal (Najm and Trussell 1999; Taylor and Wiesner 1999). Nanofiltration and reverse osmosis have the potential to remove any cyanotoxin. This technology will continue to gain in popularity as costs become more competitive and regulatory approval is assured.

Reports from the United Kingdom in 1993 and 1994 detailed treatment results from laboratory and pilot-scale tests (FWR, April 1994, March 1993a). The 1993 paper looked only at the removal of microcystin-LR from reservoir waters; whereas, the 1994 paper investigated both microcystin-LR and anatoxin-a. A summary of the conclusions follows:

- Wood-based PAC dosing effectively removes microcystin-LR but less effectively removes anatoxin-a.
- Granulated activated carbon (with a 15-minute-plus contact time) is effective for removing both toxins but better for anatoxin-a.
- Ozonation is effective—more so when applied to filtered water rather than raw water.
- Potassium permanganate is effective when applied to filtered water.
- UV radiation is effective (also reported by Alam et al. 2001 and Tsuji et al. 1995) on microcystins at doses two orders of magnitude higher than those normally used for disinfection. The organic load in the water affects the efficacy of UV (Hitzfeld, Hoger, and Dietrich 2000).
- Nanofiltration is effective on microcystin-LR with significant pretreatment.
- Chlorine is not effective on anatoxin-a or microcystin-LR, though one report suggests that for microcystin-LR alone, it is effective only at pH 5 or with long contact times at pH 7–9.

Table 9-2: Treatment options for specific toxins

	Carbon	Ozone	Chlorine	Boiling	Filtration (standard)
Anatoxin-a	x	x	–	–	–
Anatoxin-a(s)	x	–	x	–	–
Saxitoxin	x	x	–	x	–
Cylindrospermopsin					
Microcystin	x	x	–	–	x
Nodularin	x	x	x	–	x

Adapted from USEPA 2001.

Though the focus of recent research on water treatment has been the drinking water treatment process (Table 9-2), controlling the algal growth in the source water affects not only drinking water exposure but also recreational water exposure. Australia has adopted a nutrient-control policy called "Total Catchment Management." The main emphasis has been on reducing the level of nutrients (e.g., nitrogen and phosphorus), either by land use or nutrient stripping. WHO recommends keeping total phosphorus levels below 0.01 g/L in recreational waters to deter the formation of toxic blooms (WHO 1998). UK water experts are also looking at biological solutions, such as manipulating the fish community to reduce the overall phtyoplankton level, and physical solutions, such as destratification mixing systems (FWR, March 1992c). Additionally, in preliminary studies, researchers have found that decomposing barley straw using a bankside digester system produces an anti-algal compound that inhibits the growth of *M. aeruginosa* (Ball et al. 2001; FWR, March 1992c). Further experiments are under way to test the feasibility of using this method in natural water bodies, although its application to large water bodies would be impractical (Yoo et al. 1995). UK experts stated, however, that all of these methods required more research before recommendation to the water industry (FWR, March 1992a).

A common method of controlling algae growth in environmental water is to apply chemicals that inhibit growth. Copper sulfate is mentioned most often as a chemical for this control practice, but its use is very controversial and it is banned in many places around the world (Yoo et al. 1995). Copper itself is toxic and can also kill other organisms while having no impact on some algal species. In addition, the copper can accumulate in sediments, causing long-term concerns for the quality of the water ecology (Yoo et al. 1995). The biggest drawback of using copper sulfate is that it will cause the cells to lyse, resulting in a toxin release, if it is applied after blooms have already formed. Therefore, copper sulfate must be used to prevent blooms before they form, if it is used at all (Yoo et al. 1995). Other chemicals, such as potassium permanganate, lime, alum, and organic herbicides, have been used to control algal blooms, but the danger of altering a water body's ecology and health is a serious concern with these methods (Yoo et al. 1995). However, these treatments, combined with coagulation, sedimentation, and filtration, have been shown to be effective in removing microcystins (Karner et al. 2001).

See Yoo et al. (1995) or Hitzfeld and colleagues (2000) for a complete summary of algal bloom control, water treatment processes, and risk management strategies for cyanobacteria.

Conclusions

The acute health effects of cyanobacterial toxins have produced mild-to-moderate morbidity in humans, mainly associated with recreational water use. Interest in the toxins has increased since well-publicized incidents have occurred in the United Kingdom and Australia. This interest has provoked additional scientific research into the mechanisms of toxicity, the dose response, and the effectiveness of various bloom-control and water treatment measures. The emerging evidence of the microcystins' possible association with liver cancer in China is sobering. It is accepted that the microcystins are potent tumor promoters, but more research on that carcinogenic mechanism and more well-designed epidemiological studies looking at the association are needed. WHO, Australia, Canada, and the United Kingdom have all published guidelines for microcystin levels—between 0.05 and 1.0 µg/L as the maximum level allowed in drinking water. Better monitoring programs will inform regulators as to how often (if ever) those guideline levels are exceeded. The best strategy, however, is to control the conditions that allow algal blooms to form in the first place, such as land management practices and nutrient control. The potential for cyanobacterial toxins to be a serious health concern is clear.

The amount of information amassed on cyanobacteria over the past 10 years is impressive. However, more data are needed in the areas of toxicity, teratogenicity, mutagenicity, carcinogenicity, monitoring, and treatment to sufficiently evaluate cyanobacteria as a public health threat and to evaluate the best strategies for mitigation. USEPA's expert panel (USEPA 2001) concluded the highest priority toxins needing the most information are microcystin, cylindrospermopsin, and anatoxin-a.

Bibliography

Alam, Z.B., M. Otaki, H. Furumai, and S. Ohgaki. 2001. Direct and indirect inactivation of *Microcystis aeruginosa* by UV-radiation. *Water Res.*, 35:1008–1014.

Algal Toxins: National Forum. 1997. Notes of the 5th meeting held at 11 Upper Belgrave Street, London, UK; July 1, 1997 [Online]. Available: <www.atlas.co.uk/listons/index.htm>. [cited July 1999]

Ball, A.S., M. Williams, D. Vincent, and J. Robinson. 2001. Algal growth control by a barley straw extract. *Bioresour. Technol.*, 77:177–181.

Billings, W.H. 1981. Water-associated human illness in northeast Pennsylvania and its suspected association with blue-green algae blooms. In *The Water Environment Algal Toxins and Health*. Edited by W.W. Carmichael. New York: Plenum Press.

Bourke, A.T.C., and R.B. Hawes. 1983. Freshwater cyanobacteria (blue-green algae) and human health. *Med. J. Aust.*, 1:491–492.

Carmichael, W.W. 1981. Freshwater blue-green algae (cyanobacteria) toxins—A review. In *The Water Environment Algal Toxins and Health.* Edited by W.W. Carmichael. New York: Plenum Press.

Carmichael, W.W. 1994. The toxins of cyanobacteria. *Sci. Am.*, 270:78–86.

Carmichael, W.W., and J. An. 1999. Using an enzyme linked immunosorbent assay (ELISA) and a protein phosphatase inhibition assay (PPIA) for the detection of microcystins and nodularins. *Nat. Toxins.*, 7:377–385.

Codd, G.A., C.J. Ward, and S.G. Bell. 1997. Cyanobacterial toxins: occurrence, modes of action, health effects and exposure routes. *Arch. Toxicol. Suppl.*, 19:399–410.

Collins, M.D., C.S. Gowans, F. Garro, D. Estervig, and T. Swanson. 1981. Temporal association between an algal bloom and mutagenicity in a water reservoir. In *The Water Environment Algal Toxins and Health.* Edited by W.W. Carmichael. New York: Plenum Press.

Dawson, R.M. 1998. The toxicology of microcystins. *Toxicon*, 36:953–962.

Ding, W.X., H.M. Shen, H.G. Zhu, B.L. Lee, and C.N. Ong. 1999. Genotoxicity of microcystic cyanobacteria extract of a water source in China. *Mutat. Res.*, 442:69–77.

Donati, C., M. Drikas, and R. Hayes. 1994. Microcystin–LR adsorption by powdered activated carbon. *Water Res.*, 28:1735–1742.

Elder, G.H., P.R. Hunter, and G.A. Codd. 1993. Hazardous freshwater cyanobacteria (blue-green algae). *Lancet*, 341:1519–1520.

el Saadi, O.E., A.J. Esterman, S. Cameron, and D.M. Roder. 1995. Murray River water, raised cyanobacterial cell counts, and gastrointestinal and dermatological symptoms. *Med. J. Aust.*, 162:122–125.

Environmental Health Perspectives (EHP). 1999. New understanding of algae. *Environ. Health Perspect.*, 107:A13.

Falconer, I.R., and T.H. Buckley. 1989. Tumor promotion by *Microcystis* sp., a blue-green alga occurring in water supplies. *Med. J. Aust.*, 150:351–352.

Falconer, I.R., and A.R. Humpage. 2001. Preliminary evidence for in vivo tumour initiation by oral administration of extracts of the blue-green alga *Cylindrospermopsis raciborskii* containing the toxin cylindrospermopsin. *Environ. Toxicol.*, 16:192–195.

Falconer, I.R., J.V. Smith, A.R. Jackson, A. Jones, and M.T. Runnegar. 1988. Oral toxicity of a bloom of the Cyanobacterium *Microcystis aeruginosa* administered to mice over periods up to 1 year. *J. Toxicol. Environ. Health*, 24:291–305.

Fawell, J.K., R.E. Mitchell, D.J. Everett, and R.E. Hill. 1999a. The toxicity of cyanobacterial toxins in the mouse: I microcystin-LR. *Hum. Exp. Toxicol.*, 18:162–167.

Fawell, J.K., R.E. Mitchell, R.E. Hill, and D.J. Everett. 1999b. The toxicity of cyanobacterial toxins in the mouse: II anatoxin-a. *Hum. Exp. Toxicol.*, 18:168–173.

Foundation for Water Research (FWR). 1991a. Detection and removal of cyanobacterial toxins from freshwaters. Marlow, UK (June). Report No. FR0211 [Online]. Available: <www.atlast.co.uk/listons/algaltox/fr0211.htm>. [cited July 1999]

Foundation for Water Research (FWR). 1991b. Algal/bacterial toxin removal from water: a literature survey. Marlow, UK (September). Report No. FR0223 [Online]. Available: <www.atlast.co.uk/listons/algaltox/fr0223.htm>. [cited July 1999]

Foundation for Water Research (FWR). 1992a. A review of potential methods for controlling phytoplankton, with particular reference to cyanobacteria, and sampling guidelines for the water industry. Marlow, UK (March). Report No. FR0248 [Online]. Available: <www.atlast.co.uk/listons/algaltox/fr0248.htm>. [cited July 1999]

Foundation for Water Research (FWR). 1992b. An investigation of the degradation of microcystin-LR. Marlow, UK (March). Report No. FR0292 [Online]. Available: <www.atlast.co.uk/listons/algaltox/fr0292.htm>. [cited July 1999]

Foundation for Water Research (FWR). 1992c. Investigations into the use of straw to control blue-green algal growth. Marlow, UK (March). Report No. FR0285 [Online]. Available: <www.atlast.co.uk/listons/algaltox/fr0285.htm>. [cited July 1999]

Foundation for Water Research (FWR). 1992d. Levels of microcystin-LR in raw and treated waters. Marlow, UK (December). Report No. FR0337 [Online]. Available: <www.atlast.co.uk/listons/algaltox/fr0337.htm>. [cited July 1999]

Foundation for Water Research (FWR). 1993. An analytical method for anatoxin-A, a blue-green algal neurotoxin, in reservoir water. Marlow, UK (March). Report No. FR0363 [Online]. Available: <www.atlast.co.uk/listons/algaltox/fr0363.htm>. [cited July 1999]

Foundation for Water Research (FWR). 1994a. Further studies to investigate microcystin-LR and anatoxin-A removal from water. Marlow, UK (April). Report No. FR0458 [Online]. Available: <www.atlast.co.uk/listons/algaltox/fr0458.htm>. [cited July 1999]

Foundation for Water Research (FWR). 1994b. Survey of the concentration of algal toxins in water supplies: final report to the Department of the Environment. Marlow, UK (November). Report No. DWI0719 [Online]. Available: <www.atlast.co.uk/listons/waterq/dwi0719.htm>. [cited July 1999]

Foundation for Water Research (FWR). 1994c. Toxins from blue-green algae: Toxicological assessment of anatoxin-A and a method for its determination in reservoir water. Marlow, UK (November). Report No. FR0434/DoE372 [Online]. Available: <www.atlast.co.uk/listons/algaltox/fr0434.htm>. [cited July 1999]

Foundation for Water Research (FWR). 1994d. Toxins from blue-green algae: Tumor promotion by microcystin-LR – Preliminary *in vitro* studies. Marlow, UK (November). Report No. FR0493/DoE372 [Online]. Available: <www.atlast.co.uk/listons/algaltox/fr0493.htm>. [cited July 1999]

Gilroy, D.J., K.W. Kauffman, R.A. Hall, X. Huang, and F.S. Chu. 2000. Assessing potential health risks from microcystin toxins in blue-green algae dietary supplements. *Environ. Health Perspect.*, 108:435–439.

Harada, K., M. Oshikata, H. Uchida, M. Suzuki, F. Kondo, K. Sato, Y. Ueno, S.Z. Yu, G. Chen, and G.C. Chen. 1996. Detection and identification of microcystins in the drinking water of Haimen City, China. *Nat. Toxins*, 4:277–283.

Hayman, J. 1992. Beyond the Barcoo—probable human tropical cyanobacterial poisoning in outback Australia. *Med. J. Aust.*, 157:794–796.

Henriksen, P., and O. Moestrup. 1997. Seasonal variations in microcystin contents of Danish cyanobacteria. *Nat. Toxins,* 5:99–106.

Heresztyn, T., and B.C. Nicholson. 2001. Determination of cyanobacterial hepatotoxins directly in water using a protein phosphatase inhibition assay. *Water Res.,* 35:3049–3056.

Hindman, S.H., M.S. Favero, L.A. Carson, N.J. Petersen, L.B. Schonberger, and J.T. Solano. 1975. Pyrogenic reactions during haemodialysis caused by extramural endotoxin. *Lancet,* 2:732–734.

Hiripi, L., L. Nagy, T. Kalmar, A. Kovacs, and L. Voros. 1998. Insect (Locusta migratoria migratorioides) test monitoring the toxicity of cyanobacteria. *Neurotoxicology,* 19:605–608.

Hitzfeld, B.C., S.J. Hoger, and D.R. Dietrich. 2000. Cyanobacterial toxins: removal during drinking water treatment, and human risk assessment. *Environ. Health Perspect.,* 108(Suppl)1:113–122.

Humpage, A.R., S.J. Hardy, E.J. Moore, S.M. Froscio, and I.R. Falconer. 2000. Microcystins (cyanobacterial toxins) in drinking water enhance the growth of aberrant crypt foci in the mouse colon. *J. Toxicol. Environ. Health, A* 61:155–165.

James, K.J., I.R. Sherlock, and M.A. Stack. 1997. Anatoxin-a in Irish freshwater and cyanobacteria, determined using a new fluorimetric liquid chromatographic method. *Toxicon,* 35:963–971.

Jochimsen, E.M., W.W. Carmichael, J.S. An, D.M. Cardo, S.T. Cookson, C.E. Holmes, M.B. Antunes, D.A. Melo Filho, T.M. Lyra, V.S. Barreto, S.M. Azevedo, and W.R. Jarvis. 1998. Liver failure and death after exposure to microcystins at a hemodialysis center in Brazil. *N. Engl. J. Med.,* 338:873–878.

Karner, D.A., J.H. Standridge, G.W. Harrington, and R.P. Barnum. 2001. Microcystin algal toxins in source and finished drinking water. *Jour. AWWA,* 93:72–81.

Kirpenko, Y.A., L.A. Sirenko, and N.I. Kirpenko. 1981. Some Aspects Concerning Remote After-Effects of Blue-Green Algae Toxin Impact on Warm-Blooded Animals. In *The Water Environment Algal Toxins and Health.* Edited by W.W. Carmichael. New York: Plenum Press.

Lagos, N., H. Onodera, P.A. Zagatto, D. Andrinolo, S.M. Azevedo, and Y. Oshima. 1999. The first evidence of paralytic shellfish toxins in the fresh water cyanobacterium *Cylindrospermopsis raciborskii,* isolated from Brazil. *Toxicon,* 37:1359–1373.

Lahti, K., J. Ahtiainen, J. Rapala, K. Sivonen, and S.I. Niemela. 1995. Assessment of rapid bioassays for detecting cyanobacterial toxicity. *Lett. Appl. Microbiol.,* 21:109–114.

Lahti, K., J. Rapala, and D. Spink. 1997. Persistence of cyanbacteria hepatoxin, microcystin-LR in particulate material and dissolved in lake water. *Water Res.*, 31:1005–1012.

Lam, A.K.Y., E.E. Prepas, and D. Spink. 1995. Chemical control of hepatoxic phytoplankton blooms: implications for human health. *Water Res.*, 29:1845–1854.

Litvaitis, M.K. 1999. *Development of a nucleotide-based test for the detection of cyanobacteria in water supplies. New Hampshire.* Fedrip database, National Technical Information Service. Report No. FEDRIP/1999/06401477.

Long, E.G., A. Ebrahimzadeh, E.H. White, B. Swisher, and C.S. Callaway. 1990. Alga associated with diarrhea in patients with acquired immunodeficiency syndrome and in travelers. *J. Clin. Microbiol.*, 28:1101–1104.

Lopez, R.V., and E. Costas. 1999. Preference of mice to consume *Microcystis aeruginosa* (toxin-producing cyanobacteria): a possible explanation for numerous fatalities of livestock and wildlife. *Res. Vet. Sci.*, 67:107–110.

Magalhaes, V.F., R.M. Soares, and S.M. Azevedo. 2001. Microcystin contamination in fish from the Jacarepagua Lagoon (Rio de Janeiro, Brazil): ecological implication and human health risk. *Toxicon*, 39:1077–1085.

McDermott, C.M., R. Feola, and J. Plude. 1995. Detection of cyanobacterial toxins (microcystins) in waters of northeastern Wisconsin by a new immunoassay technique. *Toxicon*, 33:1433–1442.

Najm, I., and R.R. Trussell. 1999. New and Emerging Drinking Water Treatment Technologies. In *Identifying Future Drinking Water Contaminants.* Washington, DC: National Academy Press.

Negri, A.P., and G.J. Jones. 1995. Bioaccumulation of paralytic shellfish poisoning (PSP) toxins from the cyanobacterium *Anabaena circinalis* by the freshwater mussel Alathyria condola. *Toxicon*, 33:667–678.

Nicholson, B.C., J. Rositano, and M.D. Burch. 1994. Destruction of cyanobacterial peptide hepatoxins by chlorine and chloramine. *Water Res.*, 28: 1297–1303.

Ohta, T., E. Sueoka, N. Iida, A. Komori, M. Suganuma, R. Nishiwaki, M. Tatematsu, S.J. Kim, W.W. Carmichael, and H. Fujiki. 1994. Nodularin, a potent inhibitor of protein phosphatases 1 and 2A, is a new environmental carcinogen in male F344 rat liver. *Cancer Res.*, 54:6402–6406.

Park, H.D., M.F. Watanabe, K. Harda, H. Nagai, M. Suzuki, M. Watanabe, and H. Hayashi. 1993. Hepatotoxin (microcystin) and neurotoxin (anatoxin-a) contained in natural blooms and strains of cyanobacteria from Japanese freshwaters. *Nat. Toxins,* 1:353–360.

Pereira, P., H. Onodera, D. Andrinolo, S. Franca, F. Araujo, N. Lagos, and Y. Oshima. 2000. Paralytic shellfish toxins in the freshwater cyanobacterium *Aphanizomenon flos-aquae,* isolated from Montargil reservoir, Portugal. *Toxicon,* 38:1689–1702.

Pilotto, L.S., R.M. Douglas, M.D. Burch, S. Cameron, M. Beers, G.J. Rouch, P. Robinson, M. Kirk, C.T. Cowie, S. Hardiman, C. Moore, and R.G. Attewell. 1997. Health effects of exposure to cyanobacteria (blue-green algae) during recreational water-related activities. *Aust. N. Z. J. Public Health,* 21:562–566.

Pilotto, L.S., E.V. Kliewer, R.D. Davies, M.D. Burch, and R.G. Attewell. 1999. Cyanobacterial (blue-green algae) contamination in drinking water and perinatal outcomes. *Aust. N. Z. J. Public Health,* 23:154–158.

Pouria, S., A. de Andrade, J. Barbosa, R.L. Cavalcanti, V.T. Barreto, C.J. Ward, W. Preiser, G.K. Poon, G.H. Neild, and G.A. Codd. 1998. Fatal microcystin intoxication in haemodialysis unit in Caruaru, Brazil. *Lancet,* 352:21–26.

Rao, P.V., and R. Bhattacharya. 1996. The cyanobacterial toxin microcystin-LR induced DNA damage in mouse liver in vivo. *Toxicology,* 114:29–36.

Rivasseau, C., S. Martins, M.C. Hennion. 1998. Determination of some physicochemical parameters of microcystins (cyanobacterial toxins) and trace level analysis in environmental samples using liquid chromatography. *J. Chromatogr. A.,*799:155-69.

Rositano, J., G. Newcombe, B. Nicholson, and P. Sztajnbok. 2001. Ozonation of NOM and algal toxins in four treated waters. *Water Res.,* 35:23–32.

Rudi, K., F. Larsen, and K.S. Jakobsen. 1998. Detection of toxin-producing cyanobacteria by use of paramagnetic beads for cell concentration and DNA purification. *Appl. Environ. Microbiol.,* 64:34–37.

Rudi, K., O.M. Skulberg, F. Larsen, and K.S. Jakobsen. 1998. Quantification of toxic cyanobacteria in water by use of competitive PCR followed by sequence-specific labeling of oligonucleotide probes. *Appl. Environ. Microbiol.,* 64:2639–2643.

Sedmak, B., and G. Kosi. 1997. Microcystins in Slovene freshwaters (central Europe)—first report. *Nat. Toxins,* 5:64–73.

Senogles, P., G. Shaw, M. Smith, R. Norris, R. Chiswell, J. Mueller, R. Sadler, and G. Eaglesham. 2000. Degradation of the cyanobacterial toxin cylindrospermopsin, from *Cylindrospermopsis raciborskii*, by chlorination. *Toxicon*, 38:1203–1213.

Shlim, D.R., M.T. Cohen, M. Eaton, R. Rajah, E.G. Long, and B.L. Ungar. 1991. An alga-like organism associated with an outbreak of prolonged diarrhea among foreigners in Nepal. *Am. J. Trop. Med. Hyg.*, 45:383–389.

Smith, C., and A. Sutton. 1994. The persistence of anatoxin-A in reservoir water. Foundation for Water. Marlow, UK (March). Report No. FR0427 [Online]. Available: <www.atlas.co.uk/listons/waterq/fr0427.htm>. [cited July 1999]

Soong, F.S., E. Maynard, K. Kirke, and C. Lurke. 1992. Illness associated with blue-green algae. *Med. J. Aust.*, 156:67.

Standridge, J., D. Karner, and R. Barnum. 1999. Detection of microcystin algal toxins in raw and finished drinking water. Paper presented at International Symposium on Waterborne Pathogens, August 29–September 1; Milwaukee, WI.

Tao, X.G., H.G. Zhu, S.Z. Yu, Q.Y. Zhao, J.R. Wang, G.D. Wu, X.F. You, C. Li, W.L. Zhai, and J.P. Bao. 1991a. Effects of drinking water from the lower reaches of the Huangpu River on the risk of male stomach and liver cancer death. *Public Health Rev.*, 19:229–236.

Tao, X.G., H.G. Zhu, S.Z. Yu, Q.Y. Zhao, J.R. Wang, G.D. Wu, X.F. You, C. Li, W.L. Zhai, and J.P. Bao. 1991b. Pilot study on the relationship between male stomach and liver cancer death and mutagenicity of drinking water in the Huangpu River area. *Public Health Rev.*, 19:219–227.

Taylor, J.S., and M. Wiesner. 1999. Membranes. In *Water Quality and Treatment: A Handbook of Community Water Supplies.* 5th ed. Edited by American Water Works Association. New York: McGraw-Hill.

Tsuji, K., T. Watanuki, F. Kondo, M.F. Watanabe, S. Suzuki, H. Nakazawa, M. Suzuki, H. Uchida, and K.I. Harada. 1995. Stability of microcystins from cyanobacteria—II. Effect of UV light on decomposition and isomerization. *Toxicon*, 33:1619–1631.

Tsuji, K., T. Watanuki, F. Kondo, M.F. Watanabe, H. Nakazawa, M. Suzuki, H. Uchida, and K. Harada. 1997. Stability of microcystins from cyanobacteria—IV. Effect of chlorination on decomposition. *Toxicon*, 35:1033–1041.

Turner, P.C., A.J. Gammie, K. Hollinrake, and G.A. Codd. 1990. Pneumonia associated with contact with cyanobacteria. *Br. Med. J.*, 300:1440–1441.

Ueno, Y., S. Nagata, T. Tsutsumi, A. Hasegawa, M.F. Watanabe, H.D. Park, G.C. Chen, G. Chen, and S.Z. Yu. 1996. Detection of microcystins, a blue-green algal hepatotoxin, in drinking water sampled in Haimen and Fusui, endemic areas of primary liver cancer in China, by highly sensitive immunoassay. *Carcinogenesis*, 17:1317–1321.

US Environmental Protection Agency (USEPA). 2001.Creating a Cyanotoxin Target List for the Unregulated Contaminant Monitoring Rule: Meeting Summary. May 17–18, 2001. Cincinnati, OH: USEPA [Online]. Available: <http://www.epa.gov/safewater/standard/ucmr/cyanotoxinmeeting0501.pdf>. [cited August 2001]

Vasconcelos, V.M. 1999. Cyanobacterial toxins in Portugal: effects on aquatic animals and risk for human health. *Braz. J. Med. Biol. Res.*, 32:249–254.

Watanabe, M.F., H.D. Park, F. Kondo, K. Harada, H. Hayashi, and T. Okino. 1997. Identification and estimation of microcystins in freshwater mussels. *Nat. Toxins*, 5:31–35.

World Health Organization (WHO). 1998. *Guidelines for Safe Recreational Water Environments: Coastal and Fresh Waters.* Report EOS/DRAFT/98.14. Geneva: World Health Organization.

Yoo, R.S., W.W. Carmichael, R.C. Hoehn, and S.E. Hrudey. 1995. *Cyanobacterial (Blue-Green Algal) Toxins: A Resource Guide.* Denver, CO: Awwa Research Foundation and American Water Works Association.

Yoshida, T., Y. Makita, S. Nagata, T. Tsutsumi, F. Yoshida, M. Sekijima, S. Tamura, and Y. Ueno. 1997. Acute oral toxicity of microcystin-LR, a cyanobacterial hepatotoxin, in mice. *Nat. Toxins*, 5:91–95.

Yu, S.Z. 1995. Primary prevention of hepatocellular carcinoma. *J. Gastroenterol. Hepatol.*, 10:674–682.

CHAPTER 10

Echovirus in Drinking Water

by Martha A. Embrey

Executive Summary

Problem Formulation

Occurrence of Infection

- CDC surveillance reports showed that of 3,209 positive nonpolio enterovirus samples reported between 1993 and 1996, the predominant species were echovirus 9 = 12.7%; echovirus 30 = 9.5%; echovirus 6 = 5.1%; and echovirus 11 = 4.5%. WHO's surveillance found that 65% of nonpolio enterovirus infections were echovirus. On a broader scale, Canadian laboratories reported that 0.24% of all positive samples from any virus in 1993 were attributed to echovirus.
- Seroprevalence studies have shown 30% of adults positive for echovirus 30 and 60% of children under age 10 positive for echovirus 9. Occurrence varies by serotype, geography, season, and the host's age and antibody status.

Role of Waterborne Exposure

- Echoviruses are found in all types of water: lakes, rivers, marine environments, wastewater, and chlorinated drinking water. Theoretically, echovirus can have a waterborne transmission; however, little evidence supports this as the case. Most outbreaks appear to result from person-to-person spread.

Degree of Morbidity and Mortality

- Echovirus primarily causes meningitis, unspecified febrile illness, and respiratory illness. Most disease resolves on its own within a week of infection.

- Susceptibles such as newborns can have a fatality rate of up to 10%, but otherwise, fatality is uncommon.

Detection Methods in Water/Clinical Specimens

- Isolation of echoviruses in cell culture, followed by serotyping by neutralization, is still the "gold standard" of detection. Molecular methods like PCR are being used increasingly in clinical and environmental research, but although PCR is more sensitive, it is not standardly used in either type of research. PCR is appealing for clinical applications because of its rapid results; however, for environmental samples (e.g., drinking water), culturing is the only way to show infectivity.

Mechanisms of Water Contamination

- Echoviruses are shed in human fecal matter, which is the only route of water contamination.

Concentrations at Intake

- Echoviruses are found in all types of water sources—from sewage to drinking water. Concentrations reported in sewage have varied up to 900 CPU/L (echovirus 7) and up to 0.1 most probable number of infective units/L (echovirus 22) in drinking water.

Efficacy of Water Treatment

- Standard water treatment with chlorine disinfection decreases concentrations significantly but not necessarily by 100%. Free chlorine of 0.3–0.5 ppm reportedly results in rapid inactivation.
- Water temperature, contact time, and level of organic material in the source water are all related to disinfection efficacy.
- Membrane filtration technology has the potential to remove virtually all viral pathogens.

Survival/Amplification in Distribution

- Not applicable in viruses, though survival within biofilms is an untested possibility.

Routes of Exposure

- Fecal–oral: primarily person-to-person directly or via fomites.
- Theoretically, through water or food ingestion; aerosolization.

Dose Response

Infectious Dose

- A study of 149 volunteers ingesting echovirus 12 concluded that 919 pfu are necessary to infect 50% of an exposed population and an estimated 17 pfu are necessary to infect 1% of an exposed population. The infection rate was dose related up to 10,000 pfu but only for infection—not time to or length of viral shedding or the development of illness.
- Attack rates of infection have been reported variously, depending on the serotype and host characteristics: echovirus 30 showed an attack rate of 16% in a mixed-age group, 60% in an adult group, 75% in a child group (under age 5); echovirus 9 showed 17–33% in an adult group, 50–70% in a young child group; and multiple echovirus serotypes have shown 22–52% in neonatal nursery outbreaks.

Probability of Illness Based on Infection

- The probability of illness based on infection varies widely by serotype and the host's age and antibody status. Overall, a large percentage of those infected are asymptomatic. In an infective dose study, none of the 81 volunteers who were infected with echovirus 12 became ill. In a day-care center outbreak, 2 of 79 infected children developed meningitis compared with 12 of the 65 infected parents who developed meningitis. In a prospective study, 21% of neonates who were infected with nonpolio enterovirus in their first month became ill; serotype had a large influence on the incidence of illness (e.g., 0% for echovirus 22 and 54% for echovirus 5).

Efficacy of Medical Treatment

- Until recently, no antiviral drugs effectively prevented or treated echovirus infections. However, a new antipicornaviral agent, pleconaril, has shown promise in clinical trials for echovirus meningitis. It is expected to receive approval from the FDA.

Secondary Spread

- Secondary spread of echovirus infections is common. One source estimated the attack rate in family members of an index case to be 43% for echovirus; other estimates have been higher (again, depending on virus strain, etc.). Viral shedding generally lasts from a week to a month. This relatively long period increases the probability of secondary spread. Several sources report that illness in secondary cases is lessened from that of the index case; frequently secondary infection is asymptomatic.

Chronic Sequelae

- Outcomes from echovirus infections are mostly acute. Data is equivocal on whether severe central nervous system effects (e.g., seizure) in young children can cause cognitive, behavioral, or neurological sequelae later in life. The largest and most well-designed study to date showed no effect.
- Echovirus infection in children has been a suggested precursor to insulin-dependent diabetes; however, data support a stronger connection between coxsackievirus and diabetes, rather than echovirus.

Introduction

The *Picornaviridae* family comprises five genera, including enterovirus, hepatovirus, rhinovirus, cardiovirus, and aphthovirus. Along with polioviruses 1–3, coxsackieviruses A1–A22 and B1–B6, and enteroviruses 68–71, echoviruses comprise the major enterovirus phylogenetic groups. With the advent of cell culturing in the late 1940s, enteroviruses were isolated from patients suffering a wide variety of syndromes, yet it was unclear whether the association was causal. They were dubbed "enteric cytopathic human orphan" (ECHO) viruses—orphans, because they were viruses in search of a disease (Hill 1996). The echovirus serotypes include 1–9, 11–27, and 29–34. Because echoviruses 22 and 23 are now recognized as genetically different from the other echovirus serotypes (Oberste, Maher, and Pallancsh 1998), these two have been placed in their own genus, which is called human parechovirus (Ghazi et al. 1998). Echovirus 22 is now labeled human parechovirus type 1, and echovirus 23 is now labeled human parechovirus type 2. Literature before 1998 considers serotypes previously known as 22 and 23 as members of the echovirus genus and consequently reports results as such. This review will not separate these serotypes from the echovirus genus when describing studies.

Much of our knowledge of enteroviruses comes from the extensive study of poliovirus, and indeed, much of the information in the literature is presented as representative of the enterovirus genus as a whole and not specifically to virus groups, such as echovirus or coxsackievirus. Though there are many similarities, there are also significant differences among the serotypes, especially regarding pathogenesis. Though this chapter does focus on echovirus, it does present a significant amount of data on enteroviruses as a whole, which encompass all the phylogenetic groups, including echovirus.

The enteric viruses replicate in the alimentary tract (hence the name); however, they generally do not cause significant gastrointestinal symptoms. Infection is mostly asymptomatic (Minor 1998; Modlin 1986), but disease outcomes can range from fatal central nervous system complications to mild, unspecified febrile illness. Specific disease syndromes include paralytic poliomyelitis, encephalitis, aseptic meningitis, myocarditis, conjunctivitis, and respiratory illness, among others. Generally, certain serotypes are associated with particular symptoms or syndromes, although this is not exclusive. For example, echovirus is most often associated with aseptic meningitis, but poliovirus and coxsackieviruses A and B can also cause aseptic meningitis.

Epidemiology

Enterovirus infections occur both sporadically and epidemically within a population, and certain strains tend to follow one or the other pattern (CDC 2000). Some tend to show up regularly in populations (e.g., echoviruses 9 and 11), while others surface for a season or two then disappear (Strikas, Anderson, and Parker 1986). For example, echovirus 30 was reported most frequently in 1997 and 1998 in the United States (CDC 2000). Babies have a high incidence rate of echovirus serotypes that are circulating at the time of their birth. On the other hand, older children (5–10 years old) can be the most affected age group in a population when an uncommon serotype emerges because of their lack of exposure and immunity during infancy (Gondo et al. 1995; Modlin 1986; Rotbart 1995b; Yamashita et al. 1994).

Incidence and Prevalence

A large, prospective study looking at enterovirus incidence in newborns was published in 1984 (Jenista, Powell, and Menegus 1984). The researchers tested 666 newborns and 629 mothers who were not excreting nonpolio enteroviruses within the first day after delivery. After hospital discharge, the babies were tested one to four times a week for a month using culture specimens. The incidence of enterovirus infection within that first month of life was 12.8%, with 79% of the infants presenting without symptoms. The overall infection prevalence at the end of the testing period was 5.3%. Twenty-one percent of the babies who tested positive had to return to the hospital, compared with 4% of the infants overall. Risk factors associated with infection in this cohort study were low socioeconomic status and lack of breastfeeding (both $p < 0.0001$). The only factors that influenced whether or not infection produced symptoms were the virus serotype (e.g., 0% for echovirus 22 and 54% for echovirus 5) and the presence of maternal antibodies.

A Finnish study compared the seroprevalence of echovirus 22, which is now classified outside of the enterovirus group, and echovirus 30, which is a common source of meningitis outbreaks (Joki-Korpela and Hyypiä 1998). They found echovirus 22 to be very prevalent; neonates tested positive because of maternal antibodies, but later, 20% of children <1 year tested positive; 89% of children between ages 1 and 2 were positive; and 97% of adults were positive. On the other hand, the seroprevalence of echovirus 30 was much lower. Fewer neonates had neutralizing antibodies against echovirus 30; only 30% of adults were seroprevalent; overall, persons >1 year of age in the study population had a 24% seroprevalence. This difference could explain why echovirus 30 outbreaks are commonly reported—the existence of a susceptible population without antibody protection. Generalizations to populations in other countries, however, cannot be made.

A Japanese study (Gondo et al. 1995) compared the seroprevalence of echovirus 9 in children in the years before and after a large outbreak. The early samples demonstrated a seroprevalence of <20% in children <2 years, 40% in children 4–10 years, and 60% in children >10 years. After the epidemic, the same percentage of young children tested positive, but all children >2 years averaged 60%, suggesting that the epidemic increased the prevalence in the 4–10 year-old group to above-normal levels.

Enterovirus Surveillance

Surveillance systems, by nature, skew reports of disease toward including cases with more serious complications. Therefore, surveillance of enteroviral disease overestimates the true incidence of serious outcomes, especially related to the central nervous system (Morens and Pallansch 1995). Also, the detection of enterovirus in a stool sample does not necessarily correspond with disease because fecal shedding occurs for a long time after infection.

The World Health Organization (WHO) has had a viral infection surveillance program since 1963 (Minor 1998). Although the data are subject to the foibles of passive reporting, some useful information on the prevalence of various clinical outcomes emerges. WHO reported almost 60,000 nonpolio enterovirus infections between 1975 and 1983. Of the total, 65% (n = 38,191) were attributed to echovirus, 10% to coxsackievirus A, 25% to coxsackievirus B, and 0.4% to enteroviruses 68–71 (Minor 1998). Almost half of the reported infections were associated with central nervous system disease, especially aseptic meningitis (Minor 1998). The predominating echovirus serotypes were 11, 6, and 22—all commonly associated with meningitis. Almost 80% of the reported echovirus infections were spread evenly across the three youngest age groups: <1 year, 1–4 years, and 5–14 years (Minor 1998). The UK's enteroviral prevalence picture was almost identical. Between 1981 and 1991, 61% of reported nonpolio enterovirus infections were attributed to echovirus, 30% to coxsackie B, and 9% to coxsackie A (Hill 1996).

The US Centers for Disease Control and Prevention (CDC) has maintained a surveillance database for enteroviruses since 1951, and it estimates that every year in the United States, 30 million nonpolio enterovirus infections cause an unknown number of cases of aseptic meningitis, hand-foot-and-mouth disease, and upper-respiratory disease (CDC 1997). In their most recent report of nonpolio enterovirus infections (1997–1998), states reported a total of 1,672 isolates (CDC 2000). Of those isolated, the predominant echovirus serotypes were echovirus 30 (27.5%), echovirus 11 (13.8%), echovirus 9 (8.7%), and echovirus 6 (6.9%). Only 25.3% of the reports included a clinical diagnosis: aseptic meningitis (37.6%), encephalitis (4.1%), pneumonia or other respiratory disease (9.3%), paralysis (0.2%). Twelve of the 15 most frequently detected serotypes for this period were also among the 15 most common serotypes during the 1993–1998 period, on which the last CDC surveillance report was based. Unfortunately, the number of state laboratories reporting decreased significantly—from 25 in 1993 to only 8 by 1999. Clearly, this reduces the ability to generalize results across the country and accurately estimate a rate of infection. Using data on the sensitivity of cell culturing and the percentage of meningitis cases caused by enteroviruses, Rotbart (1995a) estimates that there are more than 75,000 cases of enterovirus meningitis in the United States, which is more than 10 times the number reported to the CDC.

Canada has a broader surveillance program that tracks viral and selected nonviral infections (Wilson and Weber 1995). In 1993, 38 laboratories contributed data from 66,447 virus-positive samples (from 1.44 million total viral and select antiviral samples). One hundred fifty-seven positive echovirus samples (0.24% of total positives), representing seven serotypes, were reported for 1993: 3, 4, 9, 11, 17, 25, and 30. Echovirus types 4, 9, and 11 were dominant. Though the same serotypes appeared in the surveys in 1991 and 1992, their incidence of infection varied over the years.

Outbreaks

Many outbreaks associated with enterovirus and specifically echovirus have been reported, usually in the form of aseptic meningitis. Following is a summary of several of these reports:

- Echovirus 30 was the cause of an outbreak of aseptic meningitis in Australia (Gosbell et al. 2000). The 30 cases were aged 8 months to 51 years. The primary symptoms were headache, photophobia, and fever. The water supply was ruled out as a source of transmission.
- Another echovirus 30 outbreak occurred in Switzerland, affecting 80 people with meningitis. Of the 80, 38 were children. The patients' age did not affect white blood cell count, polymorphonuclear lymphocytes, or protein in the cerebrospinal fluid (Schumacher et al. 1999).
- In the summer of 1999, a meningitis outbreak was traced to echovirus 4 in Italy (Portolani et al. 2001). An unusual aspect of this

outbreak was that 23 of 25 patients were adults (mean age 24.5). The origin of the outbreak was unknown.

- Four hundred cases of meningitis were reported in a statewide (Rhode Island) outbreak in the summer of 1991 (Rice et al. 1995). Enterovirus was isolated from 61 samples, and of the six samples that were serotyped, all were identified as echovirus 30. The patients ranged in age from 8 weeks to 74 years with 15% <2 years, 39% 2–18 years, 22% 19–28 years, and 22% >28 years. About one third of the cases reported contact with another ill person. The number of adult patients included in the report is fairly unusual; most meningitis reports focus on children. The clinical features shared by the majority of patients were headache, fever, nausea or vomiting, stiff neck, and photophobia, which are all typical of aseptic meningitis. Because of this outbreak, Rhode Island's aseptic meningitis incidence rate for 1991 was 44.5 per 100,000—10 times the national incidence.
- Seventy-nine people were diagnosed with aseptic meningitis in a 1995 outbreak in Illinois (compared with 0–7 per year in the previous 4 years) (Rodriguez et al. 1997). Echovirus 9 was isolated from nine cerebral spinal fluid or stool specimens; echovirus types 5 and 21 were each isolated from one stool specimen. The age range of patients was 0–66 years; the age groups all shared similar rates of illness. The clinical features were typical of aseptic meningitis. In terms of risk factors, investigators did not find an association between illness and recreational water use, attendance at public events, source of household drinking water, or a number of other household-related factors. The CDC speculated that the illness was transmitted from person to person rather than from a common source.
- Investigators conducted a cross-sectional and case-control study after 21 children in a German town were hospitalized with aseptic meningitis associated with echovirus 30 (Reintjes et al. 1999). They assessed every 10th child (<16 years old) from the registered 2,240 in the town. Sixty-two cases were identified from the general population and matched with other children from the same population. Results of the cross-sectional study showed an overall attack rate of 16%, with the highest rate of 24% occurring in the 6–8-year-old group. Results of the case-control study suggested the following risk factors for illness: contact with an ill household member, day-care attendance, and regular playground

use. The authors concluded that this outbreak was spread through person-to-person contact with family and community.

- A report from Western Australia described an outbreak of aseptic meningitis in 161 people (Ashwell et al. 1996). Of the cases, 64% were <15 years old, with the highest rate in the <5 years group (which could be a result of reporting bias because children are reported more frequently because they receive more health care than adults). Interestingly, this single outbreak was traced to two different echovirus serotypes—9 and 6. Forty-one percent of the cases were traced to echovirus 9, occurring mainly in the metropolitan areas; echovirus 6 caused 37% of the cases and was more widespread geographically. A risk factor for disease was recent contact with proven or likely cases. Also, the same viruses were isolated from cases and contacts. This evidence points to a person-to-person transmission instead of a single source.

- A number of parents of children in a day-care center developed meningitis from echovirus 30 (Helfand et al. 1994). To understand the extent of the outbreak, investigators surveyed and collected blood samples from the day-care families and teachers and adult and child controls. The parents and children had significantly higher infection rates (67/111 and 79/105, respectively) than the teachers (3/22) or controls. The infected parents presented with the most severe illness: 18% (n = 12) had meningitis and 11% were hospitalized. Only two of the infected children developed meningitis. The survey indicated that the good handwashing habits of teachers compared with parents resulted in their substantially lower infection rate.

- A large outbreak of "Zhi Fang" disease in China in 1985 was associated with echovirus 3 (Hill 1996). The outbreak affected approximately 20% of a village's population, and the case fatality rate was an unusually high 12%. The symptoms included very high fever, muscle spasms, dizziness, and myalgia—seemingly a different syndrome from aseptic meningitis. Again, adults were more affected and suffered the most fatalities.

- Kee et al. (1994) reported on an outbreak of 46 people with vomiting, diarrhea, and headache from exposure to a swimming pool. The outbreak was traced to one bather who vomited in the pool; that index case probably infected the rest of the cases who were swimming in the pool that day. Echovirus 30 was isolated from the index case and from six others. The chlorine levels in the pool had not been substantial enough to control the spread of infection.

- In the summer of 1998, several states, including Minnesota, Idaho, Texas, and South Dakota, reported more than 500 cases of viral meningitis, which represented "increased activity" but were not classified as full-blown outbreaks (Connolly 1998). Informally, state health officials commented that even with the increased reporting, there was still significant underreporting of the illness, partly because the symptoms can often mimic cold or flu. Several mentioned that the amount of interfamily infection was unusually high, suggesting a direct person-to-person transmission. Multiple cases in Minnesota and South Dakota were both tracked to echovirus 30.

Clinical

Echoviruses and other enteroviruses are usually acquired through the fecal–oral route, where they infect the gastrointestinal tract after ingestion. The consequent minor viremia can seed many different organ systems, resulting in a variety of clinical manifestations affecting the central nervous system, liver, lungs, or heart.

Aseptic Meningitis

Aseptic meningitis refers to meningitis of a nonbacterial nature and is most commonly caused by enterovirus infections. Meningitis is more typically seen in children <1 year old (Rorabaugh et al. 1993) and is associated with the summer months (Coyle 1999). Echovirus is more frequently associated with meningitis than the other enteroviruses. The types that are predominantly linked with this illness are 4, 6, 9, 16, and 30 (Modlin 2000; Rotbart 1995b). The symptoms are usually acute and include high fever, headache, stiff neck, and irritability. The self-limited illness typically lasts less than a week, with complete recovery in a matter of weeks. Adult patients will occasionally continue reporting symptoms for several weeks (Modlin 2000; Rotbart 1995a). Diagnosis is made through testing cerebrospinal fluid (Coyle 1999).

In a cohort study of 52 young children (< 2 years) with meningitis, 9.0% developed more advanced neurological complications such as seizures and coma; the rest recovered without incident. The complications were more commonly seen in children >3 months old (Rorabaugh et al. 1993). The neonate who develops meningitis symptoms in the first few days of life is at extreme risk for serious morbidity and death (possibly up to a 10% mortality rate) (Dagan 1996; Rotbart 1995a). Newborns who die as a result of these early echovirus infections usually succumb not to central nervous system complications but to liver failure (Modlin 2000; Rotbart 1995b). Enterovirus meningitis that occurs in patients who are beyond the neonatal stage rarely results in these types of severe complications (Rotbart 1995a). Studies have been conducted to test the long-term sequelae on children who had meningitis infection early in life (see chronic sequelae section).

Febrile Illness

The most common clinical manifestation of an enterovirus infection is unspecified febrile illness with or without rashes (Rotbart and Hayden 2000). Other symptoms that may appear are vomiting, rash, and respiratory illness. All enterovirus types can cause this type of febrile illness, but the prevalence varies among the different types and serotypes and with the presence of antibodies (Dagan 1996). Young infants are primarily affected, and differentiation between this and other, more serious illness is difficult (Rotbart and Hayden 2000). The incubation period ranges from <2 days to 2 weeks, and symptoms in young children last from <24 hours to longer than a week, though the usual course is 3–4 days (Dagan 1996; Modlin 2000). Usually more than one enterovirus serotype is circulating in the population during the season, so patients can suffer from more than one febrile episode caused by different serotypes (Dagan and Menegus 1995). Because it is rarely reported, little is known about the presentation of nonspecific febrile illness in older children and adults.

Respiratory Disease

Worldwide, enteroviruses account for approximately 2 to 15% of viral respiratory disease (Chonmaitree and Mann 1995). After unspecified febrile illness, respiratory illness is the second most common syndrome associated with the enteroviruses (Chonmaitree and Mann 1995). Most enteroviruses can cause a respiratory illness (the "common cold") that clinically cannot be distinguished from rhinovirus, respiratory syncytial virus, or adenoviruses infection (Minor 1998; Modlin 2000). One identifying feature is that respiratory disease from enterovirus occurs primarily during the summer months, whereas the others are predominantly winter illnesses (Minor 1998; Chonmaitree and Mann 1995). Most summertime colds in children are caused by the enteroviruses (Modlin 2000). Types of echoviruses associated with upper respiratory tract infection include 1–3, 5–9, 11, 14, 16, 18, 20, 22, 25, and 30. Type 11 has been used as the prototype echovirus in adult volunteer studies of respiratory infection (Chonmaitree and Mann 1995). Other clinical presentations include ear infection, tonsillitis and pharyngitis, herpangina, croup, bronchitis, and pneumonia. The successful transmission of enteroviral respiratory infection depends on the route of transmission, type of person-to-person contact, and the virus serotype (Chonmaitree and Mann 1995; Modlin 2000). Aerosolization appears to be more efficient in inducing respiratory infection than ingestion, and though contagiousness is high within households, infection after other casual contact is relatively low (Chonmaitree and Mann 1995). Respiratory diseases caused by enteroviruses are mild and require no treatment; about one half to two thirds of infected people do not develop symptoms. The percentage of asymptomatic infection is higher with echovirus than it is with coxsackievirus (Chonmaitree and Mann 1995). Respiratory illness may be less common in infected infants than in older children and adults (Dagan and Menegus 1995).

Myocarditis

The enteroviruses are the most common pathogens associated with myocarditis, and of the echoviruses, 9 and 22 have the strongest association (Rotbart and Hayden 2000). Cases most frequently occur in young adults between 20–39 years old (Rotbart and Hayden 2000). The clinical picture in older children and adults ranges from asymptomatic cardiac infection to intractable heart failure and death (Modlin 2000). Typically, an upper respiratory illness comes 1–2 weeks before cardiac symptoms of chest pain, dyspnea, and fever (Modlin 2000). In neonates, myocarditis causes heart failure, respiratory distress, tachycardia, and sometimes cyanosis and circulatory collapse. Death occurs less than 50% of the time in the newborn population (Modlin 2000).

Encephalitis

An estimated 20,000 cases of viral encephalitis occur in the United States annually, with herpes simplex being the most commonly identified source, though only accounting for about 10% (Rotbart 1995b). Arboviruses and enteroviruses (11–22% of cases) are probably the next most common causes; however, the etiology usually remains unknown (Rotbart 1995b). The primary causes of echoviral encephalitis are echoviruses 6 and 9; however, other serotypes, including echovirus 17 and 19, have also been implicated (Minor 1998; Modlin 2000). The true incidence of enterovirus-associated encephalitis is difficult to ascertain, because the virus is rarely cultured from patients with the disorder. Brain biopsy has been one of the only reliable ways to determine enterovirus as a source of encephalitis, but the advent of polymerase chain reaction (PCR) testing will undoubtedly improve the ability to gather information (Rotbart 1995a). Enterovirus encephalitis is considered a more serious illness than enterovirus meningitis (Rotbart 1995a), though patients beyond the neonatal stage usually make a full recovery (Modlin 2000).

Hepatitis

Hepatitis results from infection with coxsackievirus B and echoviruses. The infection occurs primarily in neonates but occasionally in older children or adults (Minor 1998; Modlin 2000). Hepatitis syndrome in infants causes significant morbidity and mortality. In a review of 61 cases of perinatal echovirus infections, babies who developed hepatitis as sequelae had a fatality rate of 83% (Modlin 1986).

Exanthems

Exanthems, which are usually rubelliform or maculopapular in nature, occur during enterovirus infections—usually coxsackieviruses A9, A16, and B5 and echoviruses 4, 9, 16, and 25 (Hill 1996; Minor 1998). Rashes are often transient and are accompanied by other symptoms such as fever, malaise, and aseptic meningitis (Minor 1998). In a review of 82 infants <3 months, 28% developed a rash as part of their echovirus infections (Dagan and Menegus 1995).

Paralytic Disease

Poliovirus remains the leading cause of viral paralysis in the world (Minor 1998). Sporadic cases have been reportedly linked with coxsackievirus A7 and B2–B6 and echovirus types 3, 4, 6, 11, and 19 (Minor 1998).

Other

Other syndromes rarely associated with echovirus include keratoconjunctivitis, myositis, pancreatitis, gastroenteritis, and acute arthritis (Kaye et al. 1998; Modlin 2000).

Treatment

Until recently, no antiviral drugs effectively treated or prevented nonpolio enterovirus infections. Pleconaril is a broad-spectrum, antipicornaviral agent that is close to US Food and Drug Administration approval for use in aseptic meningitis and respiratory illness. It has been used on a compassionate-release basis for patients with severe enteroviral infections and shown to have positive reactions with few adverse effects (Romero 2001; Rotbart and

Webster 2001). However, isolating ill patients to contain the spread of infection is not effective because so many asymptomatic people shed the virus and the spread from an index patient to secondary contacts is rapid. Obviously, controlling a specific environment, such as a neonatal unit in a hospital, can reduce the likelihood of an outbreak.

In very serious neonatal enterovirus infections (especially in group B coxsackievirus infections but also echoviruses), γ-globulin, maternal plasma, and exchange transfusions have been used intravenously with limited success (Dagan 1996; Minor 1998; Rotbart 1995a). Pleconaril has been used successfully in critically ill neonates (Aradottir, Alonso, and Shulman 2001). Agammaglobulinemic patients have also experienced some improvement in their conditions through γ-globulin therapy (Rotbart 1995a). In healthy patients, enteroviral meningitis can usually be treated with palliative care until recovery is complete (Dagan 1996; Rotbart 1995b).

Susceptible Subpopulations

Immunocompromised

Generally, enteroviral infections do not discriminate between healthy and immune-compromised hosts. Though severe disease can occur in the immunocompromised, enterovirus is not a major infection in people with AIDS, cancer, etc. (Morens and Pallansch 1995). There is a notable exception, however. The enteroviruses are cleared by the host by antibody-mediated mechanisms, in contrast with other viruses, which work through cellular immune mechanisms. Because of this difference, people without effective humoral immunity are sensitive to infection and present with a much different clinical expression than immunocompetent people. In many of these patients, the infection is fatal (Modlin 2000). Patients with common variable immunodeficiency or agammaglobulinemia, which are both primary immunodeficiency diseases, can develop chronic meningoencephalitis from enterovirus infection that lasts for years and can result in death (Rotbart 1995b; Sicherer and Winkelstein 1998). Sixteen patients with chronic meningoencephalitis responded positively after treatment with pleconaril (Rotbart and Webster 2001). Intravenous treatment with antibodies improves some patients; however, the virus may persist despite this type of therapy (Rotbart 1995a). Expectedly, patients undergoing allogeneic bone marrow transplants have also developed disseminated disease (in brain, lung, heart, skin, stomach, colon, and liver tissue) from echovirus infections (Schwarer et al. 1997). Reported outcomes in the affected recipients included both death and recovery without sequelae.

Children

Enterovirus infections in newborns are common; data suggest that echovirus may be the most prevalent cause of infection in neonates (Abzug 1995). Many echovirus outbreaks in newborns result in asymptomatic infections (Abzug 1995; Dagan 1996). One community-based study demonstrated a 13% enterovirus incidence in neonates <1 month old, with 21% of the infected infants becoming ill (Jenista, Powell, and Menegus 1984). Generally, serious complications from enteroviral infections are less common in young children than in older children and adults (Abzug 1995; Morens and Pallansch 1995), though the reasons are unclear. One exception to this is when fetuses are infected transplacentally (or in the first day or two after birth); this route of transmission puts them at higher risk for serious illness (Abzug 1995; Modlin 2000). Though the humoral immune system of neonates can respond to enteroviral infections, their macrophage function is not mature enough in the first few weeks to stop viral replication (Modlin 2000). Risk factors that are associated with serious clinical outcomes in neonates include infection with particular serotypes (especially echovirus 11), male gender, prematurity and low birth weight, and prepartum illness in the mother (Abzug 1995; Modlin 1986). The timing of maternal infection plays a critical role in infection outcome through the influence of maternal antibodies passively acquired by the neonate (Modlin 2000).

The most serious complication associated with echovirus infections in neonates is hepatitis. In a review of the literature, Modlin (1986) reported that 67% of children classified as having severe echovirus infection developed hepatitis, and 83% of them died. Overall, his review calculated a 3.4% fatality rate over 16 neonatal outbreaks. Meningoencephalitis from echovirus affects the central nervous system and occasionally other organs and is another serious outcome in neonates; however, hepatitis is far more serious and results in higher mortality rates (Modlin 1986). Myocarditis in the neonate results in morbidity less than 50% of the time (Modlin 2000).

Chronic Sequelae

Echovirus infection causes a variety of clinical syndromes, most commonly aseptic meningitis and respiratory infection but also rashes, encephalitis, and, in serious cases, hepatitis. Usually, these illnesses are acute, with symptoms resolving within a week of infection. Occasionally, the clinical outcome of a central nervous system-related illness will be serious and involve complex seizures, intracranial pressure, or coma (Rorabaugh et al.

1993). Evidence from several studies that have investigated the risk of long-term neurological and cognitive sequelae from these acute manifestations appears to be equivocal. An older study reported that children who had been hospitalized with enterovirus infection in their first year scored significantly lower in I.Q., language and speech, and head circumference (Pollard 1975). However, the study only included 19 children with matched controls. A similar study (Wilfert et al. 1981) found that nine children who had enteroviral meningitis in the first 3 months were significantly behind in language functioning but showed no difference in intellectual ability, hearing function, or head circumference. Bergman and colleagues (1987) compared 42 children who had enteroviral meningitis in their first year with their siblings for neurological, psychological, language, and academic indicators. They found no differences in any functioning scores between the case children and their control siblings. A large prospective cohort study is reported by Rorabaugh and coworkers (1993). They assembled a cohort of 277 children <2 years old with aseptic meningitis and tracked their acute illnesses and neurologic complications. Of the 277, neurologic signs were noted in 52 patients; 25 patients (about 10% of the total cohort) had significant neurologic complications such as seizures, coma, or intracranial pressure. These complications generally afflicted children >3 months. The children were given neurological and developmental tests four times over 42 months. The children with central nervous system complications did not score any differently from the children without complications. Of the 169 cases that had confirmed etiologies, 55% were caused by coxsackievirus group B and 38% by the echoviruses. Fewer than 5% were caused by other enteroviruses or adenoviruses. The complications occurred equally in the coxsackievirus B- and echovirus-caused cases, and no single serotype dominated.

If a neonate experiences hepatitis resulting from echovirus infection and if it survives the acute attack, the child may continue to have hepatic disease throughout infancy. However, liver function usually improves in time, and the child will continue to develop normally without long-term sequelae (Abzug 1995).

Acute myopericarditis caused by enteroviral infection can lead to chronic cardiac conditions. Electrocardiographic abnormalities (10–20%) and cardiomegaly (5–10%) are examples of permanent heart injury occurring in about one third of patients with acute myocarditis (Modlin 2000). These conditions can lead to chronic dilated cardiomyopathy, which is the second leading cause of congestive heart failure (Modlin 2000).

Various chronic neuromuscular conditions have been reported as enteroviral sequelae. Most of these associations have centered on coxsackieviruses and, of course, polioviruses. Rare case reports give accounts of paralysis, transverse myelitis, and chronic fatigue syndrome occurring after echovirus infection; however, little is known about the length of time these conditions may have persisted in affected patients (Dalakas 1995; Figueroa et al. 1989; Takahashi et al. 1995; Yamashita et al. 1994). Emerging molecular detection techniques will shine additional light on these possible associations.

Many studies have been conducted looking at the association between enteroviruses and insulin-dependent diabetes mellitus (IDDM), and data supporting both sides of an enteroviral link with IDDM have been published. The main focus has been the coxsackievirus–IDDM connection; however, echoviruses have also appeared as potential problems. Study designs have included cross-sectional, case-control, prospective cohort, retrospective cohort, and animal studies. Most epidemiological studies have centered on measuring serum antibody levels in children with IDDM, their mothers, and their siblings. The weight of evidence indicates that enteroviral infection is a risk factor for IDDM, and though echoviruses have been occasionally implicated, coxsackieviruses are implicated most frequently. Therefore, given the limited data on echovirus and IDDM, it is difficult to identify causality.

Transmission

Humans are the only known reservoir for the (human) enteroviruses. They are shed primarily in the feces but can also be shed from the throat, typically for 1–2 weeks (Rotbart 1995a). Unlike other gastrointestinal viruses, enterovirus colonization does not change the ecology of the gut; therefore, the fecal shedding period for the enteroviruses is relatively long compared with other gastrointestinal viruses, which frequently lasts only as long as the associated symptoms (Morens and Pallansch 1995). For echovirus, the shedding period in children is shorter than the other enteroviruses—seldom more than 1 month after infection (Minor 1998), while children infected with poliovirus and coxsackievirus can shed for weeks or months (Minor 1998; Rotbart 1995a). Adults shed enteroviruses for less time than children—only 1–2 weeks. Asymptomatic people frequently shed the viruses.

Transmission occurs directly through the fecal–oral route or indirectly through food or water (this is theoretical because little evidence of this type of transmission exists) (Modlin 2000; Morens and Pallansch 1995). The viruses are occasionally spread through aerosolized respiratory droplets or blood (Morens and Pallansch 1995). Though enteroviruses are often found in sewage, water, and shellfish, transmission from these sources has never been documented (Modlin 2000). Transmission can also occur vertically, from mother to child *in utero,* either transplacentally, from the amniotic fluid, or in the birth canal (Abzug 1995; Modlin 1986; Morens and Pallansch 1995; Rotbart and Hayden 2000). Vertical transmission from mother to child can have serious consequences for the infected newborn, as babies infected this way are at high risk for serious illness (Dagan 1996; Modlin 1986). In addition to the transplacental route, neonates can also become infected shortly after birth nosocomially or by family members.

Secondary Spread

Secondary spread is high in all enterovirus infections; however, echovirus has a lower secondary spread rate than the other enteroviruses. Secondary attack rates of infection in susceptible family members of as high as 92% for polioviruses, 76% for coxsackieviruses, and 43% for echoviruses have been reported (Minor 1998; Modlin 2000). The fact that the prior two viruses are shed for longer periods than echovirus may influence their attack rates. Most household members infected through secondary spread do not become sick or suffer only mild symptoms. The following factors have been found to affect virus spread to household members: (1) duration of virus excretion, (2) household size, (3) number of siblings, (4) socioeconomic status, (5) type of virus, and (6) immunity status of household members (Morens and Pallansch 1995). Infants who acquire nosocomial echovirus infections have less severe illness than the index cases (Modlin 1986). Much of the hospital-based secondary spread appears to be transmitted through health care workers (Modlin 1986).

Dose Response

Infectious Dose

Schiff et al. (1984) conducted a controlled study looking for infectious dose information on enteroviruses using echovirus 12 as a prototype. One hundred forty-nine volunteers were given 0–330,000 plaque-forming units (pfu) of echovirus 12 in nonchlorinated water. A dose response was seen in the rate of infection in doses up to 10,000 pfu. The dose, however, did not affect the time to viral shedding (the first week after inoculation in all infected volunteers) or duration of shedding (up to 26 days). None of the volunteers developed significant illness. The investigators used their results to estimate a dose of 919 pfu as necessary to infect 50% of volunteers. From that figure, they extrapolated an estimated dose necessary to infect 1% of volunteers as 17 pfu (see Table 10-1).

Table 10-1: Results of volunteer feeding studies of echovirus 12

		Number of Volunteers		
Dose (pfu)	Total	Intestinal Shedding	Seroconversion	Infection (% of group exposed)
0	34	0	0	0
330	50	14	7	30
1,000	20	8	3	45
3,300	26	18	11	73
10,000	12	11	3	100
33,000	4	2	1	50
330,000	3	2	1	67

Adapted from Schiff et al. 1984.

Previous volunteer studies using poliovirus extrapolated smaller infective doses (e.g., 4, 72, and 1 pfu), but those studies used attenuated virus in a study population of children, which may account for the discrepancy. The second part of the Schiff et al. study looked at the effect of previous infection on reinfection rate. Of 18 previously infected volunteers, 72% became reinfected with echovirus after a dose of 1,500 pfu. These results suggest that antibody status does not necessarily affect reinfection rate or number of shedding days. The paper gave no indication of how long the period was between primary infection and rechallenge. Other studies have found a link between increased antibody levels and decreased infection rates (Chonmaitree and Mann 1995).

Virulence

Generally, enteric viruses invade the intestinal epithelium, where they cause destruction of enterocytes, leading to illness. Viruses (and bacteria) use different mechanisms to evade a host's gastrointestinal defenses. Among virulence factors common to enteroviruses are a resistance to acid and bile, which helps them get through the stomach and the gut, and resistance to proteolytic enzymes, which also increases their survival and passage into the large intestine (Duncan and Edberg 1995).

Infectivity and virulence vary by genotype and phenotype. A study investigating the effect of in vivo conditions (e.g., gastric acidity) on enteroviruses found differences among the serotypes: echovirus 22 appeared more sensitive than other types to inactivation in body fluids (Piirainen, Hovi, and Roivainen 1998). All the viruses remained infective after a 2-hour exposure to pH 2 at 37° C, but after 24 hours, they were all inactivated.

In outbreaks, young children are more frequently infected by enteroviruses. The attack rates in an echovirus 9 outbreak reportedly ranged from 50 to 70% in children and 17 to 33% in adults (Morens and Pallansch 1995). A community-wide outbreak of echovirus 30 illness demonstrated an overall attack rate of 16%, with a 24% peak among the 6- to 8-year-olds (Reintjes et al. 1999). Another outbreak of echovirus 30 in a day-care center (Helfand et al. 1994) had high infection rates in both children (75%) and their parents (60%). A summary of echovirus outbreaks occurring in newborn nurseries showed an attack rate ranging from 22 to 52% with a case fatality rate of 6.3% in echovirus 11 cases and 3.4% overall (Morens and Pallansch 1995). Echovirus 11 has been strongly associated with these nursery outbreaks. The tendency of children to have higher attack rates than adults probably results from their increased exposure opportunities and susceptible immune status. In a cohort study of 75 infected babies, the incidence of symptomatic infection was 0% in those with antibodies and 38% in those without antibodies (Jenista, Powell, and Menegus 1984). However, for echoviruses especially, enteroviral illness is frequently milder in children than it is in adults (Helfand et al. 1994; Hill 1996; Morens and Pallansch 1995). (Neonates, who are discussed elsewhere in this chapter, are the exception.)

Environmental Occurrence

Enteroviruses are ubiquitous in the environment. They are found in sewage, freshwater, marine water, shrimp and shellfish, and even chlorinated water. Enteroviruses are far more prevalent in the summer and early fall seasons (in temperate climates). In tropical climates, they circulate more evenly throughout the year or are sometimes more prevalent during the rainy season (Minor 1998; Modlin 2000; Morens and Pallansch 1995).

Survival

Enteroviruses are hardy. They survive for a long time on hands, fomites, and in water (Morens and Pallansch 1995). Yates and colleagues (1985) seeded groundwater samples with three viruses, including echovirus 1, and measured the factors affecting survival. Measurements included pH, nitrates, turbidity, and hardness. They found that temperature was the only variable significantly correlated with viral decay rates and that none of the viruses tested remained viable significantly longer than the others.

The long-term survival of three strains of enterovirus (coxsackievirus B3, echovirus 7, and poliovirus 1) was tested in freshwater from five sites with different physical characteristics (Hurst, Benton, and McClellan 1989). The five sites included an artificial lake, a pond, large- and medium-sized rivers, and a small creek. The average inactivation rates were 6.5–7.0 log^{10} units over 8 weeks at 22°C; 4.0–5.0 log_{10} units over 12 weeks at 1°C; and 0.4–0.8 log_{10} units over 12 weeks at –20°C. The effect of temperature was significant ($p < 0.00001$), as were the serotypes and the water source at each of the three temperatures. Other operative factors included hardness and conductivity (which were detrimental to survival), turbidity and suspended solids (which were beneficial to survival), and, at 22° C, the number of generations of bacterial growth supported by the water sample (which was detrimental to survival).

Occurrence in Water Sources

The enteroviruses have been isolated from all types of water sources: rivers, raw sewage, wastewater, oceans, and groundwater. Table 10-2 details selected studies on echovirus occurrence in the environment.

Descriptions of enterovirus concentrations have usually not been serotype specific. Recent ability to more easily detect serotypes has resulted in more detailed analyses. Keswick and colleagues (1984) did an extensive study of the occurrence of viruses in conventionally treated drinking water coming from poor-quality source water. Seven percent of the 37 finished water samples tested positive for enterovirus (see Table 10-2 for details). Payment and colleagues (1988) reported enterovirus averages between 1 and 145 mpniu/L (most probable number of infectious units per liter) from river waters in Quebec. In a river in Japan, concentrations of enteroviruses detected over a 63-month period averaged 41 pfu/L, with a range of 1 to 190 pfu/L (Tani et al. 1995). A total of 11 echoviruses were isolated, with types 11 and 6 found most often. A 1981 study reported on the occurrence of enterovirus species, including echovirus 7, in community swimming pools (Keswick, Gerba, and Goyal 1981). Ten of 14 samples contained virus; 3 samples were positive in the presence of residual chlorine; 2 of those residual levels exceeded the 0.4-ppm standard. Seven of seven of the wading pool samples were positive, which is not surprising considering the water is usually warm and the bathers are usually toddlers.

Detection Methods

Though viruses are commonly found in water sources and cause frequent infections, they are not easy to detect in either clinical or environmental samples. This difficulty can obscure the true prevalence of enterovirus infections and estimates of concentrations in drinking water sources.

Table 10-2: Occurrence of entero- (echo-) viruses in the environment

Reference/ Location	Environmental Source	Serotype	Occurrence (%)*	Notes
Hovi, Stenvik, and Rosenlew 1996 Finland	Raw sewage	Echo 11	16.6	1,161 samples taken between 1971 and 1992; peak occurrence in early fall; cell culture
		Echo 6	15.2	
		Echo 22	2.7	
		Echo 3	2.6	
		Echo 30	1.3	
		Echo 25	0.9	
		Echo 7	0.8	
		Echo 9	0.4	
Hurst 1991 Varied	Natural freshwater	Varied	47.2 (median value of summary)	Summary of 16 studies of occurrence of enteric viruses (mainly enteroviruses) in natural freshwater
	Finished drinking water		0 (median value of summary)	Summary of nine studies of occurrence of enteroviruses in drinking water treated with coagulation, sedimentation, filtration, and disinfection; four studies did find positive samples
Keswick et al. 1984 Unknown	Raw source water	Varied (entero)	56.0	18 samples; dry season; average turbidity = 6.4; cell culture
	Clarified source water		35.0	23 samples; dry season; average turbidity = 5.6; cell culture
	Finished drinking water		41.0	39 samples; dry season; average turbidity = 1.2; cell culture
	Raw source water		7.0	14 samples; rainy season; average turbidity = 26; cell culture
	Clarified source water		29.0	7 samples; rainy season; average turbidity = 10; cell culture
	Filtered source water		15.0	15 samples; rainy season; average turbidity = 6.8; cell culture

*Percentage of samples testing positive.

Table continued on next page.

Table 10-2: Occurrence of entero- (echo-) viruses in the environment (continued)

Reference/ Location	Environmental Source	Serotype	Occurrence (%)*	Notes
	Finished drinking water		7.0	37 samples; rainy season; average turbidity = 9.6; cell culture
Krikelis et al. 1985 Greece	Raw sewage	Echo 7	23.5	191 samples taken over 15 months; peak occurrence in September; average concentration of enteroviruses = 291 CPU/L; highest = 900 CPU/L; cell culture
Payment, Trudel, and Plante 1985 Canada	Raw source water	Varied (mainly entero)	79.0	152 samples over 1 year; average concentration 3.36 mpncu/L; cell culture
	Prechlorinated source water		65.0	17 samples over 1 year; average concentration 0.072 mpncu/L; cell culture
	Sedimented source water		20.0	119 samples over 1 year; average concentration 0.016 mpncu/L; cell culture
	Filtered source water		14.0	119 samples over 1 year; average concentration 0.001 mpncu/L; cell culture
	Ozonated source water		9.0	45 samples over 1 year; average concentration 0.0003 mpncu/L; cell culture
	Finished drinking water		9.0	138 samples over 1 year; average concentration 0.0006 mpncu/L; cell culture
Payment, Fortin, and Trudel 1986 Canada	Raw sewage	Echo 3	5.3	19 samples; average concentration <0.1 mpniu/L; cell culture
		Echo 22	5.3	19 samples; average concentration 0.4 mpniu/L; cell culture
		Echo 14	0	
	Primary sewage effluent	Echo 3	5.3	19 samples; average concentration 0.7 mpniu/L; cell culture
		Echo 22	10.5	19 samples; average concentration 0.7 mpniu/L; cell culture

*Percentage of samples testing positive.

Table continued on next page.

Table 10-2: Occurrence of entero- (echo-) viruses in the environment (continued)

Reference/ Location	Environmental Source	Serotype	Occurrence (%)*	Notes
		Echo 14	0	
	Secondary sewage effluent	Echo 3	0	
		Echo 22	10.5	19 samples; average concentration <0.1 mpniu/L; cell culture
		Echo 14	0	
Puntaric et al. 1995 Croatia	Finished drinking water	Echo 6	0.7	271 samples taken from 1991–1994; cell culture
Reynolds et al. 1998 Hawaii	Primary sewage effluent	Varied	100	12 samples collected; average concentration = 3,300 pfu/L; highest = 13,000 pfu/L; cell culture
	Marine water at sewage outfall discharge site		28.6	7 samples collected; average concentration = 0.04 pfu/L; highest = 0.044 pfu/L; cell culture
	Canal freshwater		17.0	12 samples collected; average concentration = 0.035 pfu/L; highest = .045 pfu/L; cell culture
	Swimming beach water		8.0	50 samples collected; average concentration = .055 pfu/L; highest = 0.1 pfu/L; cell culture
	Stream water		0	4 samples; cell culture
van Olphen et al. 1984 The Netherlands	Raw source water	Mainly entero	67.0	45 samples; cell culture
	Partially treated drinking water		20.0	55 samples; cell culture
	Finished drinking water		0	100 samples of 500 L

*Percentage of samples testing positive.

Clinical

A problem with diagnosing enterovirus infection is that it mimics other common bacterial or viral infections. The symptoms of most enteroviral diseases are indistinguishable from symptoms of other types of infection (e.g., encephalitis/herpes simplex, respiratory illness/rhinovirus). Also, the systemic enteroviral infection developed by some neonates is clinically indistinguishable from bacterial infections and sepsis (Dagan 1996; Rotbart 1995a). Antibiotic treatment and continued in-hospital observation are standard for suspected meningitis infections until a bacterial etiology is ruled out (Dagan 1996; Rotbart and Romero 1995). Consequently, it is important for the clinician to establish etiology—or at least to rule certain things out—so that appropriate treatments can be given and unnecessary treatment averted. Sometimes the season of the year may offer hints to etiology, because enteroviral infections occur much more frequently in the summer and early fall months.

Isolation of the enterovirus in cell culture followed by serotype identification with neutralization is still the "gold standard" for diagnosis (Rigonan, Mann, and Chonmaitree 1998; Rotbart 1995a), but culturing presents numerous problems for the diagnostician. No particular cell line is appropriate for all enterovirus serotypes; however, most laboratories use a combination of monkey kidney cell lines and a human diploid fibroblast line (Dagan 1996; Modlin 2000; Rotbart 1995a). Culturing is technically challenging and labor/time intensive. Also, one fourth to two thirds of clinical samples cultured from patients with characteristic enterovirus infection are negative but are positive when tested with reverse-transcriptase PCR (RT-PCR) (Stellrecht et al. 2000; Rotbart 1995b). The echoviruses cannot be tested in any animal models (Rotbart and Romero 1995).

Because the enteroviruses do not share an antigen, immunoassays have been slow to develop as clinical detection methods (Rotbart 1995a); however, Rigonan and colleagues (1998) reported 93% sensitivity and 90% specificity compared with standard neutralization using immunofluorescent assay with commercially available antibody blends. The ability to use easily available reagents is an appealing feature of this assay, and perhaps more validation will be forthcoming. Electron microscopy has no application in detecting enteroviruses (Rotbart and Romero 1995). RT-PCR has been used broadly to detect many or all serotypes or in a more limited way to detect a few or a single serotype (Rotbart and Romero 1995). RT-PCR is primarily used to study the molecular epidemiology of enteroviruses, such as strain prominence and genetic shifts. RT-PCR is definitely a promising alternative to culturing, especially for meningitis cases for which quick diagnosis is critical. Tanel et al. (1996) reported on a comparison of RT-PCR and viral culture in the cerebrospinal fluid of meningitis cases. They found the RT-PCR assay to be 77.8% sensitive and 100% specific compared with a sensitivity of 66.7% for culture alone. Other researchers report RT-PCR's sensitivity as 85% and culture's sensitivity as 24% (Gorgievski-Hrisoho et al. 1998). Another benefit of PCR is the reduced time it takes to get results. Cell culture usually takes 3 to 15 days compared with PCR, which can be finished in the same or next day (Furione et al. 1998; Gorgievski-Hrisoho et al. 1998). In a cost–benefit analysis of using PCR as a clinical tool, Schlesinger and colleagues (1994) found PCR to be more sensitive and specific than viral culture. To analyze cost using retrospective records, they went on to estimate that if PCR had been used instead of culture for diagnosis; the time saved would have resulted in briefer hospitalizations—an average of 1.2 days fewer per child—representing a significant cost savings.

There is little need to distinguish among the specific echovirus serotypes in a clinical diagnosis, because the diseases are often not serotype specific. One important factor to consider in clinical diagnosis is the sample site. The nasopharynx and the gastrointestinal tract can be colonized for weeks or months, thereby giving a false-positive test result. The central nervous system, blood, and the genitourinary tract are infection specific—a positive result from one of these sites is likely to be associated with illness (Rotbart and Romero 1995).

Environmental

The same advantages and disadvantages of the detection methods described for clinical usage also apply for environmental usage. Cell culturing is the standard method and is the only method currently available to evaluate viral infectivity. However, as detailed above, it is slow and not nearly as specific or sensitive as newer molecular methods. PCR is especially vulnerable to the presence of inhibitory substances typically found in environmental samples and this can significantly decrease its effectiveness. From their work, Reynolds et al. (1998) concluded that RT-PCR is more useful only when the samples are either relatively free of inhibitory compounds or the viral load is so great that the inhibitors are diluted. Therefore, PCR works well for sewage but not necessarily for environmental source water. In another study, Reynolds and coworkers (2001) combined cell culture and RT-PCR to evaluate sewage, marine water, and surface drinking water sources. They found 11 positive samples where the methods were combined and only 3 when RT-PCR was used alone. Finally, when a sample contains numerous inhibitors, PCR can only test low equivalent volumes compared with cell culture. Consequently, samples with low virus concentrations but high inhibitor levels may not test positive because of the small amount of testable sample. Generally, when conditions are equivalent, PCR is a more accurate measure than conventional cell culturing.

Researchers are busy trying to simplify and improve molecular detection techniques because of their recognized potential for detecting viruses in water. Puig et al. (1994) concluded that nested PCR is even more sensitive than one-step PCR methods in detecting enteroviruses in river water and sewage samples. However, they noted that nested PCR is still susceptible to false positives from contamination with amplified DNA. Overall, they concluded that nested PCR could detect fewer than 10 particles of virus, which is 100 to 1,000 times greater than the number of particles that cell culture can detect. The ability to detect small amounts of virus is useful when evaluating the quality of different water sources, including drinking water.

With any detection method, the viruses must first be separated and concentrated from the environmental samples—sometimes from hundreds and thousands of gallons at a time. Various researchers have described particular techniques to concentrate enteroviruses using glass wool and glass powder to adsorb the viral particles (Gantzer et al. 1997; Hugues et al. 1993; Senouci, Maul, and Schwartzbrod 1996) and different types of filters and filter processes (Gilgen et al. 1995; Guttman-Bass and Nasser 1984; Li et al. 1998; Shieh, Baric, and Sobsey 1997). The results vary depending on water source, variation in chemical usage, enterovirus serotype, etc. Shieh and colleagues (1997), in their study to detect low levels of enteroviruses in sewage, point out that concentration and recovery techniques that were designed for use in cell culture are inadequate for use with PCR and, in fact, increase the natural substances in environmental samples that make PCR difficult to perform. Several researchers are working to overcome the problem of inhibition in samples; some are combining cell culture with PCR in an effort to capture the desirable qualities of each method (Reynolds et al. 2001; Straub, Pepper, and Gerba 1995).

RT-PCR has also been used to detect enteroviruses in shellfish; enteroviruses were found in 19% and 45% of oyster and mussel samples, respectively (Casas and Sunen 2001; Le Guyader et al. 2000).

Water Treatment

Enteroviruses can stay viable for years at –20° C and for weeks at –4° C, but they gradually lose their infectivity at room temperature and quickly lose infectivity at higher temperatures. Enteroviruses are easily inactivated by ultraviolet light (if there is no organic matter) and through desiccation (Minor 1998). Free residual chlorine of 0.3 to 0.5 ppm reportedly results in rapid loss of viability as well, though the water's condition (e.g., high pH, low temperature, and increased organic matter) can affect the efficacy of chlorine disinfection (Minor 1998). Water utilities maintain about 1.1 mg/L as a residual (Rice, Clark, and Johnson 1999). Hurst (1991) summarized four studies that reported on the efficiency of the entire conventional drinking water process in removing enteric viruses and coliphage. He defined conventional treatment as coagulation, sedimentation, filtration, and disinfection. The median removal efficiency after treatment was completed was ≥95.1%. The study reported on plants in Canada, Mexico, and the United States, so it is likely that treatment processes, treatment quality, and raw water quality varied. The enterovirus removal figures after full treatment in the US studies were >93.0% and >95.4% in samples from two different studies. Payment and colleagues (1985) greatly increased removal efficiency by prechlorinating the raw water, which resulted in an immediate 95.0% removal and increased to 99.97% after the rest of the full-scale treatment. The prechlorination reduction was achieved after 5 minutes at a residual chlorine level of 0.5 to 1.0 mg/L. However, the drinking water process reduced the average viral concentrations from 125 to 0.003, or 4–5 $\log_{10}$ at the maximum efficiency level. The overall virus reduction was about 4 $\log_{10}$. The treatment eliminated other indicators such as total coliforms more effectively (6 $\log_{10}$ reduction). In 138 finished water samples of 1,000 L/sample from six different plants, 12 samples contained virus (8.7%), including echovirus 7. The average concentration was 0.0006 mpncu/L and the highest concentration reported in finished water was 0.02 mpncu/L.

Researchers are busy testing ways to improve virus removal from drinking water. Efforts include filtering finished water through oxidized coal (Cloete, Da Silva, and Nel 1998) and using copper in domestic plumbing (Colquhoun et al. 1995). The latter study tested the virucidal effect of plumbing material on sample enteroviruses, including echovirus 4. In a comparison with glass, polybutylene, polyethylene, and silicone, copper was up to two orders of magnitude better at reducing viruses than the other materials, which could be significant when trying to control biofilm colonization.

Membrane filtration technology, such as micro-, ultra-, and nanofiltration, as well as reverse osmosis, has the capacity to remove water contaminants down to the ion level, which is greater than conventional water treatment. Ultrafiltration, with a nominal pore size of 0.01 μm, can achieve 4-log virus removal or greater (Najm and Trussell 1999; Taylor and Wiesner 1999). This technology will continue to gain in popularity as costs become more competitive and regulatory approval is assured.

Wastewater treatment has direct bearing on drinking water treatment because wastewater is usually discharged into drinking water sources and the amount of viral load in the source water affects the ability to treat drinking water. Irving and Smith (1981) tested wastewater for adenoviruses, reoviruses, and enteroviruses over 1 year. The average concentration of enteroviruses in raw sewage was 1,400 IU/L. Effluent after primary treatment averaged 1,250 IU/L, an 11% decrease, and secondary effluent averaged 100 IU/L, a 93% decrease. The process removed enteroviruses more efficiently than it removed adenoviruses or reoviruses. The echoviruses most often isolated in the sewage were, in descending order, 7, 30, 6, 19, 1, 11, and 20.

Payment and colleagues (1986) conducted a similar study in Quebec. When they tested raw sewage for the presence of enteric viruses, it averaged 95.1 mpniu/L. After a primary settling stage, viral reduction was 75%, and the secondary effluent was reduced by 98%. Coxsackieviruses, polioviruses, and reoviruses were most frequently isolated, while echoviruses 3, 22, and 14 were rarely encountered. As with drinking water treatment, the quality of sewage treatment facilities, procedures, and location substantially affect a plant's ability to effectively treat, making it difficult to generalize the reduction efficiency reported in studies to all wastewater.

The large number of waterborne disease outbreaks traced to groundwater (Moore et al. 1993) has sparked concerns about viral retention in groundwater. Sobsey and colleagues (1995) evaluated soil's ability to reduce the viral load in groundwater. They used four types of soil, ranging from clay loam to sand, and evaluated four viruses: echovirus 1, poliovirus 1, hepatitis A, and an indicator virus. Generally, soil filtration reduced poliovirus the most and echovirus the least. Viruses were also removed more efficiently at a higher incubation temperature. Overall, clay was the most efficient at virus reduction; it completely (>99.9% to >99.999%) removed all viruses from groundwater and primary sewage effluent at both 5° C and 25° C. Coarse sand, on the other hand, reduced viruses no more than 80% from primary effluent and no more than 98% from groundwater. Echovirus was reduced very little in coarse sand—about 50%.

Conclusions

Nonpolio enteroviruses, which include echoviruses, commonly cause infections in children and adults. Notably, echovirus is the most common cause of aseptic meningitis, though its most frequent manifestation is nonspecific febrile illness, especially in children. In all but neonates and those with agammaglobulinemia, echovirus infections are self-limited and last less than 1 week.

Because it is excreted in human fecal matter, the possibility of foodborne or waterborne transmission exists; however, this route seems to be of little or no importance in the spread of infection. Numerous outbreaks occur worldwide that are almost always caused by person-to-person transmission. Echoviruses are found in all kinds of water, from raw sewage to chlorinated drinking water. Generally, conventional water treatment with chlorine disinfection inactivates echovirus, though that cannot be assumed to be 100%.

Overall, echoviruses appear to be a minor threat as a waterborne disease.

Bibliography

Abzug, M.J. 1995. Perinatal Enterovirus Infections. In *Human Enterovirus Infections*. Edited by H.A. Rotbart. Washington, DC: ASM Press.

Aradottir, E., E.M. Alonso, and S.T. Shulman. 2001. Severe neonatal enteroviral hepatitis treated with pleconaril. *Pediatr. Infect. Dis. J.*, 20:457–459.

Ashwell, M.J.S., D.W. Smith, P.A. Phillips, and I.L. Rouse. 1996. Viral meningitis due to echovirus types 6 and 9: epidemiological data from Western Australia. *Epidemiol. Infect.*, 117:507–512.

Bergman, I., M.J. Painter, E.R. Wald, D. Chiponis, A.L. Holland, and H.G. Taylor. 1987. Outcome in children with enteroviral meningitis during the first year of life. *J. Pediatr.*, 110:705–709.

Casas, N., and E. Sunen. 2001. Detection of enterovirus and hepatitis A virus RNA in mussels (*Mytilus* spp.) by reverse transcriptase-polymerase chain reaction. *J. Appl. Microbiol.*, 90:89–95.

Centers for Disease Control and Prevention (CDC). 1997. Nonpolio enterovirus surveillance—United States, 1993–1996. *MMWR*, 46:748–750.

Centers for Disease Control and Prevention (CDC). 2000. Enterovirus surveillance—United States, 1997–1999. *MMWR*, 49:913–916.

Chonmaitree, T., and L. Mann. 1995. Respiratory Infections. In *Human Enterovirus Infections*. Edited by H.A. Rotbart. Washington, DC: ASM Press.

Cloete, T.E., E. Da Silva, and L.H. Nel. 1998. Removal of waterborne human enteric viruses and coliphages with oxidized coal. *Curr. Microbiol.*, 37:23–27.

Colquhoun, K.O., S. Timms, J. Slade, C.R. Fricker. 1995. Domestic plumbing materials and their effect on viruses. *Water Sci. Technol.*, 32:I–III.

Connolly, J. 1998. Viral meningitis cases top 500 in Texas and several Midwestern states. *Infect. Dis. Child.*, 11:8–8.

Coyle, P.K. 1999. Overview of acute and chronic meningitis. *Neurol. Clin.*, 17:691–710.

Dagan, R. 1996. Nonpolio enteroviruses and the febrile young infant: epidemiologic, clinical and diagnostic aspects. *Pediatr. Infect. Dis. J.*, 15:67–71.

Dagan, R., and M.A. Menegus. 1995. Nonpolio Enteroviruses and the Febrile Infant. In *Human Enterovirus Infections*. Edited by H.A. Rotbart. Washington, DC: ASM Press.

Dalakas, M.C. 1995. Enteroviruses and Human Neuromuscular Diseases. In *Human Enterovirus Infections*. Edited by H.A. Rotbart. Washington, DC: ASM Press.

Duncan, E.H., and S.C. Edberg. 1995. Host-microbe interaction in the gastrointestinal tract. *Crit. Rev. Microbiol.*, 21:85–100.

Figueroa, J.P., D. Ashley, D. King, and B. Hull. 1989. An outbreak of acute flaccid paralysis in Jamaica associated with echovirus type 22. *J. Med. Virol.*, 29:315–319.

Furione, M., M. Zavattoni, M. Gatti, E. Percivalle, N. Fioroni, and G. Gerna. 1998. Rapid detection of enteroviral RNA in cerebrospinal fluid (CSF) from patients with aseptic meningitis by reverse transcription-nested polymerase chain reaction. *New Microbiol.*, 21:343–351.

Gantzer, C., S. Senouci, A. Maul, Y. Levi, and L. Schwartzbrod. 1997. Enterovirus genomes in wastewater: concentration on glass wool and glass powder and detection by RT-PCR. *J. Virol. Methods*, 65:265–271.

Ghazi, F., P.J. Hughes, T. Hyypiä, and G. Stanway. 1998. Molecular analysis of human parechovirus type 2 (formerly echovirus 23). *J. Gen. Virol.*, 79:2641–2650.

Gilgen, M., B. Wegmuller, P. Burkhalter, H.P. Buhler, U. Muller, J. Luthy, and U. Candrian. 1995. Reverse transcription PCR to detect enteroviruses in surface water. *Appl. Environ. Microbiol.*, 61:1226–1231.

Gondo, K., K. Kusuhara, H. Take, and K. Heda. 1995. Echovirus type 9 epidemic in Kagoshima, Southern Japan: seroepidemiology and clinical observation of aseptic meningitis. *Pediatr. Infect. Dis. J.*, 14:787–791.

Gorgievski-Hrisoho, M., J.D. Schumacher, N. Vilimonovic, D. Germann, and L. Matter. 1998. Detection by PCR of enteroviruses in cerebrospinal fluid during a summer outbreak of aseptic meningitis in Switzerland. *J. Clin. Microbiol.*, 36:2408–2412.

Gosbell, I., D. Robinson, K. Chant, and S. Crone. 2000. Outbreak of echovirus 30 meningitis in Wingecarribee Shire, New South Wales. *Commun. Dis. Intell.*, 24:121–124.

Guttman-Bass, N., and A. Nasser. 1984. Simultaneous concentration of four enteroviruses from tap, waste, and natural waters. *Appl. Environ. Microbiol.*, 47:1311–1315.

Helfand, R.F., A.S. Khan, M.A. Pallansch, J.P. Alexander, H.B. Meyers, R.A. DeSantis, L.B. Schonberger, and L.J. Anderson. 1994. Echovirus 30 infection and aseptic meningitis in parents of children attending a child care center. *J. Infect. Dis.*, 169:1133–1137.

Hill, W.M.J. 1996. Are echoviruses still orphans? *Br. J. Biomed. Sci.*, 53:221–226.

Hovi, T., M. Stenvik, and M. Rosenlew. 1996. Relative abundance of enterovirus serotypes in sewage differs from that in patients: clinical and epidemiological implications. *Epidemiol. Infect.*, 116:91–97.

Hugues, B., M. Andre, J.L. Plantat, and H. Champsaur. 1993. Comparison of glass wool and glass powder methods for concentration of viruses from treated waste waters. *Zentralbl. Hyg. Umweltmed.*, 193:440–449.

Hurst, C.J. 1991. Presence of enteric viruses in freshwater and their removal by the conventional drinking water treatment process. *Bull. World Health Org.*, 69:113–119.

Hurst, C.J., W.H. Benton, and K.A. McClellan. 1989. Thermal and water source effects upon stability of enteroviruses in surface freshwaters. *Can. J. Microbiol.*, 35:474–480.

Irving, L.G., and F.A. Smith. 1981. One-year survey of enteroviruses, adenoviruses, and reoviruses isolated from effluent at an activated-sludge purification plant. *Appl. Environ. Microbiol.*, 41:51–59.

Jenista, J.A., K.R. Powell, and M.A. Menegus. 1984. Epidemiology of neonatal enterovirus infection. *J. Pediatr.*, 104:685–690.

Joki-Korpela, P., and T. Hyypiä. 1998. Diagnosis and epidemiology of echovirus 22 infections. *Clin. Infect. Dis.*, 26:129–136.

Kaye, S.B., C.E. Morton, C.Y. Tong, and N.P. O'Donnell. 1998. Echovirus keratoconjunctivitis. *Am. J. Ophthalmol.*, 125:187–190.

Kee, F., G. McElroy, D. Stewart, P. Coyle, and J. Watson. 1994. A community outbreak of echovirus infection associated with an outdoor swimming pool. *J. Public Health Med.*, 16:145–148.

Keswick, B.H., C.P. Gerba, H.L. DuPont, and J.B. Rose. 1984. Detection of enteric viruses in treated drinking water. *Appl. Environ. Microbiol.*, 47:1290–1294.

Keswick, B.H., C.P. Gerba, and S.M. Goyal. 1981. Occurrence of enteroviruses in community swimming pools. *Am. J. Public Health*, 71:1026–1030.

Krikelis, V., N. Spyrou, P. Markoulatos, and C. Serie. 1985. Seasonal distribution of enteroviruses and adenoviruses in domestic sewage. *Can. J. Microbiol.*, 31:24–25.

Le Guyader, F., L. Haugarreau, L. Miossec, E. Dubois, and M. Pommepuy. 2000. Three-year study to assess human enteric viruses in shellfish. *Appl. Environ. Microbiol.*, 66:3241–3248.

Li, J.W., X.W. Wang, Q.Y. Rui, N. Song, F.G. Zhang, Y.C. Ou, and F.H. Chao. 1998. A new and simple method for concentration of enteric viruses from water. *J. Virol. Methods,* 74:99–108.

Minor, P. 1998. Picornaviruses. In *Virology.* Edited by B.W.J. Mahy and L. Collier. London: Arnold.

Modlin, J.F. 1986. Perinatal echovirus infection: insights from a literature review of 61 cases of serious infection and 16 outbreaks in nurseries. *Rev. Infect. Dis.,* 8:918–926.

Modlin, J.F. 2000. Introduction to *Picornaviridae.* In *Principles and Practice of Infectious Diseases.* Edited by G.L. Mandell, J.E. Bennett, and R. Dolin. Philadelphia: Churchill Livingstone, Inc.

Moore, A.C., B.L. Herwaldt, G.F. Craun, R.L. Calderon, A.K. Highsmith, and D.D. Juranek. 1993. Surveillance for waterborne disease outbreaks—United States, 1991–1992. *MMWR,* 42:1–22.

Morens, D.M., and M.A. Pallansch. 1995. Epidemiology. In *Human Enterovirus Infections.* Edited by H.A. Rotbart. Washington, DC: ASM Press.

Najm, I., and R.R. Trussell. 1999. New and Emerging Drinking Water Treatment Technologies. In *Identifying Future Drinking Water Contaminants.* Washington, DC: National Academy Press.

Oberste, M.S., K. Maher, and M.A. Pallansch. 1998. Complete sequence of echovirus 23 and its relationship to echovirus 22 and other human enteroviruses. *Virus Res.,* 56:217–223.

Payment, P., F. Affoyon, and M. Trudel. 1988. Detection of animal and human enteric viruses in water from the Assomption River and its tributaries. *Can. J. Microbiol.,* 34:967–973.

Payment, P., S. Fortin, and M. Trudel. 1986. Elimination of human enteric viruses during conventional waste water treatment by activated sludge. *Can. J. Microbiol.,* 32:922–925.

Payment, P., M. Trudel, and R. Plante. 1985. Elimination of viruses and indicator bacteria at each step of treatment during preparation of drinking water at seven water treatment plants. *Appl. Environ. Microbiol.,* 49:1418–1428.

Piirainen, L., T. Hovi, and M. Roivainen. 1998. Variability in the integrity of human enteroviruses exposed to various simulated in vivo environments. *Microb. Pathog.,* 25:131–137.

Pollard, R.B. 1975. Inappropriate secretion of antidiuretic hormone associated with adenovirus pneumonia. *Chest,* 68:589–591.

Portolani, M., M. Pecorari, P. Pietrosemoli, A. Bartoletti, A.M. Sabbatini, M. Meacci, W. Gennari, E. Bazzani, F. Beretti, and G. Guaraldi. 2001. Outbreak of aseptic meningitis by echo 4: prevalence of clinical cases among adults. *New Microbiol.*, 24:11–15.

Puig, M., J. Jofre, F. Lucena, A. Allard, G. Wadell, and R. Girones. 1994. Detection of adenoviruses and enteroviruses in polluted waters by nested PCR amplification. *Appl. Environ. Microbiol.*, 60:2963–2970.

Puntaric, D., D. Cecuk, M. Grce, I. Vodopija, M. Ljuvicic, and Z. Baklaic. 1995. Human virus detection in drinking water in Zagreb from 1991 to 1994. *Periodicum Biologorum*, 97:347–350.

Reintjes, R., M. Pohle, U. Vieth, O. Lyytikainen, H. Timm, E. Schreier, and L. Petersen. 1999. Community-wide outbreak of enteroviral illness caused by echovirus 30: a cross-sectional survey and a case-control study. *Pediatr. Infect. Dis. J.*, 18:104–108.

Reynolds, K.A., C.P. Gerba, M. Abbaszadegan, and L.L. Pepper. 2001. ICC/PCR detection of enteroviruses and hepatitis A virus in environmental samples. *Can. J. Microbiol.*, 47:153–157.

Reynolds, K.A., K. Roll, R.S. Fujioka, C.P. Gerba, and I.L. Pepper. 1998. Incidence of enteroviruses in Mamala Bay, Hawaii using cell culture and direct polymerase chain reaction methodologies. *Can. J. Microbiol.*, 44:598–604.

Rice, E.W., R.M. Clark, and C.H. Johnson. 1999. Chlorine inactivation of *Escherichia coli* O157:H7. *Emerg. Infect. Dis.*, 5:461–463.

Rice, S.K., R.E. Heinl, L.L. Thornton, and S.M. Opal. 1995. Clinical characteristics, management strategies, and cost implications of a statewide outbreak of enterovirus meningitis. *Clin. Infect. Dis.*, 20:931–937.

Rigonan, A.L., L. Mann, and T. Chonmaitree. 1998. Use of monoclonal antibodies to identify serotypes of enterovirus isolates. *J. Clin. Microbiol.*, 36:1877–1881.

Rodriguez, A., S. Westbo, B. Adam, C. Langkop, and F.J. Francis. 1997. Outbreak of aseptic meningitis—Whiteside County, Illinois, 1995. *MMWR*, 46:221–224.

Romero, J.R. 2001. Pleconaril: a novel antipicornaviral drug. *Expert. Opin. Investig. Drugs*, 10:369–379.

Rorabaugh, M.L., L.E. Berlin, F. Heldrich, K. Roberts, L.A. Rosenberg, T. Doran, and J.F. Modlin. 1993. Aseptic meningitis in infants younger than 2 years of age: acute illness and neurologic complications. *Pediatrics*, 92:206–211.

Rotbart, H.A. 1995a. Enteroviral infections of the central nervous system. *Clin. Infect. Dis.*, 20:971–981.

Rotbart, H.A. 1995b. Meningitis and Encephalitis. In *Human Enterovirus Infections.* Edited by H.A. Rotbart. Washington, DC: ASM Press.

Rotbart, H.A., and F.G. Hayden. 2000. Picornavirus infections: a primer for the practitioner. *Arch. Fam. Med.*, 9:913–920.

Rotbart, H.A., and J.P. Romero. 1995. Laboratory Diagnosis of Enteroviral Infections. In *Human Enterovirus Infections.* H.A. Rotbart, ed., Washington, DC: ASM Press.

Rotbart, H.A., and A.D. Webster. 2001. Treatment of potentially life-threatening enterovirus infections with pleconaril. *Clin. Infect. Dis.*, 32:228–235.

Schiff, G.M., G.M. Stefanovic, E.C. Young, D.S. Sander, J.K. Pennekamp, and R.L. Ward. 1984. Studies of echovirus-12 in volunteers: determination of minimal infectious dose and the effect of previous infection on infectious dose. *J. Infect. Dis.*, 150:858–866.

Schlesinger, Y., M.H. Sawyer, and G.A. Storch. 1994. Enteroviral meningitis in infancy: potential role for polymerase chain reaction in patient management. *Pediatrics,* 94:157–162.

Schumacher, J.D., C. Chuard, F. Renevey, L. Matter, and C. Regamey. 1999. Outbreak of echovirus 30 meningitis in Switzerland. *Scand. J. Infect. Dis.*, 31:539–542.

Schwarer, A.P., S.S. Opat, A.M. Watson, D. Spelman, F. Firkin, and N. Lee. 1997. Disseminated echovirus infection after allogeneic bone marrow transplantation. *Pathology,* 29:424–425.

Senouci, S., A. Maul, and L. Schwartzbrod. 1996. Comparison study on three protocols used to concentrate poliovirus type 1 from drinking water. *Zentralbl. Hyg. Umweltmed.*, 198:307–317.

Shieh, Y.S., R.S. Baric, and M.D. Sobsey. 1997. Detection of low levels of enteric viruses in metropolitan and airplane sewage. *Appl. Environ. Microbiol.*, 63:4401–4407.

Sicherer, S.H., and J.A. Winkelstein. 1998. Primary immunodeficiency diseases in adults. *JAMA,* 279:58–61.

Sobsey, M.D., R.M. Hall, and R.L. Hazard. 1995. Comparative reductions of hepatitis A virus, enteroviruses and coliphage MS2 in miniature soil columns. *Water Sci. Technol.*, 31:203–209.

Stellrecht, K.A., I. Harding, F.M. Hussain, N.G. Mishrik, R.T. Czap, M.L. Lepow, and R.A. Venezia. 2000. A one-step RT-PCR assay using an enzyme-linked detection system for the diagnosis of enterovirus meningitis. *J. Clin. Virol.*, 17:143–149.

Straub, T.M., I.L. Pepper, and C.P. Gerba. 1995. Comparison of PCR and cell culture for detection of enteroviruses in sludge-amended field soils and determination of their transport. *Appl. Environ. Microbiol.*, 61:2066–2068.

Strikas, R.A., L.J. Anderson, and R.A. Parker. 1986. Temporal and geographic patterns of isolates of nonpolio enterovirus in the United States, 1970–1983. *J. Infect. Dis.*, 153:346–351.

Takahashi, S., A. Miyamoto, J. Oki, H. Azuma, and A. Okuno. 1995. Acute transverse myelitis caused by ECHO virus type 18 infection. *Eur. J. Pediatr.*, 154:378–380.

Tanel, R.E., S.Y. Kao, T.M. Niemiec, M.J. Loeffelholz, D.T. Holland, L.A. Shoaf, E.R. Stucky, and J.C. Burns. 1996. Prospective comparison of culture vs genome detection for diagnosis of enteroviral meningitis in childhood. *Arch. Pediatr. Adolesc. Med.*, 150:919–924.

Tani, N., Y. Dohi, N. Kurumatani, and K. Yonemasu. 1995. Seasonal distribution of adenoviruses, enteroviruses and reoviruses in urban river water. *Microbiol. Immunol.*, 39:577–580.

Taylor, J.S., and M. Wiesner. 1999. Membranes. In *Water Quality and Treatment: A Handbook of Community Water Supplies.* 5th ed. Edited by American Water Works Association. New York: McGraw-Hill.

van Olphen, M., J.G. Kapsenberg, E. van de Baan, and W.A. Kroon. 1984. Removal of enteric viruses from surface water at eight waterworks in The Netherlands. *Appl. Environ. Microbiol.*, 47:927–932.

Wilfert, C.M., R.J. Thompson Jr., T.R. Sunder, A. O'Quinn, J. Zeller, and J. Blacharsh. 1981. Longitudinal assessment of children with enteroviral meningitis during the first three months of life. *Pediatrics*, 67:811–815.

Wilson, G.A., and J.M. Weber. 1995. Laboratory reports of human viral and selected nonviral agents in Canada—1993. *Can. Med. Assoc. J.*, 153:51–53.

Yamashita, K., K. Miyamura, S. Yamadera, N. Kato, M. Akatsuka, M. Hashido, S. Inouye, and S. Yamazaki. 1994. Epidemics of aseptic meningitis due to echovirus 30 in Japan. A report of the National Epidemiological Surveillance of Infectious Agents in Japan. *Jpn. J. Med. Sci. Biol.*, 47:221–239.

Yates, M.V., C.P. Gerba, and L.M. Kelley. 1985. Virus persistence in groundwater. *Appl. Environ. Microbiol.*, 49:778–781.

CHAPTER 11

Helicobacter pylori in Drinking Water

by Martha A. Embrey

Executive Summary

Problem Formulation

Occurrence of Illness

- Approximately 50% of the world's population is infected with *H. pylori*. *H. pylori* infection is more prevalent in developing countries; the rate of infection has been decreasing steadily in developed countries over the last few decades.

Role of Waterborne Exposure

- Evidence is limited, but the waterborne route of exposure is probably important in some populations.

Degree of Morbidity and Mortality

- The majority of people infected with *H. pylori* are asymptomatic, though they may live their entire lives with the organism. *H. pylori* infection has been associated with nonulcer dyspepsia; however, the connection is equivocal. Peptic ulcer disease can result in long-term morbidity, but antibiotic treatment is effective against it. Though only a small percentage of *H. pylori*-infected people develop gastric cancer, it is the second leading cause of cancer deaths in the world and is directly related to the prevalence level of *H. pylori* in the population.

Detection Methods in Water/Clinical Specimens

- Because *H. pylori* is such a fastidious organism, culturing in environmental and clinical samples can be challenging. PCR has been used more successfully to detect *H. pylori* DNA in environmental samples. Serology and carbon urea breath tests are most often used to diagnose *H. pylori* infection in patients.

Mechanisms of Water Contamination

- Human fecal matter/sewage.

Concentrations at Intake

- Little data is available; however, of 42 surface water and 20 well water samples in the United States, 40% and 65%, respectively, tested positive using fluorescent antibody and PCR methods.

Efficacy of Water Treatment

- The efficacy of conventional treatment is unknown. Data are equivocal on the susceptibility of *H. pylori* to chlorine disinfection.
- Membrane filtration technology has the potential to remove virtually all bacterial pathogens, including *H. pylori*.

Survival/Amplification in Distribution

- Limited studies are available, indicating that *Helicobacter* spp. have been detected in water distribution system biofilm.

Routes of Exposure

- Ingestion probably through a combination of oral–oral, fecal–oral, and waterborne routes, depending on the population.

Dose Response

Infectious Dose

- 3×10^5 is the smallest documented infectious dose in an adult. The infectious dose for a child is unknown, and the majority of people are infected in childhood. The degree to which children are more susceptible than adults, in addition to having more opportunity for exposure, is unknown. Data suggest that some people are genetically more susceptible than others. Also, any element that affects the gastric ecology may be a factor (e.g., diet, nutritional status, or co-infection with another organism).

Probability of Illness Based on Infection

- Left untreated, *H. pylori* infection progresses to peptic ulcer disease in up to 20% and gastric cancer in less than 1% of those infected. The relationship between *H. pylori* infection and nonulcer dyspepsia is equivocal.

Efficacy of Medical Treatment

- Multiple therapies using a variety of antibiotics and bismuth preparations are generally 70 to 95% effective at eradicating *H. pylori* infection. Reinfection in adults seems to be a rare occurrence. The reinfection rate for children is unknown. Infection eradication's ability to improve gastric carcinoma precursors is controversial; however, antibiotic treatment can successfully cause tumor regression in gastric lymphoma cases.

Secondary Spread

- Occurrence data indicate that infection clusters in families as well as group-living situations (i.e., orphanages, mental institutions). Secondary spread, particularly among children, is likely.

Chronic Sequelae

- *H. pylori* causes chronic gastritis in all infected people, and it appears that most people acquire the infection in childhood and carry it through life. However, 80% or more of infected people remain asymptomatic. Those who develop peptic ulcer disease can eradicate it through antimicrobial therapy.

Introduction

Since it was first cultured in the early 1980s, *Helicobacter pylori* has been established as a true bacterial pathogen, meaning it is capable of causing disease in an immunocompetent host (Mobley 1997). The International Agency for Research on Cancer has declared *H. pylori* a Class 1 human carcinogen because of its association with gastric adenocarcinoma and gastric mucosa-associated lymphoid tissue (MALT) lymphoma (IARC 1994). In many parts of the world, *H. pylori* infection begins in childhood. There is some evidence that individuals with higher antibody titers and longer-standing disease are more likely to develop gastric cancer (Parsonnet et al. 1991; Nomura et al. 1991; Forman et al. 1991). Therefore, it is important to put special emphasis on this organism from the perspective of how children are affected.

Epidemiology

Chronic gastritis is a hallmark of all *H. pylori* infections, but only a small proportion of those infected (6 to 20%) go on to develop clinically significant disease at some point in their lives (Go 1997; Parsonnet 1998; Patchett 1998). In children, the infections are usually asymptomatic (Blecker et al. 1993; Lanciers et al. 1996; Rowland and Drumm 1998). Only children with duodenal ulcer disease are symptomatic, and the incidence of duodenal ulcer disease is unknown in children (Rowland and Drumm 1998). Even when children are asymptomatic, there are data suggesting that inflammation exists and progresses (Ganga-Zandzou et al. 1999).

AIDS patients are more likely to suffer from enteric infections and gastritis; however, numerous studies have not found an increased rate of *H. pylori* infection in people who are HIV^+ or who have AIDS. In Australia, 3% of 201 AIDS patients were *H. pylori* positive compared with 22% of their age-matched HIV controls (Edwards et al. 1991). Battan et al. (1990) found a 40% prevalence of infection with *H. pylori* in AIDS patients compared with 39% of age-matched controls. Several other studies have reported similar findings (Cacciarelli et al. 1996; Carella, Vallot, and Marra 1994; Marano, Smith, and Bonanno 1993; Vaira et al. 1995). Varsky and colleagues (1998) determined that cytomegalovirus was positively associated with gastric ulcers in AIDS patients but that *H. pylori* was not. In a study of 45 HIV^+ children, the rate of *H. pylori* infection was comparable with children of the same age in the normal population (Lionetti et al. 1999). It is possible that HIV-related host factors decrease *H. pylori*'s ability to colonize the gut (Carella, Vallot, and Marra 1994; Edwards et al. 1991; Marano, Smith, and

Bonanno 1993) or that AIDS patients' use of antibiotics is a contributing factor (Edwards et al. 1991).

More than 90% of duodenal ulcers and 80% of gastric ulcers are caused by *H. pylori* infection (CDC 1999). Other factors, such as reduced mucosal resistance and high acidity, may create favorable conditions for *H. pylori* to induce ulcers in some patients but not others (Tovey and Hobsley 1999; Olbe et al. 2000). *H. pylori* infection alters acid secretion in the host—either increase or decrease—and the person's preinfection acid secretion pattern may affect whether their infection outcome is ulcer disease or gastric cancer (McColl et al. 2000). Less than 1% of all infections are estimated to progress ultimately to gastric cancer (Parsonnet 1998); however, since there is such a high prevalence of infection internationally, this small percentage translates into a significant number of cases. In fact, gastric cancer is the second most common cancer killer in the world (Parkin, Pisani, and Ferlay 1999), and at least 40 to 50% of gastric cancers are related to *H. pylori* (Parkin, Pisani, and Ferlay 1999; Parsonnet 1998). Studies suggest that gastric cancer is two to three times as common in people infected with *H. pylori* (Danesh 1999). A recent study also suggested an association with pancreatic cancer (Stolzenberg-Solomon et al. 2001). Although *H. pylori* infection and its associated complications, peptic ulcer disease and gastric cancer, are decreasing in incidence, it is still a major cause of morbidity and mortality in the world (Parsonnet 1998).

Nonulcer dyspepsia is a common condition, and its association with *H. pylori* is controversial (Danesh et al. 2000; McNamara, Buckley, and O'Morain 2000). Epidemiological data support an *H. pylori* link in a certain percentage of cases but not all. Treatment therapy for the infection is beneficial to some patients.

A number of studies have looked at the association between *H. pylori* infection and anemia in both children and adults. A large evaluation of 2,794 Danish adults measured IgG antibodies against *H. pylori*, hemoglobin, and serum ferritin (Milman et al. 1998). There was no association with hemoglobin levels; however, all men and postmenopausal women who were IgG positive had significantly lower levels of serum ferritin. An Australian study of 160 women also found an association between IgG antibody and decreased serum ferritin levels (Peach, Bath, and Farish 1998).

In children, 86 Alaska natives between 2 and 6 years old were tested for anemia and IgG antibodies to *H. pylori*. Eighty percent of the anemic children and 32% of the nonanemic children were seropositive for *H. pylori* ($p = 0.02$) (CDC 1999). Another study of anemic children with IgG antibodies (mean age of 15.4 years) showed that the groups who were randomized to treatment regimens to eradicate *H. pylori* had a more efficacious response to oral iron therapy than those who had iron therapy alone (Choe et al. 1999). Even children who received treatment for *H. pylori* alone, without iron therapy, had enhanced iron absorption. However, other studies in children have not shown any association between *H. pylori* and anemia (CDC 1999; Haruma et al. 1995). In addition, a review of the literature showed that most patients (of all ages) with pernicious anemia are infected less often with *H. pylori* than their age-matched controls (Perez-Perez 1997).

Researchers have suggested several biological mechanisms for this possible association, including (1) *H. pylori*-induced hypoacidity or achlorhydia affecting iron absorption, (2) increased iron demand because of bacterial competition, or (3) gastrointestinal loss of blood from chronic inflammation (CDC 1999).

H. pylori affects people worldwide, and estimates are that more than 50% of the world's population is infected (Parsonnet 1998; Peura 1997). However, data indicate that a large discrepancy exists in the incidence of infection, the course of infection, and the manifestation of disease between populations in developed and developing countries. Incidence in adults in developed countries is 0.3 to 1% per year (Feldman, Eccersley, and Hardie 1998; Forman and Webb 1993; Lin et al. 1998; Mendall 1993; Rowland 2000). In the developed world, infection is less common in early years, with a prevalence of 60% by age 60 (Rowland 2000); 20 to 40% by age 20 (Forman and Webb 1993; Mendall 1993); and even lower rates (<10%) in young people aged 0 to 15 (Ashorn et al. 1996; Lindkvist et al. 1996; Roosendaal et al. 1997). In developing countries, infection is acquired in infancy and young childhood and can reach a prevalence of 80 to 90% by age 20, where it remains for the rest of adult life (Forman and Webb 1993; Mitchell et al. 1992; Lindkvist et al. 1996).

Childhood is the dominant period for acquisition in all populations, and adults rarely become reinfected after infection eradication (Rowland 2000). In the United States, the National Health and Nutrition Examination Survey demonstrated, using enzyme-linked immunosorbent assay (ELISA) testing for anti-*H. pylori* IgG, that 24.8% of the US population between 6 and 19 years is infected (Staat et al. 1996). This represents 16.7% of 6- to 9-year-olds, 26.2% of 10- to 14-year-olds, and 29.1% of 15- to 19-year-olds. There were no gender differences; however, there were racial differences. *H. pylori* infection was present in 40.1% of black Americans in this sample, 42% of Mexican-Americans, and 17% of European-Americans. In Nashville, Tennessee, seroprevalence was 11% among 5-year-olds and 25% among 10-year-olds (Adler-Shohet et al. 1996).

Malaty and colleagues (1999) followed a biracial cohort of 212 children for 12 years to track the seroprevalence of *H. pylori*. At the beginning of the study, 19% of 7- to 9-year-olds were seropositive; 12 years later, 33% were seropositive. There was a significant difference between black and white children: Initially, 40% of the black children compared with 11% of the whites were positive, and the disparity held after 12 years. Also, during that time, 50% of infected white children seroconverted compared with only 4% of the black children. In a recently published study (Malaty et al. 2001), researchers showed that black and Hispanic children of low socioeconomic status had a high prevalence (almost half by age 10) of *H. pylori* in Houston, Texas. A strong risk factor was how crowded the day-care center was that the children attended, showing that crowding factors outside of the home could have an effect on children's acquisition of *H. pylori* infection. This study also showed that the mother's education level was significant.

While children born to seropositive mothers will be seropositive as a result of transplacental migration of antibody, the children will not remain seropositive. Blecker et al. (1994a) showed that 100% of the children born to seropositive mothers were seropositive and 100% of the children born to seronegative mothers were seronegative. By 3 months of age, all seropositive children had become seronegative. Only 1 of 67 initially seropositive children who were given a ^{13}C-breath test at 12 to 15 months of age was positive. In Taiwanese women and infants, Gold et al. (1997) showed that 50 of 80 mothers were positive for anti-*H. pylori* IgG using ELISA testing. Forty-eight of their 80 infants were positive at birth. However, 94% of those were negative by 3 months, and 98% were negative at 6 months. By 14 months, only 13% of the infants had evidence of acquired infection. There was no association between infant seroconversion and maternal seropositivity.

Spontaneous eradication is generally accepted to be a rare occurrence (less than 1% per year) (Mendall 1993; Valle et al. 1996), though Sipponen (1997) states that spontaneous healing may occur, usually at stages when gastritis is nonatrophic. Klein et al. (1994), using ^{13}C-urea breath tests to diagnose *H. pylori* infection in children in Peru, found a decrease in the prevalence of infection from 71.4 to 47.9%. One problem with this study was that many subjects were lost to follow-up. Likewise, Granström and colleagues (1997), using ELISA testing, showed spontaneous resolution of infection among children born in Stockholm, Sweden. Among those children, the highest prevalence (10%) occurred in 2-year-olds. The prevalence was 7.5% in 4-year-olds and 3% in 11-year-olds.

In developed countries, over recent decades, the rate of infection has been decreasing in children, resulting in a birth cohort effect—older people having a higher prevalence because of the higher rate of infection when they were children (Cave 1997; Feldman, Eccersley, and Hardie 1998; Parsonnet 1998). This large decrease in incidence in children, roughly since World

War II, is probably due to improved sanitary conditions and decreased crowding, since interpersonal behavior, which would affect oral–oral spread, is unlikely to have changed enough to cause the decrease (Cave 1997). A high infection rate, however, has persisted in certain ethnic groups despite socioeconomic advancement (see Table 11-1, p. 292) (Dunn, Cohen, and Blaser 1997).

Studies from widely disparate parts of the world exemplify this birth cohort effect. Malaty et al. (1996a) looked at *H. pylori* prevalence in Korea. In older people, the prevalence of infection was high and unrelated to current socioeconomic status; however, children's rates were lower and inversely related to their family's socioeconomic status. A look at the epidemiology in Germany also supports a relationship between decreased prevalence and societal/economic changes (Forman and Webb 1993). People born before World War II have a high prevalence. Germans born after the 1950s have a much lower prevalence. These changes in Korea and Germany reflect the major societal and economic shifts that occurred after the wars in those countries. Roosendaal and colleagues (1997) have also measured a decreasing infection rate in the Netherlands. They speculate that a decline in the number of persons per household, and perhaps the need for a critical prevalence level to maintain a steady infection rate, may be responsible for their observed 52% decrease in children over a 15-year study period. Cullen et al. (1993) found similar results in Australia.

As the incidence of *H. pylori* infection decreases in developed countries, the incidence of gastric ulcer and cancer rates have also decreased. At the same time, other diseases, such as gastroesophageal reflux, Barrett's esophagus, and adenocarcinomas of the gastric cardia, have been increasing (Blaser 1999). Blaser speculates that the cag+ genotype, which is related to higher rates of *H. pylori* disease, may be protective against the gastrointestinal diseases that are now on the rise. There is also some data from a population-based study in Germany suggesting that infection with *H. pylori* protects children from diarrheal disease (Rothenbacher et al. 2000).

Medical Treatment

H. pylori can be diagnosed and successfully treated with a regimen of antibiotics; however, the high prevalence of infection without apparent adverse outcomes has led to controversy over when treatment is appropriate.

Diagnostic Methods

There are a number of ways to diagnose *H. pylori* infection. Some methods require endoscopy and others are noninvasive. If endoscopy is not required, the rapid urease test is the diagnostic method of choice, because it is inexpensive and fast (Dunn, Cohen, and Blaser 1997; Parsonnet 1998; Vaira et al. 2000). In particular cases, histology or cell culture using gastric biopsy samples may be more appropriate. Carbon-13 or -14 urea breath tests, serology, and stool tests are the methods used if endoscopy is not required, and each has its own advantages and drawbacks. Breath tests have been considered the gold standard for detecting the success of eradication chemotherapy, and even though they are expensive and more complicated to use, some epidemiological researchers have used this method to determine prevalence in a particular population (Blecker et al. 1994a; Klein et al. 1994). Serology is typically used in the clinical setting because of its low cost and simplicity (Parsonnet 1998). However, it is not appropriate for detecting infection posteradication therapy. A recent report (Corvaglia et al. 1999) compared the ^{13}C breath test with serology in children between the ages of 2.4 months and 18 years (n = 146). Results showed that the breath test was inaccurate in only four tests, compared with 24 inaccuracies and 7 indeterminate results with serology (confirmed through biopsy). The inaccuracies were most common in children under 5. The authors concluded that urea breath tests were superior to serology testing in diagnosing infection in children under 10.

Lehmann and colleagues (1999) recently reported on a new immunoassay for fecal samples. Their results show a sensitivity and specificity similar to carbon urea breath tests but with a cost and ease similar to serology. Other types of assays using other samples (e.g., urine and saliva) are also under development (Dunn, Cohen, and Blaser 1997; Lehmann et al. 1999). Polymerase chain reaction (PCR) is also seen in the research environment, but it is an unlikely candidate for routine clinical diagnosis because of expense and needed expertise. However, it is very useful in detecting genetic differences between strains in epidemiologic studies (Dunn, Cohen, and Blaser 1997).

Treatment Protocols for Adults

The US Food and Drug Administration has approved four treatment regimens combining various antibiotics with ranitidine bismuth citrate and proton-pump-inhibitor preparations, with eradication rates in adults between 70 and 90% (Peura 1997). Current recommended protocols run from 2 to 4 weeks, and though shorter treatment regimens look promising, a recent expert panel could not recommend any regimen shorter than 2 weeks (Peura 1997). Reinfection in treated adults is rare in developed countries (Parsonnet 1998), probably because infection is acquired primarily in childhood (Rowland and Drumm 1998). Though peptic ulcers can and should be treated with antimicrobial therapies, controversy surrounds whether or not treatment improves atrophic gastritis or later gastric cancer precursors (Nakajima et al. 2000; Parsonnet 1998). Also, *H. pylori* colonization could be protective against gastro-esophageal reflux disease (GERD) (Axon 2000; Perez-Perez et al. 1999). However, *H. pylori* eradication in the case of MALT lymphoma (which is rare but almost always linked to *H. pylori*) is recommended. Tumor regression occurs in almost 80% of low-grade MALT lymphoma cases after *H. pylori* eradication (Morgner et al. 2000a) and may benefit patients with the high-grade version of the disease (Morgner et al. 2000b).

Treatment Protocols for Children

The treatment of *H. pylori* infection in children is not well defined. In spite of a very large number of treatment trials reported in adults, there have been only a few controlled trials, and no double-blind, randomized, placebo-controlled trials have been reported in children. The National Institutes of Health (NIH) Consensus Conference on *H. pylori* recommends that anti-*H. pylori* therapy be reserved for adults with ulcer disease (Lee and O'Morain 1997). There is no comparable set of guidelines providing recommendations for the treatment of children, although Sherman and Hunt (1996) have pointed out the need for such a document. They recommend that children be treated only if they have clinical symptoms and complications (i.e., only in-patients who have endoscopic evidence of mucosal ulcers in the stomach or duodenum). They do not recommend treating the infection to prevent gastric carcinogenesis or to alleviate symptoms of abdominal discomfort in the absence of peptic ulcers. A European consensus group representing 11 countries gave an "uncertain" rating to the advisability of treating infected children with recurrent abdominal pain.

For children who do need to be treated, a 2-week treatment with omperazole, a proton pump inhibitor, and two antibiotics (clarithromycin and metronidazole) showed a 93% eradication rate (Dohil, Israel, and Hassall 1997). A 1-week treatment with colloidal bismuth subcitrate, clarithromycin, and metronidazole was 94% successful (Walsh et al. 1997). Others advocate dual therapy with a single antibiotic such as amoxicillin and a bismuth preparation instead of triple therapy for children because of the risk of increased side effects (Drumm 1993). In addition, some strains of *H. pylori* may be resistant to clarithromycin and metronidazole (Iovene et al. 1999). A shorter duration therapy could help improve compliance and reduce cost and side effects in children (Jones and Sherman 1998).

Treatment Debate

Who should be screened and treated is a primary question among clinicians and public health experts. The NIH recommended treating only those with active ulcer disease or those who are on continuing therapy for recurrent ulcer disease, which is the generally accepted recommendation in the health community (Lee and O'Morain 1997). Disagreement arises when considering treatment of people with gastritis, nonulcer dyspepsia, risk factors for gastric cancer, long-term proton pump inhibitor therapy, etc. A Canadian consensus group recommended treatment in patients with known peptic ulcers and also ulcer-like dyspepsia (Hunt, Fallone, and Thomson 1999). The European consensus group mentioned above arrived at a much broader advisory for treatment than the NIH, including those with a family history of gastric cancer and those on nonsteroidal anti-inflammatory therapy (Lee and O'Morain 1997). Others believe that *H. pylori* should be eradicated in all infected people, regardless of their history (Graham 1994; Moayyedi and Axon 1998). However, it is unlikely that the benefits outweigh the costs of treating such a large proportion of the population with broad-spectrum antibiotics, the majority of whom will never develop clinically significant disease (Parsonnet 1998; Patchett 1998; Stanghellini et al. 1997). Because of differences in prevalence rates, cost, and standard of care among different countries, treatment protocols will necessarily vary (Nakajima et al. 2000).

Prevention

As noted, the exact source of *H. pylori* infection is unclear in most individuals. Trying to prevent waterborne, fecal–oral, and oral–oral transmission in children would be very difficult. However, it is clear that children are not infected at birth (Blecker et al. 1994a; Gold et al. 1997). This opens the possibility of immunization as a means of preventing infection. If such immunization in childhood could prevent subsequent development of ulcer disease and gastric cancer, the reduced morbidity, mortality, and economic costs could be significant.

Lee (1996a, 1996b) and Lee and Buck (1996) reviewed the possible development of an *H. pylori* vaccine. They pointed out that one of the first questions that needs to be answered is why the natural infection can persist for life in spite of a cellular and humoral immune response. In mice, the use of an oral vaccine has been successful in eliminating an existing *Helicobacter* infection. Mice have also been protected against *de novo* infection with *H. felis* through the use of an oral vaccine consisting of sonicated *H. felis* and cholera toxin as an adjuvant. More important, mice immunized orally with *H. pylori*-produced cytotoxin VacA and *H. pylori* urease were protected from a subsequent challenge with live type I *H. pylori* organisms (Tompkins and Falkow 1995). Phase 1 clinical trials have been completed that show the safety and acceptability of urease as a vaccine antigen (Czinn 1997).

Transmission

H. pylori must gain entry to the stomach by way of the mouth (Cave 1997), so any infection must come through ingestion of the organism. Infection may be transmitted on a person-to-person basis, as a result of contact with animals (zoonotic transmission), or as a result of contact with contaminated food or water. Person-to-person spread can include gastric–oral, oral–oral, and fecal–oral spread. When multiple individuals are living together, as in a family, dormitory, barracks, or orphanage, it can be very difficult to separate person-to-person spread from common contact with a single source such as water or food. Data supporting all routes have been published, though evidence for zoonotic transmission is the weakest (Brown 2000). It is unlikely that infection is exclusive to a single route.

A number of studies support the notion of person-to-person transmission or of common-source infection with *H. pylori*. In an orphanage in Thailand, where enteric infections are endemic, 74% of 1- to 4-year-old children were seropositive for *H. pylori* antibody. A slightly older group of children in rural Thailand (who were not institutionalized) had a seroprevalence rate of 17.5% (Perez-Perez et al. 1990). Institutionalized children in Japan were significantly more likely to be seropositive than their age-matched, healthy counterparts (Yamashita et al. 2001). Malaty et al. (1991) showed that if one spouse chosen at random in a family is positive for *H. pylori*, 68% of the partners and 40% of the children will also be positive. However, if the index spouse is negative for *H. pylori*, only 9% of the spouses and 3% of the children will be positive. Blecker et al. (1994b) studied first-degree relatives who were living in the same house as children who had abdominal complaints. Forty-nine percent of relatives of patients infected with *H. pylori* had *H. pylori* antibodies, compared to 17.8% of individuals in a similar general population. More than 73% of parents of children who were positive and 81.8% of their siblings were positive for *H. pylori*.

A recent large study of preschool children in Germany (n = 1,221) compared *H. pylori* infection status of children and their parents (Rothenbacher et al. 1999). An adjusted odds ratio if the infected child's mother was also infected was 7.9 (CI [confidence interval] 95%: 4.0–15.7). The prevalence in children was 11.3% and in parents was 36.4%.

Goodman and Correa (2000) screened 684 Colombian children using the ^{13}C-urea breath test, then recorded information such as birth order and spacing and details about their siblings. They found the strongest predictors of *H. pylori* infection to be the number of 2- to 9-year-old, *H. pylori*-positive children in the family and a smaller gap to the next oldest person in the home. The authors concluded that siblings are an important factor in *H. pylori* infection in this population.

Demonstration that individuals who live together are infected with the same serotype of *H. pylori* would provide strong support for person-to-person spread or common-source infection. On the one hand, Bamford et al. (1993) showed that the same strain of organism was present in parents and their offspring using DNA fingerprinting. On the other hand, Simor et al. (1990) could not demonstrate identical strains in two sets of siblings using DNA restriction endonuclease digest patterns.

Gastric–Oral Transmission (Iatrogenic)

Case studies have confirmed that *H. pylori* has been transmitted directly from one infected person's stomach to another person via improperly cleaned endoscopes (Langenberg et al. 1990; Tytgat 1995). In addition, gastroenterologists have generally shown a higher prevalence for *H. pylori*—presumably from occupational exposure to their instruments (Chong et al. 1994; Lin et al. 1994; Mitchell, Lee, and Carrick 1989).

Oral–Oral Transmission

High incidence of infection in childhood supports both an oral–oral or fecal–oral mode of transmission. It is still unclear which mode predominates. Researchers have frequently isolated *H. pylori* DNA from saliva and dental plaque, but they have had more difficulty culturing it from these sources (Parsonnet 1998). One study in dental workers did not show a higher prevalence of infection (Luzza et al. 1995). The Chinese habit of sharing food with chopsticks may contribute to infection from an oral route, since the use of chopsticks was a significant risk factor in infected Asian immigrants to Australia (Chow et al. 1995). Leung et al. (1997) also reported the isolation of *H. pylori* DNA through polymerase chain reaction (PCR) in Chinese patients' saliva and chopsticks. In Colombian children between 2 and 9 years of age, drinking from an unwashed cup previously used by another person increased the odds ratio of *H. pylori* infection to 1.8 (CI 1.1–3.1) (Goodman et al. 1996). Based on animal studies, some authors (Lee et al. 1991) argue that oral–oral transmission is more likely. In a limited series of experiments, both *H. pylori* and *H. felis* were transmitted from infected gnotobiotic beagle puppies to uninfected animals in the same enclosure. Similar transmission did not occur in mice. The authors suggested that the difference between the murine and canine experiments was that the dogs are more likely to have oral–oral contact than rodents.

A recently published study looked at *H. pylori* infection in 242 people in Guatemala (Dowsett et al. 1999). Overall, 58% of the subjects were seropositive. Using nested PCR methods, investigators detected *H. pylori* in 87% of the positive subjects' oral cavities. They then used the same method to test underneath patients' fingernails, where they found that 58% had positive *H. pylori* fingernail samples. The positive fingernail results were significantly associated with positive tongue results ($p = 0.002$). The authors concluded that *H. pylori* may be transmitted from mouth to mouth via the hand.

Fecal–Oral Transmission

In general, it has been difficult to identify *H. pylori* in stool, which makes directly linking it to fecal transmission more difficult. It has been successfully cultured from the fecal samples of infected children in Gambia (Thomas et al. 1992) and from the freshly collected stools of half of the adults known to be positive for *H. pylori* by biopsy or ^{13}C-breath tests (Kelly et al. 1994). *H. pylori* has been detected in feces through PCR (Mapstone et al. 1993). However, technical challenges have made attempts at fecal isolation frequently unsuccessful (Cave 1997). New immunoassays developed for use with stool samples may increase the data in this area (Lehmann et al. 1999).

In a study of children living in the Colombian Andes, there were several lines of evidence linking *H. pylori* seroprevalence with the possibility of fecal–oral spread. Children whose mothers rarely used soap after cleaning up children's feces had increased odds of infection of up to 2.7 times (CI 1.1–6.6). Children whose homes had a latrine $\geq$25 meters from the house had increased odds of infection of 1.6 (CI 0.8–2.9), and children whose homes had no latrine had increased odds of infection of 2.2 (CI 0.9–5.3). Finally, children swimming in rivers, streams, or pools had increased odds of infection (OR up to 3.3 and CI 1.2–9.4) compared with children not so exposed (Goodman et al. 1996). In Chile, other studies have shown that irrigating vegetables with water contaminated with raw sewage and the subsequent consumption of those vegetables without cooking leads to the transmission of enteric pathogens and is correlated with *H. pylori* infection (Hopkins et al. 1993). Goodman and colleagues (1996) demonstrated that children consuming raw vegetables were at increased risk of infection in spite of the protective effect of the consumption of additional fruits and vegetables in the diet.

Several authors have compared the incidence of hepatitis A, which is transmitted through fecal–oral means, with the incidence of *H. pylori* to see how closely they parallel one another. Two hundred children whose serum was tested for *H. pylori* antibodies had an increase in seroprevalence with increasing age (Adler-Shohet et al. 1996). Seroprevalence for hepatitis A was higher in the *H. pylori*-positive children as well, though not significantly. The authors suggest this was because of the small sample size. A similar study with a much larger sample size (1,528 adults) was conducted in San Marino, Italy (Pretolani et al. 1997). Researchers found a significant association between the age-specific curves of *H. pylori* and hepatitis A prevalence in this population, even after controlling for socioeconomic status. In Japan, 1,015 serum samples were analyzed for both *H. pylori* and hepatitis A seropositivity in 1974, 1984, and 1994 (Fujisawa et al. 1999). Both infections followed decreasing trends over the 20 years and both had the highest infection rates in infants and young adults. However, *H. pylori* was more prevalent at all three measurements—73%, 55%, and 39%—compared with hepatitis A—58%, 42%, and 23%. The authors suggested that both infections may be spread through the fecal–oral route but that other factors must also be affecting the prevalence patterns.

However, a number of other studies do not show an association between the seroprevalence of hepatitis A and *H. pylori*. In a study of men living in England, 100 of 175 (57.1%) individuals who were *H. pylori* seropositive were also hepatitis A seropositive, while 113 of 292 (38.7%) individuals who were *H. pylori* seronegative were hepatitis A seropositive (Webb et al. 1996b). After adjustment for age group, the association between *H. pylori* seropositivity and hepatitis A seropositivity was only of borderline significance [$\chi^2 = 3.45$, $p = 0.06$]. After further adjustment for father's occupation (a surrogate for socioeconomic status), the association between *H. pylori* seropositivity and hepatitis A seropositivity was weaker still [$\chi^2 = 2.07$, $p = 0.15$]. In a study from China, cross-sectional data from rural areas supported an association between hepatitis A and *H. pylori* (Hazell et al. 1994). Although the seroprevalence of *H. pylori* was high (approximately 32%) in persons less than 10 years old in an urban area, there was no evidence of hepatitis A infection among them. In a study of 750 males from 1 to 87 years old in Italy, there was a parallel, weakly correlated ($r = 0.287$) rise in the seroprevalence of *H. pylori* and hepatitis A infections with increasing age. However, the agreement between *H. pylori* and hepatitis A seropositivity was little better than chance ($\kappa = 0.21$) and in those aged less than 20 years it was worse than chance ($\kappa = -0.064$). Furthermore, multiple logistic regression analysis did not show any risk factor shared by both infections (Luzza et al. 1997). Age prevalence curves suggest that changes responsible for reducing the prevalence of hepatitis A have not reduced the prevalence of *H. pylori* (Webb et al. 1996b).

Typically, in the developing world, the seroprevalence graphs for a fecally–orally transmitted disease like hepatitis A and an orally–orally transmitted disease like mononucleosis look identical, with everyone showing high seroprevalence in early childhood. In the developed world, childhood is still a major time of infection for both fecal and oral infections. For hepatitis A, the increase after early childhood is gradual throughout life; whereas, mononucleosis shows another jump in prevalence during teenage and young adulthood because of kissing (Feldman, Eccersley, and Hardie 1998; Mendall 1993). It is somewhat difficult to track the seroprevalence of *H. pylori* over time in developed countries because of the birth cohort effect mentioned earlier. However, it does not appear that the young adult years are important times for *H. pylori* infection, though Mendall (1993) noted an association between marital status in young adults and incidence of infection. Probably both oral–oral and fecal–oral transmission routes are likely, depending on circumstances and population.

There are some animal data that do not support the notion of fecal–oral spread (Lee et al. 1991). Germ-free mice and rats infected with *H. felis* did not transmit their infection to uninoculated mice that were housed in the same cage. Additionally, in pathogen-free mice, infected dams did not pass the *Helicobacter* to their progeny. Similarly, mice infected with a human isolate of *Gastrospirillum hominis* did not transmit the infection while in close contact with uninoculated mice. It is important to note that mice and rats are coprophagous. One would expect bacteria to spread if they were present in the feces.

Zoonotic Transmission

Though extensive data are lacking on a zoonotic transmission of *H. pylori*, based on the current evidence, it seems unlikely that animal reservoirs are a significant source of infection in humans. Handt et al. (1995) reported that domestic cats obtained from a supplier for research purposes were contaminated with *H. pylori*. Cats and other animals have been known to carry some *Helicobacter*-like organisms, though not necessarily the *pylori* species (De Groote, Ducatelle, and Haesebrouck 2000; Neiger et al. 1998). Additionally, epidemiological data do not support transmission between cats and humans. Webb et al. (1996a) looked at the *H. pylori* seropositivity of 447 factory workers and then gave them a detailed questionnaire regarding their living conditions, including pet ownership, in childhood. Though those who had pets as children were slightly more likely to be *H. pylori* positive, no differences were found between cat ownership and other pets. Fiedorek et al.

(1991) and Graham et al. (1991) showed that pet owners had a lower frequency of *H. pylori* infection. Contact with pet hamsters was significantly associated with positivity in German children (OR = 2.4) (Herbarth et al. 2001). Staat and colleagues (1996) found a significant association between exposure to cats (but not dogs) and decreased infection in a population of US children. In each of these three studies, pet ownership was seen as a proxy variable signifying higher socioeconomic status. Also, as cat ownership in the United States has increased (to overtake dogs as the most popular pet), the incidence of *H. pylori* has decreased. In a poor, Mexican population, contact with animals at home was not a risk factor for infection in children (Jimenez-Guerra et al. 2000).

Grubel et al. (1997) demonstrated that *H. pylori* was culturable from the external bodies, gut, and excreta of house flies for up to 30 hours after they had been fed *H. pylori* cultured on agar plates. The authors hypothesized that flies could pick up *H. pylori* from sewage or human waste then go on to contaminate food or even the mouths of children. Because of the fly's acidic midgut, it would seem likely that it could harbor *H. pylori* and transmit it to humans, just as it spreads other diseases (Osato et al. 1998). However, in an attempt to replicate Grubel and colleagues' study using more realistic environmental conditions, Osato and colleagues tested to see if the housefly could pick up *H. pylori* from fresh human feces. They used feces from a known *H. pylori*-infected person, a person uninfected with *H. pylori*, and human feces artificially seeded with *H. pylori* bacteria. Though other contaminating organisms were found in the flies' guts, researchers did not find any evidence of *H. pylori*. These researchers suggest that flies would have to be exposed to much higher concentrations of *H. pylori* (such as found in a pure culture) than what would likely be found in the environment and, therefore, are an unlikely source of transmission.

H. pylori is found in nonhuman primates, but these primates are probably not a source of infection in humans. Though pathogen-free pigs have been inoculated with *H. pylori*, and have been suggested as a source, there is not much other evidence to suggest that pigs are natural reservoirs (Fox 1995). Two studies have shown that meat eaters are no more likely to be infected than vegetarians are (Megraud et al. 1989; Webberley et al. 1992), and data on abattoir workers have been equivocal (Fox 1995). Despite the lack of precision in the effects, Colombian children who played with or cared for sheep were more likely to be infected with *H. pylori* (Goodman et al. 1996). That same study did not find any association between other animals and infected children (including rabbits and pigs). Fall et al. (1997) found no association between *H. pylori* seroprevalence and having lived on a farm in childhood.

Waterborne Transmission

Two important studies have been published looking at water source as a risk factor for *H. pylori* infection. Klein and coworkers (1991) looked at the prevalence of infection in 407 Peruvian children aged 2 months to 12 years. Goodman et al. (1996) conducted a population-based study of 684 children 2 to 9 years old in the Colombian Andes. In the Peruvian study, water from Lima's municipal system was either delivered by truck and stored in a cistern or tank, piped to external faucets on the street, or piped to homes with internal plumbing systems. A smaller group of high-income children had their water supplied from a community well through an internal tap. Not surprisingly, children whose homes had internal taps (and presumably better sanitation) had a lower prevalence than children with an external water source. However, when socioeconomic status was taken into account for families with internal sources of municipal water, there was no difference in risk between children of low-income and high-income families. Also, high-income children using a community well had a much lower risk than their high-income counterparts who used municipal water. In the Colombian population, drinking water came from three sources: private wells, spring water pumped into taps, and local streams. The investigators also looked at raw vegetable consumption (vegetables could be contaminated through water or sewage), swimming in rivers, and swimming in swimming pools. Children whose families reported ever drinking from streams had a higher prevalence of infection; however, 84% of the families boiled their drinking water, especially since the government had started a promotional cholera intervention campaign 3 years prior to the study. There was a stronger positive association with infection in children who swam in rivers, streams, or pools (more than three times the chance). A dose–response effect was also seen in consumption of raw vegetables, especially lettuce. Both sets of data support a waterborne transmission, though the Colombian study (Goodman et al. 1996) also found factors specifying person-to-person transmission and concluded that in that population, both routes contributed.

Experimental data back the epidemiological data in these studies. Though no one has been able to culture *H. pylori* from water, it has been detected using PCR on samples from both the Aldana (Colombia) site and from Lima (Handwerker, Fox, and Schauer 1995; Hulten et al. 1996). In the Peruvian samples, 24 of 34 of the municipal water samples were positive, and none of the community well samples were positive, which correlated with the epidemiological results (Hulten et al. 1996). Hulten et al. (1998) investigated the presence of *H. pylori* in Swedish water using PCR methods. The researchers used two PCR assays followed by a *Helicobacter* spp.-specific hybridization in a total of 74 water samples from 25 counties. Positive results using both assays were considered positive. Nine of 24 household wells, 3 of 25 municipal tap water samples, and 3 of 25 wastewater samples were positive for *H. pylori* or *H. heilmannii* DNA. The three positive tap water and wastewater samples were from the same location. Because one assay showed sensitivity to other *Helicobacter* species, the authors could not rule out the possibility that the samples may not have been positive for *H. pylori* specifically. Epidemiological data do not support a high rate of waterborne transmission in Sweden. From 1992 to 1995, surveillance statistics on worldwide stomach cancer death rates were measured in 46 countries. Of the 46, statistics showed Sweden to be ranked 42nd lowest for men and 39th lowest for women (American Cancer Society 1998). A recently published report looking at a large (n = 3,347) population of German schoolchildren found a strong association between seropositivity and nonmunicipal water supplies (OR = 16.4; 95% CI, 3.1–88.5; $p < 0.001$) (Herbarth et al. 2001).

Other studies in the United States (Elitsur, Short, and Neace 1998; Fiedorek et al. 1991), Taiwan (Teh et al. 1994), Korea (Malaty et al. 1996a), China (Mitchell et al. 1992), and elsewhere have not shown an association between infection and water supply. The Chinese study is interesting because a questionnaire and common custom indicated that most people boiled their water, no matter what the source, and stored it in a vacuum flask (Mitchell et al. 1992). However, the prevalence of infection was still relatively high (44% overall). This would suggest that a mode other than water is the primary source of transmission in this population.

Risk Factors

Increasing age, low socioeconomic status in childhood, and growing up in a crowded household are the most reliable risk factors predicting *H. pylori* infection. Table 11-1 illustrates the results of a number of studies that looked at *H. pylori* infection, mainly through cross-sectional studies, which are descriptive in nature. This type of study describes or characterizes disease occurrence and risk factors in a particular population at a particular point in time; however, cross-sectional studies are not designed to test hypotheses about cause and effect. Prevalence in this series of studies was determined by either serum antibody detection or ^{13}C breath testing, which decreases the ability to compare studies. Moreover, the studies varied in the quality of design and the numbers of participants. It is emphasized that this is not a critical review but rather a compilation and comparison of the literature on risk factors under study.

Table 11-1 Environmental and social risk factors for *H. pylori* infection

Environmental/ Social Risk Factor	Positive Association With Infection	No Association With Infection
Eating meat		Megraud et al. 1989; Webberley et al. 1992
Eating uncooked vegetables	Goodman et al. 1996; Hopkins et al. 1993	
Drinking water or the water supply	Goodman et al. 1996; Herbarth et al. 2001; Jimenez-Guerra, Shetty, and Kurpad 2000; Klein et al. 1991; Olmos, Rios, and Higa 2000 (in children)	Clemens et al. 1996; Elitsur, Short, and Neace 1998; Fiedorek et al. 1991; Malaty et al. 1996a; Mitchell et al. 1992; Nowottny and Heilmann 1990; Öztürk et al. 1996; Teh et al. 1994; Yamashita et al. 2001
Iatrogenic (endoscope) or occupational	Chong et al. 1994; Langenberg et al. 1990; Lin et al. 1994; Mitchell, Lee, and Carrick 1989; Tytgat 1995; van der Hulst et al. 1996; Wilhoite et al. 1993	Morris, Lloyd, and Nicholson 1986
Poor sanitation/no fixed hot water supply	Goodman et al. 1996; Graham et al. 1991; Mendall et al. 1992	Clemens et al. 1996; Fall et al. 1997; Mitchell et al. 1992; Oliveira et al. 1994
H. pylori-positive family member	Bamford et al. 1993; Bonamico et al. 1996; Bourke, Jones, and Sherman 1996; Dowsett et al. 1999; Drumm et al. 1990; Georgopoulos et al. 1996; Mitchell et al. 1987; Mitchell, Mitchell, and Tobias 1992; Nwokolo et al. 1992; Sakamoto 1997; Schutze et al. 1995; Tee et al. 1992	Perez-Perez et al. 1991; Sarker et al. 1997

Table continued on next page.

Table 11-1 Environmental and social risk factors for *H. pylori* infection (continued)

Environmental/ Social Risk Factor	Positive Association With Infection	No Association With Infection
Crowded environment (e.g., home, institution, nursing home, hospital)	Berkowicz and Lee 1987; Clemens et al. 1996; Drumm 1993; Drumm et al. 1990; Elitsur, Short, and Neace 1998; Fall et al. 1997; Galpin, Whitaker, and Dubiel 1992; Goodman et al. 1996; Graham et al. 1991; Herbarth et al. 2001; Jimenez-Guerra, Shetty, and Kurpad 2000; Kimura et al. 1999; Malaty and Graham 1994; Malaty et al.,1996b; McCallion et al. 1996; Mendall et al. 1992; Mitchell et al. 1992; Nowottny and Heilmann 1990; Regev et al. 1999; Robertson, Cade, and Clancy 1999; Rothenbacher et al. 1997, 1998a, 1998b; Staat et al. 1996; Yamashita et al. 2001	Lin et al. 1998; Malaty et al. 1996a
Ethnicity/race	Blecker et al. 1993; Collett et al. 1999; Dehesa et al. 1991; Fiedorek et al. 1991; Graham et al. 1991; Lanciers et al. 1996; Malaty et al. 1992, 1999; Malaty and Graham 1994; Replogle et al. 1995; Smoak, Kelley, and Taylor 1994; Staat et al. 1996; Teh et al. 1994	Adler-Shohet et al. 1996
Current low socioeconomic status	Bourke, Jones, and Sherman 1996; Collett et al. 1999; Drumm 1993; EUROGAST 1993; Fiedorek et al. 1991; Graham et al. 1991; Hopkins et al. 1993; Klein et al. 1991; Lin et al. 1998; Malaty et al. 1996a and 1996b (in children); McCallion, et al. 1996; Olmos, Rios, and Higa 2000 (in adults); Sitas et al. 1991; Staat et al. 1996	Clemens et al. 1996; Elitsur, Short, and Neace 1998; Malaty et al. 1996a and 1996b (in adults); Mendall et al. 1992
Low childhood socioeconomic status or parental education level	Adler-Shohet et al. 1996; Malaty and Graham 1994; Rothenbacher et al. 1998a; Staat et al. 1996; Webb et al. 1994	Mendall et al. 1992

Table continued on next page.

Table 11-1 Environmental and social risk factors for *H. pylori* infection (continued)

Environmental/ Social Risk Factor	Positive Association With Infection	No Association With Infection
Male gender	Fall et al. 1997; Goodman et al. 1996; Klein et al. 1994; Lin et al. 1998; Replogle et al. 1995	Adler-Shohet et al. 1996; Blecker, Mehta, and Vandenplas 1994b; Elitsur, Short, and Neace 1998; Fiedorek et al. 1991; Graham et al. 1991; Kawasaki et al. 1998; Klein et al. 1991; Megraud et al. 1989; Oliveira et al. 1994; Öztürk et al. 1996; Staat et al. 1996; Teh et al. 1994; Yamashita et al. 2001
Dyspepsia symptoms		Andersen, Elsborg, and Justesen 1988; Bode et al. 1998; Collett et al. 1999; Goodman et al. 1996; Graham et al. 1991; Hardikar et al. 1996; Kawasaki et al. 1998; Lee et al. 1994; Lin et al. 1998; Macarthur, Saunders, and Feldman 1995; O'Donohoe et al. 1996; Öztürk et al. 1996; Reifen et al. 1994
Neighborhood location (urban versus rural)	Elitsur, Short, and Neace 1998; Kawasaki et al. 1998; Mitchell et al. 1992; Teh et al. 1994	Fall et al. 1997; Fiedorek et al. 1991; Lee et al. 1994; Nowottny and Heilmann 1990

Dose Response

Though little is known about the actual number of organisms it takes to infect humans with *H. pylori,* we assume that the infectious dose must be fairly small, because the infection is so prevalent worldwide. However, known infective doses in humans have been high.

Infectious Dose

Little is known about the infectious dose of *H. pylori* required for humans. Marshall (Marshall et al. 1985) and then Morris (Morris and Nicholson 1987) ingested 1×10^7 and 3×10^5 CFU of *H. pylori,* respectively. Both needed dosing with an acid-suppressing agent to initiate infection. Therefore, 3×10^5 CFU is the smallest known infective dose in humans. A laboratory accident also provides some data on infectious dose in humans. A healthy, previously *H. pylori*-negative researcher working with the organism in the laboratory touched her finger to a filter cultured with *H. pylori* from a patient with peptic ulcer, then put her finger in her mouth (Matysiak-Budnik et al. 1995). After suffering from gastric symptoms a day later, she was diagnosed with *H. pylori.* The authors suggest that her dose was probably less than the previously measured experimental doses; however, they do not know for sure. In addition, the researcher presumably had normal gastric acidity, yet she was infected without the use of an acid suppressant. It also seems likely that cases of iatrogenic infection through improperly maintained endoscopes would involve smaller infectious doses than previously reported.

Bacterial Virulence Factors and Host Factors Influencing Infection

Most of the data related to the interaction of *H. pylori* and the human host are derived from adults. *H. pylori* is highly virulent, as evidenced by such a high infection prevalence; however, the host-specific and organism-specific factors that affect infection and disease progression are varied. In the organism, motility, adherence, and urease production are critical to its ability to infect (Atherton 1998; Mobley 1997). The body is not protected from infection by immunity (Rowland 2000). In addition, some strains produce heat-shock proteins that seem to work in conjunction with urease (Bourke, Jones, and Sherman 1996). Also, 50 to 60% of *H. pylori* strains express genes that produce a cytotoxin (Parsonnet 1998) that has been linked with the presence of peptic ulcer and other gastric diseases (Atherton 1998; Blaser 1999). However, other disease factors must come into play, because the cytotoxin-producing strain of *H. pylori* is not universally found in people with peptic ulcer disease. CagA, vacA, and iceA are all virulence factors that have been identified (Graham and Yamaoka 2000); however, none of the factors have disease specificity, though cagA does cause an increase in IL-8 and inflammation. CagA also is associated with a decreased risk of GERD (Perez-Perez et al. 1999).

Since different populations experience different incidence rates, Evans et al. (1999) tested the hypothesis that genotypes vary from population to population. They compared clinical isolates from two large heterogeneous populations (Houston, Texas, and Minas Gerais, Brazil) and a small homogenous population (Goteborg, Sweden). They found that the isolates from Texas and Brazil were very similar even though they were far apart geographically. The Swedish isolates, however, were very different from the other two, showing that genotypes vary and that population type may have a greater effect than geography.

Within the host, genetic factors appear to influence infection because *H. pylori* concordance was higher in identical twins rather than in fraternal twins tested as adults (Malaty et al. 1994). Also, several studies have shown a higher prevalence in certain racial or ethnic subgroups of a population, even after controlling for other risk factors (see Table 11-1). Other host factors, such as human leukocyte antigen genotypes and blood group antigens, may influence the development of clinically significant disease (Atherton 1998; Parsonnet 1998; Peura 1997). Gastric acid secretion as a host characteristic probably affects infection and disease development. Factors that affect the stomach's ecology, such as nutritional status and co-infection with another organism, may also contribute to the host's likelihood of infection or illness (Graham and Yamaoka 2000; McColl, el Omar, and Gillen 2000; Parsonnet 1998). In experimental infections, some subjects needed an acid suppressor to be successfully infected (Marshall et al. 1985; Morris and Nicholson 1987), but in an accidental exposure, infection occurred in a subject without adjusting normal gastric acidity (Matysiak-Budnik et al. 1995). Also, once infected, some patients have a greatly increased acid production response and others only moderately so (Go 1997). Ulcers induced by behavioral habits, including smoking, diet, and nonsteroidal anti-inflammatory use, also influence the progression to cancer (Parsonnet 1998).

Environmental Occurrence

Little is known about the occurrence of *H. pylori* in drinking water sources or how much, if any, drinking water contributes to infection. *H. pylori* is difficult to isolate in water because it changes its morphology, growth, and metabolism when exposed to varying environments (Velazquez and Feirtag 1999). Detection methods include immunoseparation, PCR, immunostaining, probe hybridization, autoradiography, and ATP bioluminescence.

H. pylori exists in two forms: a helical replicative form and a nonculturable, but possibly viable, coccoid form. Both have been shown to survive for days up to weeks in sterile river water, saline solution, and distilled water at a wide variety of pH levels and in temperatures ranging from 4°C to 15°C (Shahamat et al. 1993; West, Millar, and Thompkins 1992). Though the coccoid form fed to gnotobiotic pigs did not revert to the replicative form in vivo (Eaton et al. 1995), mice fed both the replicative and coccoid forms had both forms recovered from their stomachs later, though it took longer for the coccoid form to colonize (Cellini et al. 1994). Roe et al. (1999) compared the antigenic and morphologic profiles of *H. pylori* rods that were then converted to coccoid forms. Almost half (46.7%) of the profiles remained identical after conversion, which suggests that the coccoid form has the potential to be viable. Other reports support viability in the coccoid form (Mizoguchi, Fujioka, and Nasu 1999; Willen et al. 2000), and another does not (Enroth et al. 1999). The evidence appears to be equivocal.

The detection of *H. pylori* DNA in Peruvian water sources using PCR did not characterize the organisms' form in the samples (Hulten et al. 1996). In water, the nonculturable coccoidal form of *Campylobacter jejuni* has been used to successfully colonize chicks and mice (Fox 1995). The mechanism for infection may be the same for *H. pylori*. These limited data suggest that *H. pylori* can remain viable in water under a wide variety of conditions and can assume a form that perhaps makes it more hardy in an environmental setting.

The isolation of *H. pylori* in Peruvian drinking water using PCR detection methods confirmed the results of an epidemiological study associating water source with *H. pylori* infection (Hulten et al. 1996; Klein et al. 1991). Researchers have used PCR to detect *H. pylori* in Swedish water sources, including tap water and water from a Canadian Arctic village (Hulten et al. 1996; McKeown et al. 1999); however, PCR does not indicate the infectivity of the organism. Also, Sweden has one of the lowest death rates from stomach cancer in the world (American Cancer Society 1998), suggesting that large numbers of people are not infected with the organism. The Canadian study did not attempt to link drinking water consumption with infection.

Recently presented work has generated additional interest in the possible occurrence of *H. pylori* in water (Hegarty et al. 1999). Using a combination of fluorescent antibody and PCR assay, researchers tested 42 surface water and 20 well water samples in Ohio and Pennsylvania for *H. pylori*. They found 40% and 65% positive samples, respectively (Hegarty et al. 1999). In a very small subsample, clinical infection was significantly related to the occurrence of *H. pylori* in patients' water supplies. *H. pylori* positivity did not correlate with the occurrence of total coliforms or *Escherichia coli*—traditional indicator bacteria (Hegarty et al. 1999).

Stark and colleagues (1999) showed that *H. pylori* could form a biofilm under laboratory conditions, but Park and colleagues (2001) then used nested PCR to detect *Helicobacter* spp. from the actual biofilm of a municipal water supply system in Scotland—the first time ever reported. Other researchers have reported results indicating that *H. pylori* can survive in water under variable conditions. Ma and colleagues (1999) found that *H. pylori* levels were undetectable after 28 hours in moderately hard groundwater. The addition of acetate significantly increased its persistence, and the addition of urea significantly decreased its persistence. Difficulty in culturing the slow-growing and fastidious organism has made additional environmental data hard to generate.

The mechanism of water contamination would most likely be from human feces via sewage. The organism has been isolated from feces and sewage (Hulten et al. 1998; Thomas et al. 1992), and two studies have suggested that eating vegetables fertilized with sewage may provide a route of transmission (Goodman et al. 1996; Hopkins et al. 1993).

Water Treatment

A US Environmental Protection Agency study tested the efficacy of chlorine to inactivate *H. pylori* (Johnson, Rice, and Reasoner 1997). They tested three different strains, including two clinical, using a fecal isolate of *E. coli* as a comparison. Each isolate was tested three times at pH levels 6, 7, and 8. All experiments occurred at 5°C (chlorine is less effective at that temperature) for a maximum time exposure of 80 seconds. All samples experienced a greater than 3.5 $\log_{10}$ reduction at 80 seconds, indicating that *H. pylori* is sensitive to a chlorine concentration of 0.5 mg/L (in US water utilities, the median residual is 1.1 mg/L and the contact time is 45 minutes). When comparing the sensitivities, *E. coli* proved to be somewhat more sensitive to chlorine than *H. pylori*; however, the authors suggested that this may be related to the increased amount of particulate matter in the *H. pylori* samples compared with the *E. coli* samples.

Other data indicated that *H. pylori* was significantly more resistant to chlorine than *E. coli* (Hegarty et al. 1999a). Free chlorine at 0.2 mg/L reduced *E. coli* by four logs and *H. pylori* by less than two logs; however, there was little difference between the effectiveness when monochloramine was used. It appears that *H. pylori* may be less sensitive to chlorine than *E. coli* at low levels, but that normal residuals may be adequate to kill the organism. Further studies are necessary.

Membrane filtration technology, such as micro-, ultra-, and nanofiltration, as well as reverse osmosis, has the capacity to remove water contaminants down to the ion level. Microfiltration, with a nominal pore size of 0.2 µm, can remove bacterial pathogens better than conventional water treatment, and ultrafiltration, with a nominal pore size of 0.01 µm, can achieve up to a 6-log pathogen removal (Najm and Trussell 1999; Taylor and Wiesner 1999). This technology will continue to gain in popularity as costs become more competitive and regulatory approval is assured.

Conclusions

H. pylori is one of the most common infections in the world. The infection is acquired primarily in childhood and, if left untreated, will persist throughout life. However, the majority of infected people remain asymptomatic throughout their lives. Up to 20% will develop peptic ulcer disease, and a very small percentage will develop gastric cancer. Though the disease-to-infection ratio is small, the enormous number of infected people in the world results in significant morbidity and mortality. *H. pylori*'s exact route of transmission is unknown but it is probably person-to-person and foodborne or waterborne, depending on the exposure and risk factors in the population.

Because methods to detect *H. pylori* in environmental samples are poor, little is known about the occurrence of the organism in water. Studies on the efficacy of chlorine as a disinfectant are equivocal, and the epidemiological data pointing at a waterborne route come from developing countries in South America. Recent evidence of surface and groundwater contamination in the United States has generated much interest as well. Though it is unlikely that drinking water is a significant source of infection in countries with adequate treatment, more research into the role of water in the spread of *H. pylori* is necessary.

Bibliography

Adler-Shohet, F., P. Palmer, G. Reed, and K. Edwards. 1996. Prevalence of *Helicobacter pylori* antibodies in normal children. *Pediatr. Infect. Dis. J.*, 15:172–174.

American Cancer Society. 1998. Cancer Facts & Figures—1998. Atlanta, GA: American Cancer Society.

Andersen, L.P., L. Elsborg, and T. Justesen. 1988. *Campylobacter pylori* in peptic ulcer disease. III. Symptoms and paraclinical and epidemiologic findings. *Scand. J. Gastroenterol.*, 23:347–350.

Ashorn, M., A. Miettinen, T. Ruuska, P. Laippala, and M. Maki. 1996. Seroepidemiological study of *Helicobacter pylori* infection in infancy. *Arch. Dis. Child Fetal Neonatal Ed.*, 74:F141–2.

Atherton, J.C. 1998. *H. pylori* virulence factors. *Br. Med. Bull.*, 54:105–120.

Axon, A.T. 2000. Treatment of *Helicobacter pylori*: an overview. *Aliment. Pharmacol. Ther.*, 14(Suppl)3:1–6.

Bamford, K.B., J. Bickley, J.S. Collins, B.T. Johnston, S. Potts, V. Boston, R.J. Owen, and J.M. Sloan. 1993. *Helicobacter pylori*: comparison of DNA fingerprints provides evidence for intrafamilial infection. *Gut*, 34:1348–1350.

Battan, R., M.C. Raviglione, A. Palagiano, J.F. Boyle, M.T. Sabatini, K. Sayad, and L.J. Ottaviano. 1990. *Helicobacter pylori* infection in patients with acquired immune deficiency syndrome. *Am. J. Gastroenterol.*, 85:1576–1579.

Berkowicz, J., and A. Lee. 1987. Person-to-person transmission of *Campylobacter pylori*. *Lancet*, 2:680–681.

Blaser, M.J. 1999. Hypothesis: the changing relationships of *Helicobacter pylori* and humans: implications for health and disease. *J. Infect. Dis.*, 179:1523–1530.

Blecker, U., B. Hauser, S. Lanciers, S. Peeters, B. Suys, and Y. Vandenplas. 1993. The prevalence of *Helicobacter pylori*-positive serology in asymptomatic children. *J. Pediatr. Gastroenterol. Nutr.*, 16:252–256.

Blecker, U., S. Lanciers, E. Keppens, and Y. Vandenplas. 1994a. Evolution of *Helicobacter pylori* positivity in infants born from positive mothers. *J. Pediatr. Gastroenterol. Nutr.*, 19:87–90.

Blecker, U., S. Lanciers, D.I. Mehta, and Y. Vandenplas. 1994b. Familial clustering of *Helicobacter pylori* infection. *Clin. Pediatr.*, 3:307–308.

Blecker, U., D.I. Mehta, and Y. Vandenplas. 1994. Sex ratio of *Helicobacter pylori* infection in childhood. *Am. J. Gastroenterol.*, 89:293.

Bode, G., D. Rothenbacher, H. Brenner, and G. Adler. 1998. *Helicobacter pylori* and abdominal symptoms: a population-based study among preschool children in southern Germany. *Pediatrics*, 101:634–637.

Bonamico, M., S. Monti, I. Luzzi, F.M. Magliocca, E. Cipolletta, L. Calvani, A. Marcheggiano, R. Rossi, and P. Paoluzi. 1996. *Helicobacter pylori* infection in families of *Helicobacter pylori*-positive children. *Ital. J. Gastroenterol.*, 28:512–517.

Bourke, B., N. Jones, and P. Sherman. 1996. *Helicobacter pylori* infection and peptic ulcer disease in children. *Pediatr. Infect. Dis. J.*, 15:1–13.

Brown, L.M. 2000. *Helicobacter pylori*: epidemiology and routes of transmission. *Epidemiol. Rev.*, 22:283–297.

Cacciarelli, A.G., B.J. Marano Jr., N.M. Gualtieri, A.R. Zuretti, R.A. Torres, A.A. Starpoli, and J.G. Robilotti Jr. 1996. Lower *Helicobacter pylori* infection and peptic ulcer disease prevalence in patients with AIDS and suppressed CD4 counts. *Am. J. Gastroenterol.*, 91:1783–1784.

Carella, G., T. Vallot, and L. Marra. 1994. Role of *Helicobacter pylori* in digestive disease in AIDS. *Clin. Ter.*, 144:19–22.

Cave, D.R. 1997. How is *Helicobacter pylori* transmitted? *Gastroenterol.*, 113:S9–14.

Cellini, L., N. Allocati, D. Angelucci, T. Iezzi, E. Di Campli, L. Marzio, and B. Dainelli. 1994. Coccoid *Helicobacter pylori* not culturable in vitro reverts in mice. *Microbiol. Immunol.*, 38:843–850.

Centers for Disease Control and Prevention (CDC). [Undated.] *Helicobacter pylori*: Facts for Health Care Providers [Online]. Available: <www.cdc.gov/ncidod/dbmd/md.htm>. [cited November 11, 1999]

Centers for Disease Control and Prevention (CDC). 1999. Iron deficiency anemia in Alaska native children—Hooper Bay, Alaska, 1999. *MMWR*, 48:714–716.

Choe, Y.H., S.K. Kim, B.K. Son, D.H. Lee, Y.C. Hong, and S.H. Pai. 1999. Randomized placebo-controlled trial of *Helicobacter pylori* eradication for iron-deficiency anemia in preadolescent children and adolescents. *Helicobacter*, 4:135–139.

Chong, J., B.J. Marshall, J.S. Barkin, R.W. McCallum, D.K. Reiner, S.R. Hoffman, and C. O'Phelan. 1994. Occupational exposure to *Helicobacter pylori* for the endoscopy professional: a sera epidemiological study. *Am. J. Gastroenterol.*, 89:1987–1992.

Chow, T.K., J.R. Lambert, M.L. Wahlqvist, and B.H. Hsu-Hage. 1995. *Helicobacter pylori* in Melbourne Chinese immigrants: evidence for oral–oral transmission via chopsticks. *J. Gastroenterol. Hepatol.*, 10:562–569.

Clemens, J., M.J. Albert, M. Rao, S. Huda, F. Qadri, F.P. Van Loon, B. Pradhan, A. Naficy, and A. Banik. 1996. Sociodemographic, hygienic and nutritional correlates of *Helicobacter pylori* infection of young Bangladeshi children. *Pediatr. Infect. Dis. J.*, 15:1113–1118.

Collett, J.A., M.J. Burt, C.M. Frampton, K.H. Yeo, T.M. Chapman, R.C. Buttimore, H.B. Cook, and B.A. Chapman. 1999. Seroprevalence of *Helicobacter pylori* in the adult population of Christchurch: risk factors and relationship to dyspeptic symptoms and iron studies. *N. Z. Med. J.*, 112:292–295.

Corvaglia, L., P. Bontems, J.M. Devaster, P. Heimann, Y. Glupczynski, E. Keppens, and S. Cadranel. 1999. Accuracy of serology and ^{13}C-urea breath test for detection of *Helicobacter pylori* in children. *Pediatr. Infect. Dis. J.*, 18:976–979.

Cullen, D.J., B.J. Collins, K.J. Christiansen, J. Epis, J.R. Warren, I. Surveyor, and K.J. Cullen. 1993. When is *Helicobacter pylori* infection acquired? *Gut*, 34:1681–1682.

Czinn, S.J. 1997. What is the role for vaccination in *Helicobacter pylori*? *Gastroenterol.*, 113:S149–S153.

Danesh, J. 1999. *Helicobacter pylori* infection and gastric cancer: systematic review of the epidemiological studies. *Aliment. Pharmacol. Ther.*, 13:851–856.

Danesh, J., M. Lawrence, M. Murphy, S. Roberts, and R. Collins. 2000. Systematic review of the epidemiological evidence on *Helicobacter pylori* infection and nonulcer or uninvestigated dyspepsia. *Arch. Intern. Med.*, 160:1192–1198.

De Groote, D., R. Ducatelle, and F. Haesebrouck. 2000. *Helicobacters* of possible zoonotic origin: a review. *Acta. Gastroenterol. Belg.*, 63:380–387.

Dehesa, M., C.P. Dooley, H. Cohen, P.L. Fitzgibbons, G.I. Perez-Perez, and M.J. Blaser. 1991. High prevalence of *Helicobacter pylori* infection and histologic gastritis in asymptomatic Hispanics. *J. Clin. Microbiol.*, 29:1128–1131.

Dohil, R., D.M. Israel, and E. Hassall. 1997. Effective 2-wk therapy for *Helicobacter pylori* disease in children. *Am. J. Gastroenterol.*, 92:244–247.

Dowsett, S.A., L. Archila, V.A. Segreto, C.R. Gonzalez, A. Silva, K.A. Vastola, R.D. Bartizek, and M.J. Kowolik. 1999. *Helicobacter pylori* infection in indigenous families of Central America: serostatus and oral and fingernail carriage. *J. Clin. Microbiol.*, 37:2456–2460.

Drumm, B. 1993. *Helicobacter pylori* in the pediatric patient. *Gastroenterol. Clin. North Am.*, 22:169–182.

Drumm, B., G.I. Perez-Perez, M.J. Blaser, and P.M. Sherman. 1990. Intrafamilial clustering of *Helicobacter pylori* infection. *N. Engl. J. Med.*, 322:359–363.

Dunn, B.E., H. Cohen, and M.J. Blaser. 1997. *Helicobacter pylori*. *Clin. Microbiol. Rev.*, 10:720–741.

Eaton, K.A., C.E. Catrenich, K.M. Makin, and S. Krakowka. 1995. Virulence of coccoid and bacillary forms of *Helicobacter pylori* in gnotobiotic piglets. *J. Infect. Dis.*, 171:459–462.

Edwards, P.D., J. Carrick, J. Turner, A. Lee, H. Mitchell, and D.A. Cooper. 1991. *Helicobacter pylori*-associated gastritis is rare in AIDS: antibiotic effect or a consequence of immunodeficiency? *Am. J. Gastroenterol.*, 86:1761–1764.

Elitsur, Y., J.P. Short, and C. Neace. 1998. Prevalence of *Helicobacter pylori* infection in children from urban and rural West Virginia. *Dig. Dis. Sci.*, 43:773–778.

Enroth, H., K. Wreiber, R. Rigo, D. Risberg, A. Uribe, and L. Engstrand. 1999. In vitro aging of *Helicobacter pylori*: changes in morphology, intracellular composition and surface properties. *Helicobacter*, 4:7–16.

EUROGAST Study Group. 1993. Epidemiology of, and risk factors for, *Helicobacter pylori* infection among 3194 asymptomatic subjects in 17 populations. *Gut*, 34:1672–1676.

Evans, D.G., D.M. Queiroz, E.N. Mendes, A.M. Svennerholm, and D.J. Evans Jr. 1999. Differences among *Helicobacter pylori* strains isolated from three different populations and demonstrated by restriction enzyme analysis of an internal fragment of the conserved gene hpaA. *Helicobacter*, 4:82–88.

Fall, C.H., P.M. Goggin, P. Hawtin, D. Fine, and S. Duggleby. 1997. Growth in infancy, infant feeding, childhood living conditions, and *Helicobacter pylori* infection at age 70. *Arch. Dis. Child.*, 77:310–314.

Feldman, R.A., A.J. Eccersley, and J.M. Hardie. 1998. Epidemiology of *Helicobacter pylori*: acquisition, transmission, population prevalence and disease-to-infection ratio. *Br. Med. Bull.*, 54:39–53.

Fiedorek, S.C., H.M. Malaty, D.L. Evans, C.L. Pumphrey, H.B. Casteel, D.J. Evans, and D.Y. Graham. 1991. Factors influencing the epidemiology of *Helicobacter pylori* infection in children. *Pediatrics,* 88:578–582.

Forman, D., D.G. Newell, F. Fullerton, J.W. Yarnell, A.R. Stacey, N. Wald, and F. Sitas. 1991. Association between infection with *Helicobacter pylori* and risk of gastric cancer: evidence from a prospective investigation. *Br. Med. J.*, 302:1302–1305.

Forman, D., and P. Webb. 1993. Geographic Distribution and Association with Gastric Cancer. In *Helicobacter pylori Infection: Pathophysicology, Epidemiology and Management.* Edited by T.C. Northfield, M. Mendall, and P.M. Goggin. Boston: Kluwer Academic Publishers.

Fox, J.G. 1995. Non-human reservoirs of *Helicobacter pylori. Aliment. Pharmacol. Ther.*, 9(Suppl)2:93-103:93-103.

Fujisawa, T., T. Kumagai, T. Akamatsu, K. Kiyosawa, and Y. Matsunaga. 1999. Changes in seroepidemiological pattern of *Helicobacter pylori* and hepatitis A virus over the last 20 years in Japan. *Am. J. Gastroenterol.*, 94:2094–2099.

Galpin, O.P., C.J. Whitaker, and A.J. Dubiel. 1992. *Helicobacter pylori* infection and overcrowding in childhood. *Lancet,* 339:619.

Ganga-Zandzou, P.S., L. Michaud, P. Vincent, M.O. Husson, N. Wizla-Derambure, E.M. Delassalle, D. Turck, and F. Gottrand. 1999. Natural outcome of *Helicobacter pylori* infection in asymptomatic children: a two-year follow-up study. *Pediatrics,* 104:216–221.

Georgopoulos, S.D., A.F. Mentis, C.A. Spiliadis, L.S. Tzouvelekis, E. Tzelepi, A. Moshopoulos, and N. Skandalis. 1996. *Helicobacter pylori* infection in spouses of patients with duodenal ulcers and comparison of ribosomal RNA gene patterns. *Gut,* 39:634–638.

Go, M.F. 1997. What are the host factors that place an individual at risk for *Helicobacter pylori*-associated disease? *Gastroenterol.*, 113:S15–20.

Gold, B.D., B. Khanna, L.M. Huang, C.Y. Lee, and N. Banatvala. 1997. *Helicobacter pylori* acquisition in infancy after decline of maternal passive immunity. *Pediatr. Res.*, 41:641–646.

Goodman, K.J., and P. Correa. 2000. Transmission of *Helicobacter pylori* among siblings. *Lancet,* 355:358–362.

Goodman, K.J., P. Correa, H.J. Tengana Aux, H. Ramirez, J.P. DeLany, O. Guerrero Pepinosa, M. Lopez Quinones, and T. Collazos Parra. 1996. *Helicobacter pylori* infection in the Colombian Andes: a population-based study of transmission pathways. *Am. J. Epidemiol.*, 144:290–299.

Graham, D.Y. 1994. Benefits from elimination of *Helicobacter pylori* infection include major reduction in the incidence of peptic ulcer disease, gastric cancer, and primary gastric lymphoma. *Prev. Med.*, 23:712–716.

Graham, D.Y., H.M. Malaty, D.G. Evans, D.J. Evans, P.D. Klein, and E. Adam. 1991. Epidemiology of *Helicobacter pylori* in an asymptomatic population in the United States. Effect of age, race, and socioeconomic status. *Gastroenterol.*, 100:1495–1501.

Graham, D.Y., and Y. Yamaoka. 2000. Disease-specific *Helicobacter pylori* virulence factors: the unfulfilled promise. *Helicobacter*, 5(Suppl)1: S3–S9.

Granström, M., Y. Tindberg, and M. Blennow. 1997. Seroepidemiology of *Helicobacter pylori* infection in a cohort of children monitored from 6 months to 11 years of age. *J. Clin. Microbiol.*, 35:468–470.

Grubel, P., J.S. Hoffman, F.K. Chong, N.A. Burstein, C. Mepani, and D.R. Cave. 1997. Vector potential of houseflies (Musca domestica) for *Helicobacter pylori*. *J. Clin. Microbiol.*, 35:1300–1303.

Handt, L.K., J.G. Fox, I.H. Stalis, R. Rufo, G. Lee, J. Linn, X. Li, and H. Kleanthous. 1995. Characterization of feline *Helicobacter pylori* strains and associated gastritis in a colony of domestic cats. *J. Clin. Microbiol.*, 33:2280–2289.

Handwerker, J., J.G. Fox, and D.B. Schauer. 1995. Detection of *Helicobacter pylori* in drinking water using polymerase chain reaction amplification, Abstr. Q-203. In *Abstracts of the 95th General Meeting of the American Society for Microbiology 1995*. Washington, DC: American Society for Microbiology. p. 435.

Hardikar, W., C. Feekery, A. Smith, F. Oberklaid, and K. Grimwood. 1996. *Helicobacter pylori* and recurrent abdominal pain in children. *J. Pediatr. Gastroenterol. Nutr.*, 22:148–152.

Haruma, K., K. Komoto, H. Kawaguchi, S. Okamoto, M. Yoshihara, K. Sumii, and G. Kajiyama. 1995. Pernicious anemia and *Helicobacter pylori* infection in Japan: evaluation in a country with a high prevalence of infection. *Am. J. Gastroenterol.*, 90:1107–1110.

Hazell, S.L., H.M. Mitchell, M. Hedges, X. Shi, P.J. Hu, Y.Y. Li, A. Lee, and E. Reiss-Levy. 1994. Hepatitis A and evidence against the community dissemination of *Helicobacter pylori* via feces. *J. Infect. Dis.*, 170:686–689.

Hegarty, J.P., M.T. Dowd, and K.H. Baker. 1999. Occurrence of *Helicobacter pylori* in surface water in the United States. *J. Appl. Microbiol.*, 87:697–701.

Hegarty, J.P., D.S. Herson, B.W. Redmond, N.A. Reed, and K.H. Baker. 1999a. Effect of oxidizing disinfectants on *Helicobacter pylori*—A comparative study. Abstr. Q-132. In *Abstracts of the 99th General Meeting of the American Society for Microbiology 1999*. Washington, DC: American Society for Microbiology. p. 558.

Herbarth, O., P. Krumbiegel, G.J. Fritz, M. Richter, U. Schlink, D.M. Muller, and T. Richter. 2001. *Helicobacter pylori* prevalences and risk factors among school beginners in a german urban center and its rural county. *Environ. Health. Perspect.*, 109:573–577.

Hopkins, R.J., P.A. Vial, C. Ferreccio, J. Ovalle, P. Prado, V. Sotomayor, R.G. Russell, S.S. Wasserman, and J.G. Morris Jr. 1993. Seroprevalence of *Helicobacter pylori* in Chile: vegetables may serve as one route of transmission. *J. Infect. Dis.*, 168:222–226.

Hulten, K., H. Enroth, T. Nystrom, and L. Engstrand. 1998. Presence of Helicobacter species DNA in Swedish water. *J. Appl. Microbiol.*, 85:282–286.

Hulten, K., S.W. Han, H. Enroth, P.D. Klein, A.R. Opekun, R.H. Gilman, D.G. Evans, L. Engstrand, D.Y. Graham, and F.A. El Zaatari. 1996. *Helicobacter pylori* in the drinking water in Peru. *Gastroenterol.*, 110:1031–1035.

Hunt, R.H., C.A. Fallone, and A.B. Thomson. 1999. Canadian *Helicobacter pylori* Consensus Conference update: infections in adults. Canadian *Helicobacter* Study Group. *Can. J. Gastroenterol.*, 13:213–217.

IARC. 1994. Schistosomes, liver flukes and *Helicobacter pylori*. IARC Working Group on the Evaluation of Carcinogenic Risks to Humans. Lyon, 7–14 June 1994. *IARC Monogr. Eval. Carcinog. Risks Hum.*, 61:1-241.

Iovene, M.R., M. Romano, A.P. Pilloni, B. Giordano, F. Montella, S. Caliendo, and M.A. Tufano. 1999. Prevalence of antimicrobial resistance in eighty clinical isolates of *Helicobacter pylori*. *Chemotherapy*, 45:8–14.

Jimenez-Guerra, F., P. Shetty, and A. Kurpad. 2000. Prevalence of and risk factors for *Helicobacter pylori* infection in school children in Mexico. *Ann. Epidemiol.*, 10:474.

Johnson, C.H., E.W. Rice, and D.J. Reasoner. 1997. Inactivation of *Helicobacter pylori* by chlorination. *Appl. Environ. Microbiol.*, 63:4969–4970.

Jones, N.L., and P.M. Sherman. 1998. *Helicobacter pylori* infection in children. *Curr. Opin. Pediatr.*, 10:19–23.

Kawasaki, M., T. Kawasaki, T. Ogaki, K. Itoh, S. Kobayashi, Y. Yoshimizu, K. Aoyagi, A. Iwakawa, S. Takahashi, S. Sharma, and G.P. Acharya. 1998. Seroprevalence of *Helicobacter pylori* infection in Nepal: low prevalence in an isolated rural village. *Eur. J. Gastroenterol. Hepatol.*, 10:47–50.

Kelly, S.M., M.C. Pitcher, S.M. Farmery, and G.R. Gibson. 1994. Isolation of *Helicobacter pylori* from feces of patients with dyspepsia in the United Kingdom. *Gastroenterol.*, 107:1671–1674.

Kimura, A., T. Matsubasa, H. Kinoshita, N. Kuriya, Y. Yamashita, T. Fujisawa, H. Terakura, and M. Shinohara. 1999. *Helicobacter pylori* seropositivity in patients with severe neurologic impairment. *Brain Dev.*, 21:113–117.

Klein, P.D., R.H. Gilman, R. Leon-Barua, F. Diaz, E.O. Smith, and D.Y. Graham. 1994. The epidemiology of *Helicobacter pylori* in Peruvian children between 6 and 30 months of age. *Am. J. Gastroenterol.*, 89:2196–2200.

Klein, P.D., D.Y. Graham, A. Gaillour, A.R. Opekun, and E.O. Smith. 1991. Water source as risk factor for *Helicobacter pylori* infection in Peruvian children. Gastrointestinal Physiology Working Group. *Lancet*, 337:1503–1506.

Lanciers, S., B. Hauser, Y. Vandenplas, and U. Blecker. 1996. The prevalence of *Helicobacter pylori* positivity in asymptomatic children of different ethnic backgrounds living in the same country. *Ethn. Health*, 1:169–173.

Langenberg, W., E.A. Rauws, J.H. Oudbier, and G.N. Tytgat. 1990. Patient-to-patient transmission of *Campylobacter pylori* infection by fiberoptic gastroduodenoscopy and biopsy. *J. Infect. Dis.*, 161:507–511.

Lee, A. 1996a. Prevention of *Helicobacter pylori* infection. *Scand. J. Gastroenterol. Suppl.*, 215:11–15.

Lee, A. 1996b. Vaccination against *Helicobacter pylori*. *J. Gastroenterol.*, 31(Suppl)9:69–74.

Lee, A., and F. Buck. 1996. Vaccination and mucosal responses to *Helicobacter pylori* infection. *Aliment. Pharmacol. Ther.*, 10(Suppl)1:129–138.

Lee, A., J.G. Fox, G. Otto, E.H. Dick, and S. Krakowka. 1991. Transmission of Helicobacter spp. A challenge to the dogma of fecal–oral spread. *Epidemiol. Infect.*, 107:99–109.

Lee, J., and C. O'Morain. 1997. Who should be treated for *Helicobacter pylori* infection? A review of consensus conferences and guidelines. *Gastroenterol.*, 113:S99–106.

Lee, M.G., M. Arthurs, S.I. Terry, E. Donaldson, P. Scott, F. Bennett, B. Hanchard, and P.N. Levett. 1994. *Helicobacter pylori* in patients undergoing upper endoscopy in Jamaica. *West Indian Med. J.*, 43:84–86.

Lehmann, F., J. Drewe, L. Terracciano, R. Stuber, R. Frei, and C. Beglinger. 1999. Comparison of stool immunoassay with standard methods for detecting *Helicobacter pylori* infection. *Br. Med. J.*, 319:1409.

Leung, W.K., J.J. Sung, T.K. Ling, K.L. Siu, and A.F. Cheng. 1997. Does the use of chopsticks for eating transmit *Helicobacter pylori*? [letter]. *Lancet*, 350:31–31.

Lin, S.K., J.R. Lambert, L. Nicholson, W. Lukito, and M. Wahlqvist. 1998. Prevalence of *Helicobacter pylori* in a representative Anglo-Celtic population of urban Melbourne. *J. Gastroenterol. Hepatol.*, 13:505–510.

Lin, S.K., J.R. Lambert, M.A. Schembri, L. Nicholson, and M.G. Korman. 1994. *Helicobacter pylori* prevalence in endoscopy and medical staff. *J. Gastroenterol. Hepatol.*, 9:319–324.

Lindkvist, P., D. Asrat, I. Nilsson, E. Tsega, G.L. Olsson, B. Wretlind, and J. Giesecke. 1996. Age at acquisition of *Helicobacter pylori* infection: comparison of a high and a low prevalence country. *Scand. J. Infect. Dis.*, 28:181–184.

Lionetti, P., S. Amarri, F. Silenzi, L. Galli, M. Cellini, M. de Martino, and A. Vierucci. 1999. Prevalence of *Helicobacter pylori* infection detected by serology and ^{13}C-urea breath test in HIV-1 perinatally infected children. *J. Pediatr. Gastroenterol. Nutr.*, 28:301–306.

Luzza, F., M. Imeneo, M. Maletta, G. Paluccio, A. Giancotti, F. Perticone, A. Foca, and F. Pallone. 1997. Seroepidemiology of *Helicobacter pylori* infection and hepatitis A in a rural area: evidence against a common mode of transmission. *Gut*, 41:164–168.

Luzza, F., M. Maletta, M. Imeneo, E. Fabiano, P. Doldo, L. Biancone, and F. Pallone. 1995. Evidence against an increased risk of *Helicobacter pylori* infection in dentists: a serological and salivary study. *Eur. J. Gastroenterol. Hepatol.*, 7:773–776.

Ma, Y.L., J.P. Hegarty, K.H. Baker. 1999. Persisence of *Helicobacter pylori*, *Campylobacter jejuni*, and *Escherichia coli* in culturable form in groundwater. Abstr. Q-170. In *Abstracts of the 99th General Meeting of the American Society for Microbiology 1999*. Washington, DC: American Society for Microbiology. p. 565.

Macarthur, C., N. Saunders, and W. Feldman. 1995. *Helicobacter pylori*, gastroduodenal disease, and recurrent abdominal pain in children. *JAMA*, 273:729–734.

Malaty, H.M., L. Engstrand, N.L. Pedersen, and D.Y. Graham. 1994. *Helicobacter pylori* infection: genetic and environmental influences. A study of twins. *Ann. Intern. Med.*, 120:982–986.

Malaty, H.M., D.G. Evans, D.J. Evans, and D.Y. Graham. 1992. *Helicobacter pylori* in Hispanics: comparison with blacks and whites of similar age and socioeconomic class. *Gastroenterol.*, 103:813–816.

Malaty, H.M., and D.Y. Graham. 1994. Importance of childhood socioeconomic status on the current prevalence of *Helicobacter pylori* infection. *Gut*, 35:742–745.

Malaty, H.M., D.Y. Graham, P.D. Klein, D.G. Evans, E. Adam, and D.J. Evans. 1991. Transmission of *Helicobacter pylori* infection. Studies in families of healthy individuals. *Scand. J. Gastroenterol.*, 26:927–932.

Malaty, H.M., D.Y. Graham, W.A. Wattigney, S.R. Srinivasan, M. Osato, and G.S. Berenson. 1999. Natural history of *Helicobacter pylori* infection in childhood: 12-year follow-up cohort study in a biracial community. *Clin. Infect. Dis.*, 28:279–282.

Malaty, H.M., J.G. Kim, S.D. Kim, and D.Y. Graham. 1996a. Prevalence of *Helicobacter pylori* infection in Korean children: inverse relation to socioeconomic status despite a uniformly high prevalence in adults. *Am. J. Epidemiol.*, 143:257–262.

Malaty, H.M., N.D. Logan, D.Y. Graham, and J.E. Ramchatesingh. 2001. *Helicobacter pylori* infection in preschool and school-aged minority children: effect of socioeconomic indicators and breast-feeding practices. *Clin. Infect. Dis.*, 32:1387–2.

Malaty, H.M., V. Paykov, O. Bykova, A. Ross, D.P. Graham, J.F. Anneger, and D.Y. Graham. 1996b. *Helicobacter pylori* and socioeconomic factors in Russia. *Helicobacter*, 1:82–87.

Mapstone, N.P., D.A. Lynch, F.A. Lewis, A.T. Axon, D.S. Tompkins, M.F. Dixon, and P. Quirke. 1993. PCR identification of *Helicobacter pylori* in faeces from gastritis patients. *Lancet*, 341:447.

Marano, B.J., Jr., F. Smith, and C.A. Bonanno. 1993. *Helicobacter pylori* prevalence in acquired immunodeficiency syndrome. *Am. J. Gastroenterol.*, 88:687–690.

Marshall, B.J., J.A. Armstrong, D.B. McGechie, and R.J. Glancy. 1985. Attempt to fulfil Koch's postulates for pyloric *Campylobacter. Med. J. Aust.*, 142:436–439.

Matysiak-Budnik, T., F. Briet, M. Heyman, and F. Megraud. 1995. Laboratory-acquired *Helicobacter pylori* infection [letter]. *Lancet,* 346:1489–1490.

McCallion, W.A., L.J. Murray, A.G. Bailie, A.M. Dalzell, D.P. O'Reilly, and K.B. Bamford. 1996. *Helicobacter pylori* infection in children: relation with current household living conditions. *Gut,* 39:18–21.

McColl, K.E., E. el Omar, and D. Gillen. 2000. *Helicobacter pylori* gastritis and gastric physiology. *Gastroenterol. Clin. North Am.*, 29:687–703, viii.

McKeown, I., P. Orr, S. Macdonald, A. Kabani, R. Brown, G. Coghlan, M. Dawood, J. Embil, M. Sargent, G. Smart, and C.N. Bernstein. 1999. *Helicobacter pylori* in the Canadian arctic: seroprevalence and detection in community water samples. *Am. J. Gastroenterol.*, 94:1823–1829.

McNamara, D.A., M. Buckley, and C.A. O'Morain. 2000. Nonulcer dyspepsia. Current concepts and management. *Gastroenterol. Clin. North Am.*, 29:807–818.

Megraud, F., M.P. Brassens-Rabbe, F. Denis, A. Belbouri, and D.Q. Hoa. 1989. Seroepidemiology of *Campylobacter pylori* infection in various populations. *J. Clin. Microbiol.*, 27:1870–1873.

Mendall, M. 1993. Natural history and mode of transmission. In *Helicobacter pylori Infection: Pathophysicology, Epidemiology and Management.* Edited by T.C. Northfield, M. Mendall, and P.M. Goggin. Boston: Kluwer Academic Publishers.

Mendall, M.A., P.M. Goggin, N. Molineaux, J. Levy, T. Toosy, D. Strachan, and T.C. Northfield. 1992. Childhood living conditions and *Helicobacter pylori* seropositivity in adult life [see comments]. *Lancet,* 339:896–897.

Milman, N., S. Rosenstock, L. Andersen, T. Jorgensen, and O. Bonnevie. 1998. Serum ferritin, hemoglobin, and *Helicobacter pylori* infection: a seroepidemiologic survey comprising 2794 Danish adults. *Gastroenterol.*, 115:268–274.

Mitchell, H.M., T.D. Bohane, J. Berkowicz, S.L. Hazell, and A. Lee. 1987. Antibody to *Campylobacter pylori* in families of index children with gastrointestinal illness due to *C. pylori*. *Lancet*, 2:681–682.

Mitchell, H.M., A. Lee, and J. Carrick. 1989. Increased incidence of *Campylobacter pylori* infection in gastroenterologists: further evidence to support person-to-person transmission of *C. pylori*. *Scand. J. Gastroenterol.*, 24:396–400.

Mitchell, H.M., Y.Y. Li, P.J. Hu, Q. Liu, M. Chen, G.G. Du, Z.J. Wang, A. Lee, and S.L. Hazell. 1992. Epidemiology of *Helicobacter pylori* in southern China: identification of early childhood as the critical period for acquisition. *J. Infect. Dis.*, 166:149–153.

Mitchell, J.D., H.M. Mitchell, and V. Tobias. 1992. Acute *Helicobacter pylori* infection in an infant, associated with gastric ulceration and serological evidence of intra-familial transmission. *Am. J. Gastroenterol.*, 87:382–386.

Mizoguchi, H., T. Fujioka, and M. Nasu. 1999. Evidence for viability of coccoid forms of *Helicobacter pylori*. *J. Gastroenterol.*, 34(Suppl)11:32–36.

Moayyedi, P., and A.T. Axon. 1998. Is there a rationale for eradication of *Helicobacter pylori*? Cost-benefit: the case for. *Br. Med. Bull.*, 54:243–250.

Mobley, H.L. 1997. *Helicobacter pylori* factors associated with disease development [see comments]. *Gastroenterol.*, 113:S21–8.

Morgner, A., E. Bayerdorffer, A. Neubauer, and M. Stolte. 2000a. Gastric MALT lymphoma and its relationship to *Helicobacter pylori* infection: management and pathogenesis of the disease. *Microsc. Res. Tech.*, 48:349–356.

Morgner, A., E. Bayerdorffer, A. Neubauer, and M. Stolte. 2000b. Malignant tumors of the stomach. Gastric mucosa-associated lymphoid tissue lymphoma and *Helicobacter pylori*. *Gastroenterol. Clin. North Am.*, 29:593–607.

Morris, A., G. Lloyd, and G. Nicholson. 1986. *Campylobacter pyloridis* serology among gastroendoscopy clinic staff [letter]. *Lancet*, 339:619.

Morris, A., and G. Nicholson. 1987. Ingestion of *Campylobacter pyloridis* causes gastritis and raised fasting gastric pH. *Am. J. Gastroenterol.*, 82:192–199.

Najm, I., and R.R. Trussell. 1999. New and Emerging Drinking Water Treatment Technologies. In *Identifying Future Drinking Water Contaminants*. Washington, DC: National Academy Press.

Nakajima, S., D.Y. Graham, T. Hattori, T. Bamba. 2000. Strategy for treatment of *Helicobacter pylori* infection in adults. I. Updated indications for test and eradication therapy suggested in 2000. *Curr. Pharm. Des.*, 6:1503–1514.

Neiger, R., C. Dieterich, A. Burnens, A. Waldvogel, I. Corthesy-Theulaz, F. Halter, B. Lauterburg, and A. Schmassmann. 1998. Detection and prevalence of *Helicobacter* infection in pet cats. *J. Clin. Microbiol.*, 36:634–637.

Nomura, A., G.N. Stemmermann, P.H. Chyou, I. Kato, G.I. Perez-Perez, and M.J. Blaser. 1991. *Helicobacter pylori* infection and gastric carcinoma among Japanese Americans in Hawaii. *N. Engl. J. Med.*, 325:1132–1136.

Nowottny, U., and K.L. Heilmann. 1990. Epidemiology of *Helicobacter pylori* infection. *Leber Magen Darm.*, 20:180, 183–180, 186.

Nwokolo, C.U., J. Bickley, A.R. Attard, R.J. Owen, M. Costas, and I.A. Fraser. 1992. Evidence of clonal variants of *Helicobacter pylori* in three generations of a duodenal ulcer disease family. *Gut*, 33:1323–1327.

O'Donohoe, J.M., P.B. Sullivan, R. Scott, T. Rogers, M.J. Brueton, and D. Barltrop. 1996. Recurrent abdominal pain and *Helicobacter pylori* in a community-based sample of London children. *Acta. Paediatr.*, 85:961–964.

Olbe, L., L. Fandriks, A. Hamlet, and A.M. Svennerholm. 2000. Conceivable mechanisms by which *Helicobacter pylori* provokes duodenal ulcer disease. *Baillieres Best Pract. Res. Clin. Gastroenterol.*, 14:1–12.

Oliveira, A.M., D.M. Queiroz, G.A. Rocha, and E.N. Mendes. 1994. Seroprevalence of *Helicobacter pylori* infection in children of low socioeconomic level in Belo Horizonte, Brazil. *Am. J. Gastroenterol.*, 89:2201–2204.

Olmos, J.A., H. Rios, and R. Higa. 2000. Prevalence of *Helicobacter pylori* infection in Argentina: results of a nationwide epidemiologic study. Argentinean Hp Epidemiologic Study Group. *J. Clin. Gastroenterol.*, 31:33–37.

Osato, M.S., K. Ayub, H.H. Le, R. Reddy, and D.Y. Graham. 1998. Houseflies are an unlikely reservoir or vector for *Helicobacter pylori*. *J. Clin. Microbiol.*, 36:2786–2788.

Öztürk, H., M.E. Senocak, B. Uzunalimoglu, G. Hascelik, N. Buyukpamukcu, and A. Hicsonmez. 1996. *Helicobacter pylori* infection in symptomatic and asymptomatic children: a prospective clinical study. *Eur. J. Pediatr. Surg.*, 6:265–269.

Park, S.R., W.G. Mackay, and D.C. Reid. 2001. *Helicobacter* sp. recovered from drinking water biofilm sampled from a water distribution system. *Water Res.*, 35:1624–1626.

Parkin, D.M., P. Pisani, and J. Ferlay. 1999. Global cancer statistics. *CA Cancer J. Clin.* 49:33–64, 1.

Parsonnet, J. 1998. *Helicobacter pylori*. *Infect. Dis. Clin. North Am.*, 12:185–197.

Parsonnet, J., G.D. Friedman, D.P. Vandersteen, Y. Chang, J.H. Vogelman, N. Orentreich, and R.K. Sibley. 1991. *Helicobacter pylori* infection and the risk of gastric carcinoma. *N. Engl. J. Med.*, 325:1127–1131.

Patchett, S.E. 1998. *Helicobacter pylori* eradication cost-benefit: the case against. *Br. Med. Bull.*, 54:251–257.

Peach, H.G., N.E. Bath, and S.J. Farish. 1998. *Helicobacter pylori* infection: an added stressor on iron status of women in the community. *Med. J. Aust.*, 169:188–190.

Perez-Perez, G.I. 1997. Role of *Helicobacter pylori* infection in the development of pernicious anemia. *Clin. Infect. Dis.*, 25:1020–1022.

Perez-Perez, G.I., R.M. Peek, A.J. Legath, P.R. Heine, and L.B. Graff. 1999. The role of CagA status in gastric and extragastric complications of *Helicobacter pylori*. *J. Physiol. Pharmacol.*, 50:833–845.

Perez-Perez, G.I., D.N. Taylor, L. Bodhidatta, J. Wongsrichanalai, W.B. Baze, B.E. Dunn, P.D. Echeverria, M.J. Blaser. 1990. Seroprevalence of *Helicobacter pylori* infections in Thailand. *J. Infect. Dis.*, 161:1237–1241.

Perez-Perez, G.I., S.S. Witkin, M.D. Decker, and M.J. Blaser. 1991. Seroprevalence of *Helicobacter pylori* infection in couples. *J. Clin. Microbiol.*, 29:642–644.

Peura, D.A. 1997. The report of the Digestive Health Initiative International Update Conference on *Helicobacter pylori*. *Gastroenterol.*, 113:S4–S8.

Pretolani, S., T. Stroffolini, M. Rapicetta, F. Bonvicini, L. Baldini, F. Megraud, G.C. Ghironzi, F. Sampogna, U. Villano, F. Cecchetti, G. Giulianelli, M.L. Stefanelli, A. Armuzzi, F. Miglio, and G. Gasbarrini. 1997. Seroprevalence of hepatitis A virus and *Helicobacter pylori* infections in the general population of a developed European country (the San Marino study): evidence for similar pattern of spread. *Eur. J. Gastroenterol. Hepatol.*, 9:1081–1084.

Regev, A., G.M. Fraser, M. Braun, E. Maoz, L. Leibovici, and Y. Niv. 1999. Seroprevalence of *Helicobacter pylori* and length of stay in a nursing home. *Helicobacter*, 4:89–93.

Reifen, R., I. Rasooly, B. Drumm, K. Murphy, and P. Sherman. 1994. *Helicobacter pylori* infection in children. Is there specific symptomatology? *Dig. Dis. Sci.*, 39:1488–1492.

Replogle, M.L., S.L. Glaser, R.A. Hiatt, and J. Parsonnet. 1995. Biologic sex as a risk factor for *Helicobacter pylori* infection in healthy young adults. *Am. J. Epidemiol.*, 142:856–863.

Robertson, M.S., J.F. Cade, and R.L. Clancy. 1999. *Helicobacter pylori* infection in intensive care: increased prevalence and a new nosocomial infection. *Crit. Care Med.*, 27:1276–1280.

Roe, I.H., S.H. Son, H.T. Oh, J. Choi, J.H. Shin, J.H. Lee, and Y.C. Hah. 1999. Changes in the evolution of the antigenic profiles and morphology during coccoid conversion of *Helicobacter pylori*. *Korean J. Intern. Med.*, 14:9–14.

Roosendaal, R., E.J. Kuipers, J. Buitenwerf, C. van Uffelen, S.G. Meuwissen, G.J. van Kamp, and C.M. Vandenbroucke-Grauls. 1997. *Helicobacter pylori* and the birth cohort effect: evidence of a continuous decrease of infection rates in childhood. *Am. J. Gastroenterol.*, 92:1480–1482.

Rothenbacher, D., M.J. Blaser, G. Bode, and H. Brenner. 2000. Inverse relationship between gastric colonization of *Helicobacter pylori* and diarrheal illnesses in children: results of a population-based cross-sectional study. *J. Infect. Dis.*, 182:1446–1449.

Rothenbacher, D., G. Bode, G. Berg, R. Gommel, T. Gonser, G. Adler, and H. Brenner. 1998a. Prevalence and determinants of *Helicobacter pylori* infection in preschool children: a population-based study from Germany. *Int. J. Epidemiol.*, 27:135–141.

Rothenbacher, D., G. Bode, G. Berg, U. Knayer, T. Gonser, G. Adler, and H. Brenner. 1999. *Helicobacter pylori* among preschool children and their parents: evidence of parent-child transmission. *J. Infect. Dis.*, 179:398–402.

Rothenbacher, D., G. Bode, F. Peschke, G. Berg, G. Adler, and H. Brenner. 1998b. Active infection with *Helicobacter pylori* in an asymptomatic population of middle-aged to elderly people. *Epidemiol. Infect.*, 120:297–303.

Rothenbacher, D., G. Bode, T. Winz, G. Berg, G. Adler, and H. Brenner. 1997. *Helicobacter pylori* in out-patients of a general practitioner: prevalence and determinants of current infection. *Epidemiol. Infect.*, 119:151–157.

Rowland, M. 2000. Transmission of *Helicobacter pylori*: is it all child's play? *Lancet*, 355:332–333.

Rowland, M., and B. Drumm. 1998. Clinical significance of *Helicobacter* infection in children. *Br. Med. Bull.*, 54:95–103.

Sakamoto, K. 1997. Investigation of infection routes of *Helicobacter pylori* in health checks of residents by random sampling. *Kurume Med. J.*, 44:273–280.

Sarker, S.A., D. Mahalanabis, P. Hildebrand, M.M. Rahaman, P.K. Bardhan, G. Fuchs, C. Beglinger, and K. Gyr. 1997. *Helicobacter pylori*: prevalence, transmission, and serum pepsinogen II concentrations in children of a poor periurban community in Bangladesh. *Clin. Infect. Dis.*, 25:990–995.

Schutze, K., E. Hentschel, B. Dragosics, and A.M. Hirschl. 1995. *Helicobacter pylori* reinfection with identical organisms: transmission by the patients' spouses. *Gut*, 36:831–833.

Shahamat, M., U. Mai, C. Paszko-Kolva, M. Kessel, and R.R. Colwell. 1993. Use of autoradiography to assess viability of *Helicobacter pylori* in water. *Appl. Environ. Microbiol.*, 59:1231–1235.

Sherman, P.M., and R.H. Hunt. 1996. Why guidelines are required for the treatment of *Helicobacter pylori* infection in children. *Clin. Invest. Med.*, 19:362–367.

Simor, A.E., B. Shames, B. Drumm, P. Sherman, D.E. Low, and J.L. Penner. 1990. Typing of *Campylobacter pylori* by bacterial DNA restriction endonuclease analysis and determination of plasmid profile. *J. Clin. Microbiol.*, 28:83–86.

Sipponen, P. 1997. *Helicobacter pylori* gastritis—epidemiology. *J. Gastroenterol.*, 32:273–277.

Sitas, F., D. Forman, J.W. Yarnell, M.L. Burr, P.C. Elwood, S. Pedley, and K.J. Marks. 1991. *Helicobacter pylori* infection rates in relation to age and social class in a population of Welsh men. *Gut*, 32:25–28.

Smoak, B.L., P.W. Kelley, and D.N. Taylor. 1994. Seroprevalence of *Helicobacter pylori* infections in a cohort of US Army recruits. *Am. J. Epidemiol.*, 139:513–519.

Staat, M.A., D. Kruszon-Moran, G.M. McQuillan, and R.A. Kaslow. 1996. A population-based serologic survey of *Helicobacter pylori* infection in children and adolescents in the United States. *J. Infect. Dis.*, 174:1120–1123.

Stanghellini, V., L. Cogliandro, R. Cogliandro, R. De Giorgio, and R. Corinaldesi. 1997. Widespread eradication of *Helicobacter pylori*: a debate. *Helicobacter,* 2(Suppl)1:S77–80:S77–80.

Stark, R.M., G.J. Gerwig, R.S. Pitman, L.F. Potts, N.A. Williams, J. Greenman, I.P. Weinzweig, T.R. Hirst, and M.R. Millar. 1999. Biofilm formation by *Helicobacter pylori. Lett. Appl. Microbiol.*, 28:121–126.

Stolzenberg-Solomon, R.Z., M.J. Blaser, P.J. Limburg, G. Perez-Perez, P.R. Taylor, J. Virtamo, and D. Albanes. 2001. *Helicobacter pylori* seropositivity as a risk factor for pancreatic cancer. *J. Natl. Cancer. Inst.*, 93:937–941.

Taylor, J.S., and M. Wiesner. 1999. Membranes. In *Water Quality and Treatment: A Handbook of Community Water Supplies.* 5th ed. Edited by American Water Works Association. New York: McGraw-Hill.

Tee, W., J. Lambert, R. Smallwood, M. Schembri, B.C. Ross, and B. Dwyer. 1992. Ribotyping of *Helicobacter pylori* from clinical specimens. *J. Clin. Microbiol.*, 30:1562–1567.

Teh, B.H., J.T. Lin, W.H. Pan, S.H. Lin, L.Y. Wang, T.K. Lee, and C.J. Chen. 1994. Seroprevalence and associated risk factors of *Helicobacter pylori* infection in Taiwan. *Anticancer Res.*, 14:1389–1392.

Thomas, J.E., G.R. Gibson, M.K. Darboe, A. Dale, and L.T. Weaver. 1992. Isolation of *Helicobacter pylori* from human faeces. *Lancet,* 340:1194–1195.

Tompkins, L.S., and S. Falkow. 1995. The new path to preventing ulcers. *Science,* 267:1621–1622.

Tovey, F.I., and M. Hobsley. 1999. Is *Helicobacter pylori* the primary cause of duodenal ulceration? *J. Gastroenterol. Hepatol.*, 14:1053–1056.

Tytgat, G.N. 1995. Endoscopic transmission of *Helicobacter pylori. Aliment. Pharmacol. Ther.*, 9(Suppl)2:105–110.

Vaira D., J. Holton, M. Menegatti, C. Ricci, L. Gatta, A. Geminiani, M. Miglioli. 2000. Review article:invasive and non-invasive tests for *Helicobacter pylori* infection. *Aliment. Pharmacol. Ther.*, 14 Suppl 3:13-22.

Vaira, D., M. Miglioli, M. Menegatti, J. Holton, A. Boschini, M. Vergura, C. Ricci, P. Azzarone, P. Mule, and L. Barbara. 1995. *Helicobacter pylori* status, endoscopic findings, and serology in HIV-1-positive patients. *Dig. Dis. Sci.*, 40:1622–1626.

Valle, J., M. Kekki, P. Sipponen, T. Ihamaki, and M. Siurala. 1996. Long-term course and consequences of *Helicobacter pylori* gastritis. Results of a 32-year follow-up study. *Scand. J. Gastroenterol.*, 31:546–550.

van der Hulst, R.W.M., B. Koycu, J.J. Keller, M. Feller, E.A. Rauws, J. Dankert, and G.N.J. Tytgat. 1996. *H. pylori* reinfection after successful eradication analysed by RAPD or RFLP [Abstract]. *Gastroenterol.*, 110:A284.

Varsky, C.G., M.C. Correa, N. Sarmiento, M. Bonfanti, G. Peluffo, A. Dutack, O. Maciel, P. Capece, G. Valentinuzzi, and D. Weinstock. 1998. Prevalence and etiology of gastroduodenal ulcer in HIV-positive patients: a comparative study of 497 symptomatic subjects evaluated by endoscopy. *Am. J. Gastroenterol.*, 93:935–940.

Velazquez, M., and J.M. Feirtag. 1999. *Helicobacter pylori*: characteristics, pathogenicity, detection methods and mode of transmission implicating foods and water. *Int. J. Food Microbiol.*, 53:95–104.

Walsh, D., N. Goggin, M. Rowland, M. Durnin, S. Moriarty, and B. Drumm. 1997. One week treatment for *Helicobacter pylori* infection. *Arch. Dis. Child.*, 76:352–355.

Webb, P.M., T. Knight, J.B. Elder, D.G. Newell, and D. Forman. 1996a. Is *Helicobacter pylori* transmitted from cats to humans? *Helicobacter*, 1:79–81.

Webb, P.M., T. Knight, S. Greaves, A. Wilson, D.G. Newell, J. Elder, and D. Forman. 1994. Relation between infection with *Helicobacter pylori* and living conditions in childhood: evidence for person to person transmission in early life. *Br. Med. J.*, 308:750–753.

Webb, P.M., T. Knight, D.G. Newell, J.B. Elder, and D. Forman. 1996b. *Helicobacter pylori* transmission: evidence from a comparison with hepatitis A virus. *Eur. J. Gastroenterol. Hepatol.*, 8:439–441.

Webberley, M.J., J.M. Webberley, D.G. Newell, P. Lowe, and V. Melikian. 1992. Seroepidemiology of *Helicobacter pylori* infection in vegans and meat-eaters. *Epidemiol. Infect.*, 108:457–462.

West, A.P., M.R. Millar, and D.S. Tompkins. 1992. Effect of physical environment on survival of *Helicobacter pylori*. *J. Clin. Pathol.*, 45:228–231.

Wilhoite, S.L., D.A. Ferguson, D.R. Soike, J.H. Kalbfleisch, and E. Thomas. 1993. Increased prevalence of *Helicobacter pylori* antibodies among nurses. *Arch. Intern. Med.*, 153:708–712.

Willen, R., B. Carlen, X. Wang, N. Papadogiannakis, R. Odselius, and T. Wadstrom. 2000. Morphologic conversion of *Helicobacter pylori* from spiral to coccoid form. Scanning (SEM) and transmission electron microscopy (TEM) suggest viability. *Ups. J. Med. Sci.*, 105:31–40.

Yamashita, Y., T. Fujisawa, A. Kimura, and H. Kato. 2001. Epidemiology of *Helicobacter pylori* infection in children: a serologic study of the Kyushu region in Japan. *Pediatr. Int.*, 43:4–7.

CHAPTER 12

Microsporidia in Drinking Water

by Martha A. Embrey

Executive Summary

Problem Formulation

Occurrence of Illness

- Much knowledge available on microsporidia-related illness is in the form of case reports. Serological testing is still being developed; however, there is one study that reported a seroprevalence for *Encephalitozoon* spp. as 8% in Dutch blood donors and 5% in French pregnant women. Microsporidial infections have been estimated to affect between 7 and 50% of AIDS patients, though a seroprevalence of *E. bieneusi* in children with AIDS was shown to be 2%.

Role of Waterborne Exposure

- Microsporidia have been loosely associated with one outbreak from water. They have been isolated from surface water and groundwater, raw sewage, and sewage effluent, so the risk of waterborne exposure exists.

Degree of Morbidity and Mortality

- In the immunocompromised (especially AIDS patients), microsporidiosis causes mainly chronic diarrhea with wasting syndrome. However, microsporidial infection can occur any place in the body (e.g., eyes, sinuses, and kidneys). Mortality is difficult to measure because those affected are generally end-stage AIDS patients, who have a myriad of other opportunistic infections. Patients with disseminated disease almost always die, though the exact cause may be unclear. The disease appears to be uncommon in the immunocompetent, but infections in this group have been associated with self-limited diarrhea and keratoconjunctivitis that is not severe.

Detection Methods in Water/Clinical Specimens

- Light microscopy and transmission electron microscopy (the gold standard necessary for speciation) are used for clinical samples; PCR is most often used for water samples. PCR, immunofluorescence techniques, enzyme-linked immunosorbent assays, and Western blot assays are under research for clinical use. Cell culture, though possible for some species, is not practical.

Mechanisms of Water Contamination

- Human fecal matter/sewage and probably animal fecal matter.

Concentrations at Intake

- Dependent on level of human and possibly animal fecal contamination in the source water. In published reports, 22 of 39 samples of surface water and 7 of 14 samples of surface, groundwater, and sewage effluent samples were positive for pathogenic microsporidia; the concentrations were not given.

Efficacy of Water Treatment

- One study showed that standard chlorine disinfection was effective against *E. intestinalis*; however, their small size may decrease the efficacy of sand filtration. Coagulation and flocculation may decrease their concentration. Membrane filtration technology has the potential to remove virtually all microbial pathogens.

Survival/Amplification in Distribution

- Unknown.

Routes of Exposure

- Ingestion (fecal–oral) and probably inhalation, inoculation, and sexual contact.

Dose Response

Infectious dose

- In cell culture, *E. intestinalis* was shown to have a $TCID_{50}$ of between 800 and 900 spores. It is not known how that extrapolates to human infectious doses.

Probability of Illness Based on Infection

- Asymptomatic infection occurs in immunocompetent and immunosuppressed hosts, but the rate is unknown.

Efficacy of Medical Treatment

- No standard treatment exists for any microsporidial infection. Treatments have been based on those used for protozoal infections; the best results in AIDS patients have come from antiretroviral therapy.

Secondary Spread

- Unknown but likely a possibility.

Chronic Sequelae

- Chronic diarrhea and wasting disease is common in immunosuppressed patients with microsporidiosis.

Introduction

"Microsporidia" is a nontaxonomic designation for the phyla Microspora, which are obligate intracellular parasites. Microsporidia had been classified as ancient eukaryotes, because some RNA sequences are smaller than those found in prokaryotes and because they lack mitochondria but they also display features of prokaryotes (Didier, Snowden, and Shadduck 1998; Mathis 2000; Weiss and Keohane 1997). However, recent analyses suggest that microsporidia should be reclassified from primitive protozoa to highly evolved organisms related to fungi (Mathis 2000; Vivares and Metenier 2000; Weiss 2000). With the new knowledge regarding this classification, microsporidia may serve as new intracellular model organisms. There are more than 140 genera and 1,200 species of microsporidia. To date, these genera have been demonstrated to be pathogenic in human hosts: *Enterocytozoon* (including the most common, *E. bieneusi*), *Encephalitozoon* (including *E. intestinalis,* formerly *Septata intestinalis*), *Nosema, Trachipleistophora, Brachiola,* and *Vittaforma* (Cali, Kotler, and Orenstein 1993; Mathis 2000; Weber et al. 1994; Weiss 2000; Yachnis et al. 1996). New microsporidial pathogens such as *Brachiola vesicularum* are just beginning to be recognized and classified as human pathogens (Cali et al. 1998). Unclassified or incompletely characterized microsporidial organisms are placed in the universal genus *Microsporidium.*

Microsporidia usually range in size from 1–3 μm, compared with *Giardia* cysts that are 9–12 μm long, and *Cryptosporidia* oocysts at 2–4 microns (Collins and Wright 1997; Smith 1997). Microsporidia have three stages in their life cycle: (1) the environmentally resistant spore; (2) merogony, asexual reproduction characterized by repeated karyokinesis prior to cytokinesis; and (3) the sporogony stage, in which it produces additional infectious spores (Haas, Rose, and Gerba 1999). The high environmental resistance of the spore stage may be attributable to a chitinous outer layer surrounding the cyst (Didier, Snowden, and Shadduck 1998). Once the spore, or cyst, enters the host, either through ingestion or possibly inhalation or inoculation, the organism enters the merogony stage. The best data on life cycles and infection sites are on *E. bieneusi* and *E. intestinalis,* because they are the most common infectious species. The spore infects the epithelial cells of the intestinal lining via a polar filament that pierces the new host cell, then enters the cell and completes its merogony and sporogony cycles (Magaud, Achbarou, and Desportes-Livage 1997; Mathis 2000). Infection then spreads to neighboring cells or to other areas, possibly by infecting macrophage cells (Kotler and Orenstein 1998; Orenstein 1991).

Epidemiology

Microsporidial spores have been isolated from many parts of the body. Several species have been found in sputum, bronchoalveolar lavage fluid, and epithelium tissue, making infection through inhalation a distinct possibility (Orenstein 1991; Weber et al. 1994). Spores are most often isolated in stool and intestinal tissue, typical of the horizontal infection path common to parasites that are ingested and infect the gastrointestinal tract. *E. bieneusi* and *E. intestinalis* are the most commonly identified species in the intestinal tract. Spores have also been detected in urine, which has not been reported with protozoal infections but is consistent with finding microsporidial infections in the kidney and the genitourinary tract (Schwartz et al. 1994; Wittner, Tanowitz, and Weiss 1993).

Immunocompromised Hosts

Microsporidia have numerous clinical manifestations in humans, and some species disseminate into systemic infection. Infections occur in patients with and without immunocompromising conditions, though the preponderance of disease is associated with patients with a severely depleted immune system (Navin et al. 1999). Data suggest that the incidence of microsporidial infections is increasing, but this information could be due to greater awareness of the disease and the refinement of diagnostic techniques. Microsporidia were first associated with diarrheal syndromes in AIDS patients, as reported by Desportes and colleagues in 1985. Microsporidiosis may be one of the most common causes of AIDS-related diarrhea (Orenstein 1991).

Information concerning the incidence of microsporidial infections examines cases either at endemic levels or on a case-specific basis. Some of the species identified as human pathogens, such as *Trachipleistophora hominis* and *Vittaforma corneae,* have only been reported in a few patients. The literature has generally reported the most common microsporidial infections, *E. bieneusi* and *E. intestinalis,* in adults with severely depleted CD4 counts. The standard detection method using transmission electron microscopy (TEM) on biopsy tissue is not practical for epidemiological studies. As a consequence, the development of the polymerase chain reaction (PCR) methodology should be valuable in learning about the epidemiology of microsporidiosis (Coyle et al. 1994; Talal et al. 1998).

Two of the earliest case reports of microsporidiosis occurred in severely immunodeficient children (Margileth et al. 1973; Weber and Bryan 1994). A later study examined the seroprevalence of microsporidiosis in HIV-positive children in Spain and found only a 2% seroprevalence, compared with 14.4% of children positive for *Cryptosporidium* and 8.4% for *Giardia* (del Aguila et al. 1997). (Seroprevalence methodologies are still being developed.) The species of microsporidia found was *E. bieneusi,* consistent with the species most commonly identified in adult patients with HIV-related disease. The mean CD4 count in these study children was 504.7/mm^3, which is significantly higher than that described for adults with microsporidiosis. The authors of the study called for new epidemiological studies to further determine prevalence and risk factors in this population.

Microsporidiosis is not limited to people whose immune status is diminished by HIV. Several of the most disseminated infections have occurred in patients with severely compromised immune systems due to other conditions (Wanke, DeGirolami, and Federman 1996). Examples of this include a woman who had an allogeneic marrow transplant (Kelkar et al. 1997), an athymic child (Margileth et al. 1973), and an immune-deficient male patient who developed one of the rare cases of myositis (Cali et al. 1997). Kelkar and colleagues (1997) concluded that transplant patients must also be considered at risk for microsporidiosis, and supporting reports have been published (Cotte et al. 1999; Goetz et al. 2001; Metge et al. 2000). No studies to date have explored the specific susceptibilities of immunocompromised groups other than patients with HIV infection.

Immunocompetent Hosts

For humans with intact immune systems, the most common form of microsporidiosis seems to be persistent traveler's diarrhea (Albrecht and Sobottka 1997; Raynaud et al. 1998; Sowerby 1995). Though ocular infections from *V. corneae, Nosema* spp., and *Microsporidium* spp. have also been reported in the literature (Font et al. 2000; Weber and Bryan 1994). Including diagnostic tests for microsporidia with those for coccidia (*Cryptosporidium* and *Cyclospora*) would help confirm the role of microsporidia as pathogenic agents in traveler's diarrhea (Deluol et al. 1994).

A study by van Gool and colleagues in 1997 looked at the seroprevalence of Dutch blood donors and pregnant French women for *E. intestinalis.* Enzyme-linked immunosorbent assay (ELISA), immunofluorescence assay (IFA) technique, and counterimmunoelectrophoresis (CIE) methods were used with *E. intestinalis* antigen. The results showed that 8% of the Dutch and 5% of the French were seropositive for *Encephalitozoon* spp. The results showed no differences between men and women, and the prevalence was highest in people in their forties and fifties, which suggests that the disease may not be common in childhood. The study was, however, unable to draw further epidemiological conclusions due to the sample sizes and limits of the methods used. The techniques were not specific enough to differentiate between *E. hellem* and *E. intestinalis*. From the data collected, the researchers could not draw conclusions as to the incidence of disease—only seroprevalence. Though research is under way to improve methods, serological testing for microsporidia has a history of unreliability (van Gool et al. 1997; Weber et al. 1994). The development and continuing refinement of serological testing for microsporidia is needed to provide data that are more reliable for study.

A year-long study in Sweden compared 851 adults with diarrhea and 203 healthy controls (Svenungsson et al. 2000). Microsporidia was isolated from stools less than 1% of the time, but PCR was not used as the detection method. In 1,454 patients whose stool samples were tested for microsporidia, of the 338 who had positive stool samples, 261 patients were HIV positive, 16 were transplant patients, and 61 were assumed immunocompetent (Cotte et al. 1999). Again, staining with standard microscopy was the detection method, so a more sensitive methodology like PCR would have likely yielded more positive results.

Clinical

Microsporidiosis is known primarily as an opportunistic infection in patients with AIDS. Microsporidial infections are estimated to affect between 2 and 50% of all AIDS patients (Weiss and Keohane 1997). Though infections are usually localized, *Encephalitozoon* spp. and *E. bieneusi* can cause infections that disseminate widely throughout a host (Didier et al. 1996a). Microsporidial infections have been found in almost every system of the human body (Orenstein et al. 1997). The dissemination possibly occurs as macrophages are infected and then circulate through the system, or possibly in free blood (Shadduck and Orenstein 1993). Based on this information, it is not surprising that many clinical syndromes are associated with microsporidiosis including keratoconjunctivitis, sinusitis, encephalitis, nephritis, hepatitis, osteomyelitis, and myositis. However, chronic diarrhea and wasting disease (i.e., malabsorption, weight loss, and fever) are the primary conditions associated with the disease. Chronic diarrhea is usually defined as an episode that lasts more than 28 days (Thielman and Guerrant 1998).

The major risk factor for severe acute and chronic microsporidial infections is an extreme depletion of CD4 cells, often as low as 15–30 cells/mm^3 (Kotler and Orenstein 1994). AIDS patients with higher CD4 counts have also demonstrated asymptomatic infections (Deluol et al. 1994). A study of 106 HIV-infected patients showed a 33% occurrence rate for *E. bieneusi* in intestinal biopsies regardless of CD4 count (Rabeneck et al. 1993). Sowerby et al. (1995) performed a small study examining patients with chronic diarrhea who had CD4 counts greater than 200/mm^3, and three of the 31 patients had microsporidiosis. Any patient with chronic, unexplained diarrhea should consider microsporidiosis as an etiology.

Finally, data on how microsporidia affect the immunocompetent are almost nonexistent; therefore, references to clinical disease are limited mainly to reports on the immunocompromised, particularly AIDS patients. Also, based on few seroprevalence data and case reports, we know that microsporidial infections can be asymptomatic in both immunocompetent and immunocompromised hosts. What percentage of the infected in either group goes on to develop clinical symptoms is unknown.

Enteric Infections

Coyle et al. (1996) used PCR in a study of 111 AIDS patients with and without diarrhea. Of the patients with diarrhea, 44% were positive for *E. bieneusi* or *E. intestinalis,* compared with 2.3% of patients without diarrhea. The mean CD4 lymphocyte count of the patients with diarrhea was 21/mm^3, compared with 60/mm^3 in the patients without diarrhea. This confirms earlier findings that most patients with microsporidiosis have a severely depressed immune system. This study was one of the first to demonstrate that rRNA primers used in PCR were sensitive and specific when used in a clinical setting. Microsporidia accounted for 50% of nine AIDS patients with unexplained diarrhea and D-xylose malabsorption in a study by Molina et al. (1993). Careful analysis by Kotler and Orenstein (1994), using intestinal biopsies, demonstrated an incidence of 33% in AIDS patients with chronic diarrhea.

E. bieneusi may play a roll in cholangitis, another common late complication of HIV infection. Cholangitis in patients with AIDS can be severe and is often associated with biliary tract infection by *Cryptosporidium* and cytomegalovirus. However, in about 40% of cases, no etiologic agent is identified. Pol et al. (1993) found that all eight patients in their study with unexplained cholangitis demonstrated biliary tract infection with microsporidia. The cases of cholangitis were all associated with severe immune deficiency.

In the immunocompetent host, intestinal microsporidial infections can result in self-limiting diarrheal illness that may occasionally become prolonged (Fournier et al. 1998; Gainzarain et al. 1998; Wanke, DeGirolami, and Federman 1996). Standard diarrheal treatments are used to control symptoms, and the primary complication associated with the infection—dehydration—is treated with standard practices (DuPont and Ericsson 1993).

Ocular Infections

Ocular infections of microsporidia usually remain confined to conjunctive tissue; however, there was a case report of a corneal infection (Font et al. 2000). *V. corneae* and *Nosema* spp. are the species reportedly most responsible for ocular infections. These infections have occurred mostly in the immunocompromised but also in some immunocompetent hosts. As previously mentioned, infections in immunocompetent hosts have usually involved trauma to the eye. Infections may occur more readily in this area due to a lack of localized immune response. Ocular infections due to *Encephalitozoon* spp. occur primarily in patients infected with HIV and are usually found in ocular epithelial tissue (Weber and Bryan 1994). Concomitant urinary tract infections often accompany ocular *Encephalitozoon* infections (Schwartz et al. 1994).

Systemic and Other Infections

Infections of the liver, biliary tract, kidneys, and respiratory system have been detected through thorough microscopic examinations of samples from the respiratory system, the genitourinary system, and the conjunctiva, as well as postmortem examinations of severely immunocompromised patients (Gatti et al. 1997; Schwartz et al. 1992; Tanowitz et al. 1996). These infections have included not only *E. bieneusi* and *E. intestinalis* but also *E. hellem* and *E. cuniculi*. It is presumed that these infections begin in the gastrointestinal tract, or perhaps through aerosol contact and then disseminate widely through the body (Weiss and Keohane 1997).

Mortality

Mortality rates for microsporidiosis are not well documented. The muscular and disseminated microsporidiosis almost always result in death, and intestinal infections are often co-factors in final illnesses for people with severely compromised immune systems (Weber et al. 1994; Yachnis et al. 1996). The problems with quantifying the rates of mortality are twofold: (1) The lack of epidemiology for microsporidiosis makes it inherently difficult to track, and (2) patients with severe microsporidiosis have such crippled immune systems that they often have systemic shutdowns and other opportunistic infections, making it difficult to determine their exact cause of death (Dunand et al. 1997).

Medical Treatment

No standard treatment exists for any of the microsporidial infections (USPHS/IDSA 1999). Current treatment protocols have been based primarily on those used for protozoal infections. Reports examining the use of antiretroviral therapies in patients with HIV to treat persistent cases of microsporidiosis have been positive; however, it appeared that the parasites were not completely eliminated in all patients and relapses occurred as soon as patients' CD4 counts declined (Carr et al. 1998).

Enteric Infections

Microsporidiosis symptoms occur because of the injury and atrophy to the villius cells in the epithelial portion of the small intestine. Microsporidia inject a polar tube into the cell to enter. It is the trauma of the continuing infection that is believed to cause chronic diarrhea. In addition to chronic diarrhea, D-xylose, fat, and sometimes B-12 malabsorption are prevalent in these patients, and progressive weight loss is often noted in association with the above symptoms (Kotler and Orenstein 1998).

In immunocompetent patients, the gastrointestinal illness is usually self-limiting and is allowed to run its course. Most studies indicate a "spontaneous" clearance of the organism (Fournier et al. 1998). Antidiarrheal medications, such as loperamide or bismuth, may relieve symptoms (DuPont and Ericsson 1993). Immunocompromised patients face a much greater challenge. Treatment of symptoms does not rid the system of the organisms for the majority of microsporidia species. Albendazole, a microtubule inhibitor, has been used successfully with some microsporidian species, most notably *E. intestinalis* and *E. cuniculi* (Didier 1998; Weiss et al. 1994). Patients who have successfully eliminated symptoms with treatment, however, may still excrete spores. A double-blind, placebo-controlled trial of albendazole (400 mg twice daily for 3 weeks) cleared *E. intestinalis* infection and, when taken as a prophylaxis, reduced the risk of relapses (Molina et al. 1998). *E. hellem* and *E. cuniculi* have demonstrated in vitro sensitivity to albendazole as well (Ditrich, Kucerova, and Koudela 1994; Weiss et al. 1994). The primary treatment for microsporidial infections is albendazole, with concurrent therapies to treat symptoms such as parenteral nutrition to combat malnutrition and octreotide to treat chronic diarrhea (Didier 1998; Kotler and Orenstein 1998).

E. bieneusi has been generally unresponsive to any of the antimicrobial therapies used to date. Dieterich et al. (1994) reported that albendazole did reduce the symptoms of *E. bieneusi* infection in some patients but did not eliminate the infection. Fumagillin has recently been used in HIV-infected patients with *E. bieneusi*. Molina et al. (2000) found that 72% (8/11) HIV-positive patients treated with 60 mg/day of fumagillin for 14 days cleared their *E. bieneusi* infections and had not relapsed after almost a year.

Antiretroviral therapy has been shown to be effective against many opportunistic infections in AIDS patients, including microsporidiosis, and is probably the best treatment available for AIDS-related microsporidiosis. Maggi and colleagues (2000) found that therapy resulting in an increased CD4 cell count resolved diarrhea in 50 patients positive for cryptosporidiosis or microsporidiosis. Other similar studies of nine and six patients had the same results (Carr et al. 1998; Miao et al. 2000).

Ocular Infections

Antimicrobial drugs and sulfonamides have had some success in clearing infections. Debulking the cornea in addition to chemotherapy has also been effective. Clearance of *E. hellem* infection has been achieved using 200 mg/day of itraconazole or 400 mg/day of albendazole (Rossi et al. 1999). Similar treatments are used for the other species affecting the eye. Topical fumagillin has also been used but may need to be followed by systemic albendazole (Weber et al. 1994).

Systemic and Other Infections

Limited treatment is available for most systemic and disseminated microsporidiosis infections, and the majority of them have been found through postmortem examinations. Molina et al. (1995) did find that five patients with systemic *E. (Septata) intestinalis* infection responded well to albendazole (400 mg twice a day) but encountered a problem with recurrence. Only a few patients have presented with myositis and they had been infected by a number of different microsporidian species. Several of these species, including *Nosema* spp. and *Trachipleistophora* spp., have been described. Myositis occurring locally and in conjunction with systemic infections has been described as well (Cali et al. 1997). It might be inferred that aggressive chemotherapy with albendazole or other antimicrobials could be used if a diagnosis was made soon enough. No data have been published regarding antiretroviral therapy and systemic disease.

Albendazole is also the treatment of choice for sinusitis due to microsporidia (Kotler and Orenstein 1998).

Chronic Sequelae

Microsporidiosis is a newly identified infection, and there is scarce literature available on the long-term effects of the disease. Most patients with known microsporidiosis have had severely depleted immune systems, thus high rates of morbidity and mortality are associated with the disease. Microsporidiosis is associated with patients with chronic, rather than acute, diarrheal episodes (Gainzarain et al. 1998; Navin et al. 1999). In a retrospective look at 73 HIV-positive patients infected with enteric microsporidiosis, after 6 months, 54.8% had diarrhea persisting and 51.2% reported weight loss (Dascomb et al. 1999). Risk factors included high HIV viral load without any antiretroviral therapy. In patients with HIV, and even for a very few without, other chronic sequelae in addition to diarrhea involve wasting and D-xylose malabsorption (Weiss and Keohane 1997). This syndrome results in substantial weight loss, which can last over a period of several months to more than a year (Shadduck and Orenstein 1993). In immunocompetent patients, microsporidial diarrhea appears to be self-limiting (Lopez-Velez et al. 1999).

Transmission

There are no known exposure or transmission risk factors for microsporidia (Didier 1998). Because they are classified as enteric, obligate parasites, researchers theorized that many infections are transmitted through the same pathways as protozoa: ingestion being the most common, direct oral–fecal contamination, and possibly by aerosol inhalation (Hutin et al. 1998; Weber et al. 1993; Wittner, Tanowitz, and Weiss 1993). How their reclassification to a fungi or type of organism other than protozoa changes these interpretations is yet to be seen.

There has been one reported outbreak of microsporidiosis (Cotte et al. 1999). An ongoing French surveillance study claimed to uncover an outbreak by reviewing retrospective data, but the data's association with drinking water exposure was weak (Hunter 2000). Sexual transmission may also be another mode of infection (Weiss and Keohane 1997). In corneal infections, trauma often precedes the infection in immunocompetent patients (Didier, Snowden, and Shadduck 1998), which suggests inoculation as a transmission pathway. It has also been suggested that some microsporidia, such as *E. bieneusi*, might be a natural human parasite; if the immune system weakens, the infection may be "activated" to cause disease (Mathis 2000; Weber and Bryan 1994). In animals, some species infect offspring vertically through the placenta (Didier, Snowden, and Shadduck 1998; Mathis 2000).

Data for sources and modes of transmission are insufficient to draw specific conclusions regarding the risk factors for acquiring microsporidial infection. A US Public Health Service report on the prevention of opportunistic infections for people with HIV (USPHS/IDSA 1999) gives no specific guidance on the prevention of microsporidiosis. However, in a single study, swimming was identified as a risk factor for people with HIV (Hutin et al. 1998). Some humans have asymptomatic microsporidial infections, even those with suppressed immune systems (Deluol et al. 1994; Rabeneck et al. 1995), though the rates are unknown. Data collected thus far are insufficient to estimate the extent of second-degree microsporidial infections.

Many vertebrates and invertebrates harbor microsporidia, and some of the species can infect humans. Although the possibility and extent of zoonotic transmission is not known, data continue to support certain species as zoonotic (see Table 12-1). All four of the major pathogenic microsporidian species have animal hosts (Mathis 2000). *E. bieneusi* was recently discovered in the feces of pigs, cattle, rabbits, dogs, and cats, and *E. cuniculi* is known to infect a number of vertebrate species as well as humans (Cali et al. 1998; Deplazes et al. 1996; Mathis 2000). Domestic dogs and cats have also been found to excrete human pathogenic microsporidia (Didier et al. 1996b; Rinder et al. 2000). For protozoa, the zoonotic reservoirs in the environment are widely known, including, but not limited to, raccoons, muskrats, and beaver for *Giardia* and cattle, other ruminants, and birds for *Cryptosporidium*, as well as house pets for both (Smith 1997). Extrapolating from data collected on protozoa, it is likely that animal reservoirs for microsporidia exist in the environment and may play a role in the transmission of disease.

Table 12-1: Microsporidia infection of animal and human hosts

Species	Hosts	Infection Sites
Enterocytozoon bieneusi	Humans, mammals	Small intestine, biliary tract, respiratory system
Encephalitozoon intestinalis (previously *Septata intestinalis*)	Humans, mammals	Small intestine, respiratory system, disseminated
Encephalitozoon hellem	Humans, birds	Disseminated, ocular epithelial tissue
Encephalitozoon cuniculi	Humans, mammals	Liver, pancreas, brain
Nosema algerae	Humans	Corneal stroma
Nosema ocularum	Humans	Corneal stroma
Vittaforma corneae (previously *Nosema corneum*)	Humans	Corneal stroma, disseminated
Pleistophora spp.	Humans	Skeletal muscle
Trachipleistophora spp.	Humans	Skeletal muscle, nasal tissue, disseminated

Adapted from Mathis 2000; Didier, Snowden, and Shaddock 1998.

Travel to tropical or developing regions has been cited as a potential exposure risk, specifically for immunocompetent people (Fournier et al. 1998; Sandfort et al. 1994). In a 1983 study, the World Health Organization tested 115 travelers to the tropics and 49 nontraveling controls for seroprevalence of *E. cuniculi*. Fourteen of the travelers had positive results for microsporidia, compared with none of the controls. Ten percent of travelers to the tropics were diagnosed with microsporidiosis after suffering persistent diarrhea (Lopez-Velez et al. 1999); Muller and colleagues (2001) also found that 9 of 148 travelers returned from the tropics with microsporidial infections.

Dose Response

Little information is available on the dose response of microsporidia in humans.

Infectious Dose

The only infectious dose data available are estimated from a rabbit kidney cell-culture system. The 50% tissue culture infective dose for *E. intestinalis* is approximately 800 to 900 spores (Wolk et al. 2000). How this data extrapolates to infectivity in humans is unknown. Sources and modes of transmission need further investigation and study before this can be determined. For comparison, the infectious dose for *Giardia* is estimated to be 10–100 cysts (Rose, Haas, and Regli 1991). The infectious dose for *Cryptosporidia* is estimated to be 132 oocysts based on volunteer feeding studies (DuPont et al. 1995).

Environmental Occurrence

Researchers assume that microsporidia are waterborne pathogens similar in behavior and resistance to *Cryptosporidium* and *Giardia*. There is an abundance of data regarding the presence of *Giardia* and *Cryptosporidium* in the environment, but to date, only a few studies have been published regarding the presence of microsporidia in water or the environment. Their study in water is made more difficult by their small size, as well as the difficulty of sorting out the human pathogenic genera from all of the others, some of which may only affect fish or insects. The need for testing mechanisms that do not involve TEM was identified as a key factor for possibly identifying human pathogenic microsporidia in water supplies (Dowd et al. 1999).

A small study in France found human pathogenic microsporidia, namely *E. bieneusi*, but also *V. cornea* and *Pleistophora* spp. in surface water from locations on two separate rivers (Sparfel et al. 1997). Samples of spiked filters, using known species of microsporidia, were first used to test the efficacy of the sampling method. Three sequential filters were used, the smallest having a 0.4-μm pore size. The extracted material was then examined using light microscopy and PCR testing using known microsporidial DNA sequences. The test was able to detect as few as 1–3 spores/2 L of water. Five 200- to 400-L samples were then taken from different surface water sources near Paris. Using the same filtration method described earlier, they examined the field samples. Though the detection of spores using light microscopy was unsuccessful, PCR amplification positively identified several microsporidia species. This was the first study to document the presence of human pathogenic microsporidia in the environment. In a follow-up study, Fournier and colleagues (2000) collected water samples from the Seine River over 1 year. Each of the 25 samples was analyzed using the methods described earlier. Using PCR, 64% of the samples were positive; however, only one of those was identified as *E. bieneusi*. Eight other samples were similar genetically to *V. corneae* or *Pleistophora* spp. The rest were unidentifiable microsporidial species.

In a study by Dowd and colleagues (1998), raw sewage and raw and treated water were tested for the presence of *E. intestinalis*, *E. bieneusi*, and *V. corneae* using PCR. They used methods similar to those used in the Sparfel study. Water samples from 14 locations were concentrated, then evaluated using DNA extraction, followed by PCR amplification and sequencing for specific microsporidia species. The sequence analysis was completed using a National Center for Biotechnology database. The test found *E. intestinalis* in 5 of the 14 samples—in surface water and groundwater, raw sewage, and tertiary effluent. *V. corneae* was found in one sample of tertiary effluent and *E. bieneusi* was found in one sample of surface water. The significance of finding microsporidia in the tertiary effluent sample may be an indication that microsporidia can survive water treatment practices—though PCR would not be able to measure viability. Finding *E. intestinalis* in surface water may also lend further credence to the idea of animal reservoirs for microsporidia. Overall, the findings of these studies take a step closer to linking human pathogenic microsporidia with waterborne disease.

Detection Methods

Because microsporidia have only recently been identified as a possibly waterborne pathogen, the detection methods in both patients and in the environment are still evolving.

Clinical

Diagnosis for microsporidia, as with protozoa, is difficult (Collins and Wright 1997). Microsporidia require specific tests, because the standard parasite test protocols do not screen for these organisms (Goodgame 1996). Furthermore, identification can be approached on two levels: (1) the general diagnosis of microsporidiosis and (2) a species-specific identification (Cali 1994). Information such as the site of infections, spore size, and ultrastructural details of the spore morphology can facilitate diagnosis (Cali et al. 1994); however, it should also be noted that at this time, the diagnosis of microsporidiosis is labor intensive, regardless of the method used (Owen 1997).

For gastrointestinal infections, the first step is to take a stool sample from patients with long-term, chronic diarrhea (Bryan and Weber 1993). Chromotrope, Giemsa, or fluorescent chitin staining and examination by light microscopy can be an efficient first step in finding the etiologic agent (Conteas et al. 1996; Weber et al. 1992). These techniques may be used on tissue samples as well (Giang et al. 1993; Weber and Bryan 1994). New techniques are being developed in order to improve quick, efficient light microscopic diagnosis of microsporidiosis (Thielman and Guerrant 1998; Velasquez et al. 1999). TEM on tissue samples is the gold standard for diagnosis and is necessary to determine speciation (Conteas, Didier, and Berlin 1997). The absence of organisms from an initial screening may indicate the need for more invasive diagnostic procedures, such as biopsy (Orenstein 1991).

The protocol for using endoscopy for patients with chronic diarrhea is controversial, but for patients with CD4 counts less than 200 cells/mm^3 and with diarrhea that may or may not be controlled with medication, biopsy is a consideration (Simon, Kotler, and Brandt 1996). While looking at tissue samples from intestinal biopsy, one of the clearest indications of the infection (other than the organism itself) is the villius flattening (Shadduck and Orenstein 1993). Biopsy is used not only for the diagnosis of enteric microsporidia but for ocular, muscular, and other infection sites. Lavage and ocular scrapings can also be used for initial screenings. For patients with extraintestinal microsporidiosis, lavage and biopsies of the respiratory system or other affected areas are key to diagnosis. Gatti et al.

(1997) demonstrated that of nine patients with confirmed extraintestinal microsporidiosis, all had stool samples negative for microsporidia. Some species have been found in urine samples using light microscopy, but electron microscopy is required for speciation (Cali et al. 1994). As with other "difficult" organisms, the success rate for diagnosis is highly dependent on the skill and experience of the lab doing the testing (Collins and Wright 1997).

Serological tests that can accurately indicate disease or the speciation for microsporidiosis have not been developed. The *Encephalitozoon* species often cross-react with one another (van Gool et al. 1997). Further development of monoclonal antibodies for use in Western blot testing may make serological testing a better possibility in the future (Croppo et al. 1994). Many species of microsporidia can be cultured, including *E. hellem*, *N. corneum*, and *E. intestinalis*, but *E. bieneusi* has no standard cell culture system (Snowden et al. 1998). This barrier poses further difficulty to developing serological tests. However, Weiss et al. (1992) did propose that the cross-reactivity of the *Encephalitozoon* spp. could allow the development of a Western blot test that might be able to detect *E. bieneusi*. Seroprevalence studies are not feasible at this time for immunodeficient patients, because these people are often unable to generate antibody responses (Didier 1998).

The advent of PCR tests will further advance the detection and recognition of microsporidia. DNA and RNA sequences collected for the human pathogenic species of microsporidia will allow for specific speciation, which is key to both identifying the correct treatment and for epidemiologic studies (Kock et al. 1996). DNA and RNA sequences are now available for most of the microsporidia species known to affect humans (Dowd et al. 1999; Weiss and Vossbrinck 1998). Both stool samples and biopsies have been examined using PCR, and the detection and speciation have been effective and accurate, having been compared with the results from electron microscopy of the same biopsy samples (Delbac and Vivares 1997; Muller et al. 2001). Researchers have developed an extraction-free, filter-based PCR protocol for microsporidia spores that requires less preparation and, therefore, less time and expense (Orlandi and Lampel 2000). Using this methodology, they were able to detect 10–50 *E. intestinalis* spores in a seeded stool sample. The ability to do accurate PCR testing on stool is a crucial step in finding noninvasive techniques for diagnosing species-specific microsporidiosis (Ligoury et al. 1997; Talal et al. 1998).

Environmental

Dowd and colleagues (1999) refined and evaluated detection methods of human pathogenic microsporidia species in water. This study compared IFA techniques for detection with those of PCR and also described two methods for isolating microsporidia DNA for PCR amplification. The IFA technique had only a 4.8% recovery rate for large volumes of water, comparable to what a water treatment plant would use. Furthermore, the IFA technique had many false-positive and false-negative results and was unable to react with *E. bieneusi*, which is one of the most important human pathogenic microsporidia. Finally, the light microscopy used with this technique was unable to confirm species, which is a crucial aspect of testing. IFA was deemed to have an unacceptable methodology and recovery rate.

In contrast, the PCR method could detect very small numbers of organisms in purified samples with either of the DNA isolation technique used. The problems encountered were the lack of extensive assessment for species determination, difficulty in making the determination, and the inability to demonstrate the viability of the isolated microsporidia (a problem common to any microbial detection using this method). The authors also discussed lack of available data on the efficacy of disinfection and survival rates for microsporidia. Dowd and coworkers (1999) demonstrated that detection and speciation of microsporidia from water samples can be done with PCR testing and that PCR could be used to assess the effectiveness of removal, if not inactivation. However, the study brings to light the need for more research in order to begin to answer these questions on environmental detection.

It is difficult to make suppositions based on one study. However, Dowd and colleagues base some of their conclusions on their test results in combination with the larger body of data found on protozoa such as *Cryptosporidia* and *Giardia* (Brasseur 1997). Their detection in groundwater may indicate that because their size is similar to enteric bacteria and viruses, microsporidia species may be capable of travelling through substrate to infect groundwater sources. The presence of the organism in surface water may also indicate that, similar to protozoa, microsporidia have animal reservoirs that may serve to contaminate surface water supplies.

Water Treatment

The microsporidian chitinous outer membrane provides excellent protection against the environment (Didier 1998); also, the size of microsporidia is similar to bacteria that "slip" through sand filters, though sedimentation and flocculation may reduce the number of organisms. Spores remain viable after 10 years in distilled water (Shadduck and Polley 1978). Koudela and coworkers (1999) found no significant reduction in infectivity of microsporidial spores stored in water at 4°C or frozen at –12°C and –24°C for up to 24 hours. However, heating the spores to 60°C for 5 minutes and 70°C for 1 minute completely deactivated them. Researchers conducted chlorine disinfection studies on *E. intestinalis* using spectrophotometric and hemacytometric counting methods, complemented by a cell-culture model (Wolk et al. 2000). Using these systems, they calculated the effectiveness of free chlorine at concentrations ranging from 0 to 10 mg/L at concentration times up to 80 minutes. They observed a 99.9% (3-log) reduction in infectivity at concentrations of 2 mg/L or higher for a minimum of 16 minutes. Based on these results, the authors concluded that the majority of water utilities provide abundant treatment to inactivate *E. intestinalis* spores. We still do not know the differences among types of microsporidia and whether some strains are more resistant than others. Further studies on the prevalence and possible reservoirs for microsporidia in raw and finished drinking water are needed before appropriate treatment techniques can be determined, but the current evidence supporting the efficacy of chlorine disinfection is compelling.

Membrane filtration technology, such as micro-, ultra-, and nanofiltration, as well as reverse osmosis, has the capacity to remove water contaminants down to the ion level. Microfiltration, with a nominal pore size of 0.2 μm, can remove bacterial and protozoal pathogens better than conventional water treatment, and ultrafiltration, with a nominal pore size of 0.01 μm, can achieve up to a 6-log pathogen removal (Najm and Trussell 1999; Taylor and Wiesner 1999). This technology will continue to gain in popularity as costs become more competitive and regulatory approval is assured.

Conclusions

Microsporidiosis is a newly identified disease that is prevalent primarily in people with AIDS. It most commonly manifests itself as chronic diarrhea and wasting disease in patients who have severely depleted immune systems (i.e., end-stage immunocompromising disease). In addition to the gastrointestinal system, microsporidial infection can spread throughout the body, causing any number of disease syndromes. Mortality is difficult to evaluate because the people who are most severely affected by infection have many other co-infections that are capable of causing death. Though immunocompetent people can become infected, little data exist on the course of infection and disease in them.

Microsporidia have been isolated in surface water and groundwater, raw sewage, and sewage effluent using PCR methods, though we have very little other information on its occurrence in the environment. Little is known about the effect of drinking water treatment on microsporidia, but one study indicates that microsporidia may be susceptible to chlorine disinfection. However, its relatively small size may make it more difficult to remove through filtration.

At this point, not enough data exist to define the role of water exposure in the transmission of microsporidia. In addition to ingestion, reports suggest that the organism can be transmitted through inhalation, sexual contact, and inoculation. It is not known which route is primary to infection in which populations.

Much more basic knowledge about the health effects (especially in the immunocompetent host), prevalence of infection, and occurrence in the environment is necessary before any conclusions can be drawn about the risk of microsporidia as a waterborne pathogen.

Bibliography

Albrecht, H., and I. Sobottka. 1997. *Enterocytozoon bieneusi* infection in patients who are not infected with human immunodeficiency virus. *Clin. Infect. Dis.*, 25:344.

Brasseur, P. 1997. Waterborne cryptosporidiosis: a major environmental risk. *J. Eukaryot. Microbiol.*, 44:67S–68S.

Bryan, R.T., and R. Weber. 1993. Microsporidia. Emerging pathogens in immunodeficient persons. *Arch. Pathol. Lab. Med.*, 117:1243–1245.

Cali, A. 1994. Human microsporidiosis: past and present, an overview of the Cleveland workshop. *J. Eukaryot. Microbiol.*, 41:69S.

Cali, A., D.P. Kotler, and J.M. Orenstein. 1993. *Septata intestinalis* N. G., N. sp., an intestinal microsporidian associated with chronic diarrhea and dissemination in AIDS patients. *J. Eukaryot. Microbiol.*, 40:101–112.

Cali, A., P.M. Takvorian, E.M. Keohane, and L.M. Weiss. 1997. Opportunistic microsporidian infections associated with myostitis. *J. Eukaryot. Microbiol.*, 44:86S–86S.

Cali, A., P.M. Takvorian, S. Lewin, M. Rendel, C.S. Sian, M. Wittner, H.B. Tanowitz, E. Keohane, and L.M. Weiss. 1998. *Brachiola vesicularum*, n. g., n. sp., a new microsporidium associated with AIDS and myositis. *J. Eukaryot. Microbiol.*, 45:240–251.

Cali, A., L. Weiss, P. Takvorian, H. Tanowitz, and M. Wittner. 1994. Ultrastructural identification of AIDS associated microsporidiosis. *J. Eukaryot. Microbiol.*, 41:24S.

Carr, A., D. Marriott, A. Field, E. Vasak, and D.A. Cooper. 1998. Treatment of HIV-1-associated microsporidiosis and cryptosporidiosis with combination antiretroviral therapy. *Lancet*, 351:256–261.

Collins, P.A., and M.S. Wright. 1997. Emerging intestinal protozoa: a diagnostic dilemma. *Clin. Lab. Sci.*, 10:273–278.

Conteas, C.N., E.S. Didier, and O.G. Berlin. 1997. Workup of gastrointestinal microsporidiosis. *Dig. Dis.*, 15:330–345.

Conteas, C.N., T. Sowerby, G.W. Berlin, F. Dahlan, A. Nguyen, R. Porschen, J. Donovan, M. LaRiviere, and J.M. Orenstein. 1996. Fluorescence techniques for diagnosing intestinal microsporidiosis in stool, enteric fluid, and biopsy specimens from acquired immunodeficiency syndrome patients with chronic diarrhea. *Arch. Pathol. Lab. Med.*, 120:847–853.

Cotte, L., M. Rabodonirina, F. Chapuis, F. Bailly, F. Bissuel, C. Raynal, P. Gelas, F. Persat, M.A. Piens, and C. Trepo. 1999. Waterborne outbreak of intestinal microsporidiosis in persons with and without human immunodeficiency virus infection. *J. Infect. Dis.*, 180:2003–2008.

Coyle, C.M., H.B. Tanowitz, D.P. Kotler, J.M. Orenstein, M. Wittner, A. Cali, and L.M. Weiss. 1994. Microsporidial diarrhea in AIDS. The use of the polymerase chain reaction for epidemiology. *Clin. Infect. Dis.*, 19:581–581.

Coyle, C.M., M. Wittner, D.P. Kotler, C. Noyer, J.M. Orenstein, H.B. Tanowitz, and L.M. Weiss. 1996. Prevalence of microsporidiosis due to *Enterocytozoon bieneusi* and *Encephalitozoon (Septata) intestinalis* among patients with AIDS-related diarrhea: determination by polymerase chain reaction to the microsporidian small-subunit rRNA gene. *Clin. Infect. Dis.*, 23:1002–1006.

Croppo, G.P., G.J. Leitch, S. Wallace, and G.S. Visvesvara. 1994. Immunofluorescence and western blot analysis of microsporidia using anti-*Encephalitozoon hellem* immunoglobulin G monoclonal antibodies. *J. Eukaryot. Microbiol.*, 41:31S.

Dascomb, K., R. Clark, J. Aberg, J. Pulvirenti, R.G. Hewitt, P. Kissinger, and E.S. Didier. 1999. Natural history of intestinal microsporidiosis among patients infected with human immunodeficiency virus. *J. Clin. Microbiol.*, 37:3421–3422.

del Aguila, C., R. Navajas, D. Gurbindo, J.T. Ramos, M.J. Mellado, S. Fenoy, M.A. Munoz Fernandez, M. Subirats, J. Ruiz, and N.J. Pieniazek. 1997. Microsporidiosis in HIV-positive children in Madrid (Spain). *J. Eukaryot. Microbiol.*, 44:84S–85S.

Delbac, F., and C. Vivares. 1997. A PCR technique for detecting and differentiating known microsporidia in AIDS patients. *J. Eukaryot. Microbiol.*, 44:75S.

Deluol, A.M., J.L. Poirot, F. Heyer, P. Roux, D. Levy, and W. Rozenbaum. 1994. Intestinal microsporidiosis: about clinical characteristics and laboratory diagnosis. *J. Eukaryot. Microbiol.*, 41:33S.

Deplazes, P., A. Mathis, C. Muller, and R. Weber. 1996. Molecular epidemiology of *Encephalitozoon cuniculi* and first detection of *Enterocytozoon bieneusi* in faecal samples of pigs. *J. Eukaryot. Microbiol.*, 43:93S.

Desportes, I., Y. Le Charpentier, A. Galian, F. Bernard, B. Cochand-Priollet, A. Lavergne, P. Ravisse, and R. Modigliani. 1985. Occurrence of a new microsporidan: *Enterocytozoon bieneusi* n.g., n. sp., in the enterocytes of a human patient with AIDS. *J. Protozool.*, 32(2):250-254.

Didier, E.S. 1998. Microsporidiosis. *Clin. Infect. Dis.*, 27:1–7.

Didier, E.S., L.B. Rogers, J.M. Orenstein, M.D. Baker, C.R. Vossbrinck, T. Van Gool, R. Hartskeerl, R. Soave, and L.M. Beaudet. 1996a. Characterization of *Encephalitozoon (Septata) intestinalis* isolates cultured from nasal mucosa and bronchoalveolar lavage fluids of two AIDS patients. *J. Eukaryot. Microbiol.*, 43:34–43.

Didier, E.S., K.F. Snowden, and J.A. Shadduck. 1998. Biology of microsporidian species infecting mammals. *Adv. Parasitol.*, 40:283–320.

Didier, E.S., G.S. Visvesvara, M.D. Baker, L.B. Rogers, D.C. Bertucci, M.A. De Groote, and C.R. Vossbrinck. 1996b. A microsporidian isolated from an AIDS patient corresponds to *Encephalitozoon cuniculi* III, originally isolated from domestic dogs. *J. Clin. Microbiol.*, 34:2835–2837.

Dieterich, D.T., E.A. Lew, D.P. Kotler, M.A. Poles, and J.M. Orenstein. 1994. Treatment with albendazole for intestinal disease due to *Enterocytozoon bieneusi* in patients with AIDS. *J. Infect. Dis.*, 169:178–183.

Ditrich, O., Z. Kucerova, and B. Koudela. 1994. In vitro sensitivity *of Encephalitozoon cuniculi* and *E. hellem* to albendazole. *J. Eukaryot. Microbiol.*, 41:37S.

Dowd, S.E., C.P. Gerba, M. Kamper, and I.L. Pepper. 1999. Evaluation of methodologies including immunofluorescent assay (IFA) and the polymerase chain reaction (PCR) for detection of human pathogenic microsporidia in water. *J. Microbiol. Methods*, 35:43–52.

Dowd, S.E., C.P. Gerba, and I.L. Pepper. 1998. Confirmation of the human-pathogenic microsporidia *Enterocytozoon bieneusi, Encephalitozoon intestinalis,* and *Vittaforma corneae* in water. *Appl. Environ. Microbiol.*, 64:3332–3335.

Dunand, V.A., S.M. Hammer, R. Rossi, M. Poulin, M.A. Albrecht, J.P. Doweiko, P.C. DeGirolami, E. Coakley, E. Piessens, and C.A. Wanke. 1997. Parasitic sinusitis and otitis in patients infected with human immunodeficiency virus: report of five cases and review. *Clin. Infect. Dis.*, 25:267–272.

DuPont, H.L., C.L. Chappell, C.R. Sterling, P.C. Okhuysen, J.B. Rose, and W. Jakubowski. 1995. The infectivity of *Cryptosporidium parvum* in healthy volunteers. *N. Engl. J. Med.*, 332:855–859.

DuPont, H.L., and C.D. Ericsson. 1993. Prevention and treatment of traveler's diarrhea. *N. Engl. J. Med.*, 328:1821–1827.

Font, R.L., A.N. Samaha, M.J. Keener, P. Chevez-Barrios, and J.D. Goosey. 2000. Corneal microsporidiosis. Report of case, including electron microscopic observations. *Ophthalmology,* 107:1769–1775.

Fournier, S., O. Ligoury, V. Garrait, J.P. Gagneux, C. Sarfati, F. Derouin, and J.M. Molina. 1998. Microsporidiosis due to *Enterocytozoon bieneusi* infection as a possible cause of traveller's diarrhea. *Eur. J. Clin. Microbiol. Infect. Dis.,* 17:743–744.

Fournier, S., O. Liguory, M. Santillana-Hayat, E. Guillot, C. Sarfati, N. Dumoutier, J. Molina, and F. Derouin. 2000. Detection of microsporidia in surface water: a one-year follow-up study. *FEMS Immunol. Med. Microbiol.,* 29:95–100.

Gainzarain, J.C., A. Canut, M. Lozano, A. Labora, F. Carreras, S. Fenoy, R. Navajas, N.J. Pieniazek, A.J. Da Silva, and C. del Aguila. 1998. Detection of *Enterocytozoon bieneusi* in two human immunodeficiency virus-negative patients with chronic diarrhea by polymerase chain reaction in duodenal biopsy specimens and review. *Clin. Infect. Dis.,* 27:394–398.

Gatti, S., L. Sacchi, S. Novati, S. Corona, A.M. Bernuzzi, H. Moura, N.J. Pieniazek, G.S. Visvesvara, and M. Scaglia. 1997. Extraintestinal microsporidiosis in AIDS patients: clinical features and advanced protocols for diagnosis and characterization of the isolates. *J. Eukaryot. Microbiol.,* 44:79S.

Giang, T.T., D.P. Kotler, M.L. Garro, and J.M. Orenstein. 1993. Tissue diagnosis of intestinal microsporidiosis using the Chromotrope-2R modified Tri-Chrome stain. *Arch. Pathol. Lab. Med.,* 117:1249–1251.

Goetz, M., S. Eichenlaub, G.R. Pape, and R.M. Hoffmann. 2001. Chronic diarrhea as a result of intestinal microsposidiosis in a liver transplant recipient. *Transplantation,* 71:334–337.

Goodgame, R.W. 1996. Understanding intestinal spore-forming protozoa: cryptosporidia, microsporidia, isospora, and cyclospora. *Ann. Intern. Med.,* 124:429–441.

Haas, C.N., J.B. Rose, and C.P. Gerba. 1999. *Quantitative Microbial Risk Assessment.* New York: John Wiley & Sons.

Hunter, P.R. 2000. Waterborne outbreak of microsporidiosis. *J. Infect. Dis.,* 182:380–381.

Hutin, Y.J.F., M.N. Sombardier, O. Ligoury, C. Sarfati, F. Derouin, J. Modai, and J.M. Molina. 1998. Risk factors for intestinal microsporidiosis in patients with human immunodeficiency virus infection: A case control study. *J. Infect. Dis.,* 178:904–907.

Kelkar, R., P. Sastry, S.S. Kulkarni, T. Saikia, P.M. Parikh, and S.H. Advani. 1997. Pulmonary microsporidial infection in a patient with CML undergoing allogenic marrow transplant. *Bone Marrow Transplant.*, 19:179–182.

Kock, N.P., H. Petersen, T. Fenner, I. Sobottka, C. Schmetz, P. Deplazes, N.J. Pieniazek, H. Albrect, and J. Schottelius. 1996. Species-specific identification of microsporidia in stool and intestinal biopsy specimens by the polymerase chain reaction. *Eur. J. Clin. Microbiol. Infect. Dis.*, 16:369–376.

Kotler, D.P., and J.M. Orenstein. 1994. Prevalence of intestinal microsporidiosis in HIV-infected individuals referred for gastroenterological evaluation. *Am. J. Gastroenterol.*, 89:1998–2002.

Kotler, D.P., and J.M. Orenstein. 1998. Clinical syndromes associated with microsporidiosis. *Adv. Parasitol.*, 40:321–349.

Koudela, B., S. Kucerova, and T. Hudcovic. 1999. Effect of low and high temperatures on infectivity of *Encephalitozoon cuniculi* spores suspended in water. *Folia Parasitol. (Praha)*, 46:171–174.

Ligoury, O., F. David, C. Sarfati, A. Schuitema, R. Hartskeerl, F. Derouin, J. Modai, and J.M. Molina. 1997. Diagnosis of infections caused by *Enterocytozoon bieneusi* and *Encephalitozoon intestinalis* using polymerase chaing reaction in stool specimens. *AIDS*, 11:723–726.

Lopez-Velez, R., M.C. Turrientes, C. Garron, P. Montilla, R. Navajas, S. Fenoy, and C. del Aguila. 1999. Microsporidiosis in travelers with diarrhea from the tropics. *J. Travel. Med.*, 6:223–227.

Magaud, A., A. Achbarou, and I. Desportes-Livage. 1997. Cell invasion by the microsporidium *Encehpalitozoon intestinalis*. *J. Eukaryot. Microbiol.*, 44:81S–81S.

Maggi, P., A.M. Larocca, M. Quarto, G. Serio, O. Brandonisio, G. Angarano, and G. Pastore. 2000. Effect of antiretroviral therapy on cryptosporidiosis and microsporidiosis in patients infected with human immunodeficiency virus type 1. *Eur. J. Clin. Microbiol. Infect. Dis.*, 19:213–217.

Margileth, A.M., A.J. Strano, R. Chandra, R. Neafie, M. Blum, and R.M. McClully. 1973. Disseminated nosematosis in an immunologically compromised infant. *Arch. Pathol.*, 95:145–150.

Mathis, A. 2000. Microsporidia: emerging advances in understanding the basic biology of these unique organisms. *Int. J. Parasitol.*, 30:795–804.

Metge, S., J.T. Van Nhieu, D. Dahmane, P. Grimbert, F. Foulet, C. Sarfati, and S. Bretagne. 2000. A case of *Enterocytozoon bieneusi* infection in an HIV-negative renal transplant recipient. *Eur. J. Clin. Microbiol. Infect. Dis.*, 19:221–223.

Miao, Y.M., F.M. Awad-El-Kariem, C. Franzen, D.S. Ellis, A. Muller, H.M. Counihan, P.J. Hayes, and B.G. Gazzard. 2000. Eradication of *Cryptosporidia* and microsporidia following successful antiretroviral therapy. *J. Acquir. Immune Defic. Syndr.*, 25:124–129.

Molina, J.M., C. Chastang, J.M. Goguel, C. Sarfati, I. Desportes-Livage, J. Horton, F. Derouin, and J. Modai. 1998. Albendazole for treatment and prophylaxis of microsporidiosis due to *Encephalitozoon intestinalis* in patients with AIDS: A radomized double-blind controlled trial. *J. Infect. Dis.*, 177:1373–1377.

Molina, J.M., J. Goguel, C. Sarfati, J.F. Michiels, I. Desportes-Livage, S. Balkan, C. Chastang, L. Cotte, C. Maslo, A. Struxiano, F. Derouin, and J.M. Decazes. 2000. Trial of oral fumagillin for the treatment of intestinal microsporidiosis in patients with HIV infection. ANRS 054 Study Group. Agence Nationale de Recherche sur le SIDA. *AIDS*, 14:1341–1348.

Molina, J.M., E. Oksenhendler, B. Beauvais, C. Sarfati, A. Jaccard, and J. Modai. 1995. Disseminated microsporidosis due to *Septata intestinalis* in patients with AIDS: Clincial features and response to Albendazole therapy. *Clin. Infect. Dis.*, 171:245–249.

Molina, J.M., C. Sarfati, B. Beauvis, M. Lemann, A. Lesourd, F. Ferchal, I. Casin, P. Lagrange, R. Modigliani, F. Derouin, and J. Modai. 1993. Intestinal microsporidiosis in human immunodeficiency virus-infected patients with chronic unexplained diarrhea: Prevalence and clinical and biologic features. *J. Infect. Dis.*, 167:217–221.

Muller, A., R. Bialek, A. Kamper, G. Fatkenheuer, B. Salzberger, and C. Franzen. 2001. Detection of microsporidia in travelers with diarrhea. *J. Clin. Microbiol.*, 39:1630–1632.

Najm, I., and R.R. Trussell. 1999. New and Emerging Drinking Water Treatment Technologies. In *Identifying Future Drinking Water Contaminants.* Washington, DC: National Academy Press.

Navin, T.R., R. Weber, D.J. Vugia, D. Rimland, J.M. Roberts, D.G. Addiss, G.S. Visvesvara, S.P. Wahlquist, S.E. Hogan, L.E. Gallagher, D.D. Juranek, D.A. Schwartz, C.M. Wilcox, J.M. Stewart, S.E. Thompson III, and R.T. Bryan. 1999. Declining CD4+ T-lymphocyte counts are associated with increased risk of enteric parasitosis and chronic diarrhea: results of a 3-year longitudinal study. *J. Acquir. Immune Defic. Syndr. Hum. RetroVirol.*, 20:154–159.

Orenstein, J.M. 1991. Microsporidiosis in the acquired immunodeficiency syndrome. *J. Parasitol.*, 77:843–864.

Orenstein, J.M., H.P. Gaetz, A.T. Yachnis, S.S. Frankel, R.B. Mertens, and E.S. Didier. 1997. Disseminated microsporidiosis in AIDS: are any organs spared? *AIDS*, 11:385–386.

Orlandi, P.A., and K.A. Lampel. 2000. Extraction-free, filter-based template preparation for rapid and sensitive PCR detection of pathogenic parasitic protozoa. *J. Clin. Microbiol.*, 38:2271–2277.

Owen, R.L. 1997. Polymerase chain reaction of stool: a powerful tool for specific diagnosis and epidemiologic investigation of enteric microsporidial infections. *AIDS*, 11:817–818.

Pol, S., C.A. Romana, S. Richard, P. Amouyal, I. Desportes-Livage, F. Carnot, J.F. Pays, and P. Berthelot. 1993. Microsporidia infections in patients with the human immunodeficiency virus and unexplained cholangitis. *N. Engl. J. Med.*, 328:95–99.

Rabeneck, L., R.M. Genta, F. Gyorkey, P. Gyorkey, J.E. Clarridge, and L.W. Foote. 1995. Oberservations on the pathological spectrum and clinical course of microsporidiosis in men infected with the human immunodeficiency virus: Follow-up study. *Clin. Infect. Dis.*, 20:1229–1235.

Rabeneck, L., F. Gyorkey, R.M. Genta, P. Gyorkey, L.W. Foote, and J.M.H. Risser. 1993. The role of microsporidia in the pathogenesis of HIV-related chronic diarrhea. *Ann. Intern. Med.*, 119:895–899.

Raynaud, L., F. Delbac, V. Broussolle, M. Rabodonirina, V. Girault, M. Wallon, G. Cozon, C.P. Vivares, and F. Peyron. 1998. Identification of *Encephalitozoon intestinalis* in travelers with chronic diarrhea by specific PCR amplification. *J. Clin. Microbiol.*, 36:37–40.

Rinder, H., A. Thomschke, B. Dengjel, R. Gothe, T. Loscher, and M. Zahler. 2000. Close genotypic relationship between *Enterocytozoon bieneusi* from humans and pigs and first detection in cattle. *J. Parasitol.*, 86:185–188.

Rose, J.B., C.N. Haas, and S. Regli. 1991. Risk assessment and control of waterborne giardiasis. *Am. J. Public Health*, 81:709–713.

Rossi, P., C. Urbani, G. Donelli, and E. Pozio. 1999. Resolution of microsporidial sinusitis and keratoconjuntivitis by itraconazole treatment. *Am. J. Ophthalmol.*, 127:210–212.

Sandfort, J., A. Hannemann, H. Gelderblom, K. Stark, R.L. Owen, and B. Ruf. 1994. *Enterocytozoon bieneusi* infection in an immunocompetent patient who had acute diarrhea and who was not infected with the human immunodefiency virus. *Clin. Infect. Dis.*, 19:514–516.

Schwartz, D.A., R.T. Bryan, K.O. Hewan-Lowe, G.S. Visvesvara, R. Weber, A. Cali, and P. Angritt P. 1992. Disseminated microsporidiosis (*Encephalitozoon hellem*) and acquired immunodeficiency syndrome. Autopsy evidence for respiratory acquisition. *Arch. Pathol. Lab. Med.*, 116:660-668.

Schwartz, D.A., R.T. Bryan, and G.S. Visvesvara. 1994. Diagnostic approaches for *Encephalitozoon* infections in patients with AIDS. *J. Eukaryot. Microbiol.*, 41:59S–60S.

Schwartz, D.A., G. Visvesvara, R. Weber, and R.T. Bryan. 1994. Male genital tract microsporidiosis and AIDS: prostatic abscess due to *Encephalitozoon hellem*. *J. Eukaryot. Microbiol.*, 41:61S.

Shadduck, J.A., and J.M. Orenstein. 1993. Comparative pathology of microsporidiosis. *Arch. Pathol. Lab. Med.*, 117:1215–1219.

Shadduck, J.A., and M.B. Polley. 1978. Some factors influencing the *in vitro* infectivity and replication of *Encephalitozoon cuniculi*. *J. Protozool.*, 25:491–496.

Simon, D., D.P. Kotler, and L.J. Brandt. 1996. Chronic unexplained diarrhea in human immunodeficiency virus infection: Determination of the best diagnostic approach. *Gastroenterol.*, 111:269–271.

Smith, L.A. 1997. Still around and dangerous: *Giardia lamblia* and *Entamoeba histolytica*. *Clin. Lab. Sci.*, 10:279–286.

Snowden, K.F., E.S. Didier, J.M. Orenstein, and J.A. Shadduck. 1998. Animal models of human microsporidial infections. *Lab. Anim. Sci.*, 48:589–592.

Sowerby, T.M., C.N. Conteas, O.G. Berlin, and J. Donovan. 1995. Microsporidiosis in patients with relatively preserved CD4 counts. *AIDS*, 9:975.

Sparfel, J.M., C. Sarfati, O. Ligoury, B. Caroff, N. Dumontier, B. Gueglio, E. Billaud, F. Raffi, J.M. Molina, M. Miegeville, and F. Derouin. 1997. Detection of microsporidia and indentification of *Enterocytozoon bieneusi* in surface water by filtration followed by specfic PCR. *J. Eukaryot. Microbiol.*, 44:78S.

Svenungsson, B., A. Lagergren, E. Ekwall, B. Evengard, K.O. Hedlund, A. Karnell, S. Lofdahl, L. Svensson, and A. Weintraub. 2000. Enteropathogens in adult patients with diarrhea and healthy control subjects: a 1-year prospective study in a Swedish clinic for infectious diseases. *Clin. Infect. Dis.*, 30:770–778.

Talal, A.H., D.P. Kotler, J.M. Orenstein, and L.M. Weiss. 1998. Detection of *Enterocytozoon bieneusi* in fecal specimens by polymerase chain reaction analysis with primers to the small-subunit rRNA. *Clin. Infect. Dis.*, 26:673–675.

Tanowitz, H.B., D. Simon, L.M. Weiss, C. Noyer, C.M. Coyle, and M. Wittner. 1996. Management of the HIV infected patient, part 1: Gastrointestinal manifestations. *Med. Clin. North Am.*, 80:1395–1415.

Taylor, J.S., and M. Wiesner. 1999. Membranes. In *Water Quality and Treatment: A Handbook of Community Water Supplies.* 5th ed. Edited by American Water Works Association. New York: McGraw-Hill.

Thielman, N.M., and R.L. Guerrant. 1998. Persistent diarrhea in the returned traveler. *Infect. Dis. Clin. North Am.*, 12: 489–501.

USPHS/IDSA (US Public Health Service/Infectious Diseases Society of America). 1999. 1999 USPHS/IDSA Guidelines for the prevention of opportunistic infections in persons infected with the human immunodeficiency virus. *MMWR*, 48:1–59.

van Gool, T., J.C.M. Vetter, B. Weinmayr, A. VanDam, F. Derouin, and J. Dankert. 1997. High seroprevalence rates of *Enterocytozoon* species in immunocompetent subjects. *J. Infect. Dis.*, 175:1020–1024.

Velasquez, J.N., S. Carnevale, J.H. Labbe, A. Chertcoff, M.G. Cabrera, and W. Oelemann. 1999. In situ hybridization: A molecular approach for the diagnosis of the microsporidian parasite *Enterocytozoon bieneusi*. *Hum. Pathol.*, 30:54–58.

Vivares, C.P., and G. Metenier. 2000. Towards the minimal eukaryotic parasitic genome. *Curr. Opin. Microbiol.*, 3:463–467.

Wanke, C.A., P.C. DeGirolami, and M. Federman. 1996. *Enterocytozoon bieneusi* infection and diarrheal disease in patients who were not infected with the human immunodeficiency virus: Case report and review. *Clin. Infect. Dis.*, 23:816–818.

Weber, R., and R.T. Bryan. 1994. Microsporidial infections in immunodeficient and immunocompetent patients. *Clin. Infect. Dis.*, 19:517–521.

Weber, R., R.T. Bryan, R.L. Owen, C.M. Wilcox, L. Gorelkin, and G.S. Visvesvara. 1992. Improved light-microscopical detection of microsporidia spores in stool and duodenal aspirates. The Enteric Opportunistic Infections Working Group. *N. Engl. J. Med.*, 326:161–166.

Weber, R., R.T. Bryan, D.A. Schwartz, and R.L. Owen. 1994. Human microsporidial infections. *Clin. Microbiol. Rev.*, 7:426–461.

Weber, R., H. Kuster, G.S. Visvesvara, R.T. Bryan, D.A. Schwartz, and R. Luthy. 1993. Disseminated microsporidiosis due to *Encephalitozoon hellem*: pulmonary colonization, microhematuria, and mild conjunctivitis in a patient with AIDS. *Clin. Infect. Dis.*, 17:415–419.

Weiss, L.M. 2000. Molecular phylogeny and diagnostic approaches to microsporidia. *Contrib. Microbiol.*, 6:209–235.

Weiss, L.M., A. Cali, E. Levee, D. Laplace, H. Tonowitz, D. Simon, and M. Wittner. 1992. Diagnosis of *Encephlitozoon cuniculi* infection by Western Blot and the use of cross-reactive antigens for the possible detection of microsporidiosis in humans. *Am. J. Trop. Med. Hyg.*, 47:456–462.

Weiss, L.M., and E.M. Keohane. 1997. The uncommon gastrointestinal protozoa: microsporidia, blastocystis, isospora, dientamoeba, and balantidium. *Curr. Clin. Top. Infect. Dis.*, 17:147–187.

Weiss, L.M., E. Michalakakis, C.M. Coyle, H.B. Tanowitz, and M. Wittner. 1994. The in vitro activity of albendazole against *Encephalitozoon cuniculi. J. Eukaryot. Microbiol.*, 41:65S.

Weiss, L.M., and C.R. Vossbrinck. 1998. Microsporidiosis: molecular and diagnositic aspects. *Adv. Parasitol.*, 40:354–395.

Wittner, M., H.B. Tanowitz, and L.M. Weiss. 1993. Parasitic infections in AIDS patients. *Infect. Dis. Clin. North Am.*, 7:569–586.

Wolk, D.M., C.H. Johnson, E.W. Rice, M.M. Marshall, K.F. Grahn, C.B. Plummer, and C.R. Sterling. 2000. A spore counting method and cell culture model for chlorine disinfection studies of *Encephalitozoon syn. Septata intestinalis. Appl. Environ. Microbiol.*, 66:1266–1273.

Yachnis, A.T., J. Berg, A. Martinez-Salazar, B.S. Bender, L. Diaz, A.M. Rojiana, T.A. Eskin, and J.M. Orenstein. 1996. Disseminated microsporidiosis especially infecting the brain, heart, and kidneys: Report of a newly recognized pansporoblastic species in two symptomatic AIDS patients. *Am. J. Clin. Pathol.*, 106:535–543.

CHAPTER 13

Mycobacterium avium Complex in Drinking Water

by Martha A. Embrey

Executive Summary

Problem Formulation

Occurrence of Illness

- In P-MAC disease, estimates of 0.98 cases/100,000 and 1.28 cases/100,000 have been made for this country. P-MAC occurs mostly in the elderly. In AIDS patients, disseminated MAC occurs in 15 to 40% of the population, that number is decreasing though with the advent of retroviral therapy. MAC-related lymphadenitis, which is rare, occurs mainly in children.

Role of Waterborne Exposure

- Important. Many studies have implicated water as the transmission route for MAC infection. However, MAC is ubiquitous throughout the environment, and infection undoubtedly occurs through other routes of environmental exposure. No evidence supports any person-to-person transmission.

Degree of Morbidity and Mortality

- Generally, MAC is opportunistic—occurring in those who already have underlying disease. P-MAC can cause significant pulmonary and systemic symptoms; disseminated MAC commonly causes fever, anemia, night sweats, and weight loss. Both can decrease expected lifespan in patients, but the mortality rate is unknown.

Detection Methods in Water/Clinical Specimens

- MAC strains need dedicated growth media and extended incubation. Because they grow slowly, isolation is made difficult by other organisms' overgrowth. Samples are usually treated to remove other organisms and physically concentrated to improve isolation. These steps may decrease the rate of recovery and viability of MAC strains.

Mechanisms of Water Contamination

- Mycobacteria are common and ubiquitous in the environment.

Concentrations at Intake

- Variable. Concentrations measured in raw water have ranged from 0 to 4,800 CFU/100 mL.

Efficacy of Water Treatment

- Chlorine is a relatively ineffective disinfectant, especially at residual levels. Physical treatment, such as sand filtration and coagulation–sedimentation, is effective at removing most, but not all, of the organisms. Membrane filtration technology has the potential to remove virtually all bacterial pathogens.

Survival/Amplification in Distribution

- Major problem. Chlorine residual is not an effective disinfectant and MAC can colonize distribution pipes. The organism appears to be thermophilic, and MAC has been frequently isolated from showerheads and recirculating hot water systems in institutions.

Routes of Exposure

- Ingestion and inhalation of aerosols. Some believe the latter is responsible for more transmission than the former.

Dose Response

Infectious Dose

- 7.5×10^6 in beige mice. Human dose unknown.

Probability of Illness Based on Infection

- MAC is generally an opportunistic infection; however, the incidence of P-MAC in people (especially elderly women) without risk factors and lymphadenitis in children has been increasing. Data from the late 1970s and early 1980s showed an annual rate of 1.3 cases/100,000. Disseminated MAC occurs in approximately 15–40% of AIDS patients.

Efficacy of Medical Treatment

- P-MAC and disseminated MAC are difficult to treat and are resistant to antimicrobials; symptomatic infection can be highly problematic and result in lengthy and complicated chemotherapy. Relapse is relatively common.

Secondary Spread

- None.

Chronic Sequelae

- Untreated, MAC can cause numerous respiratory and systemic symptoms. Available treatment is difficult and sometimes ineffective and can last for months or years. Permanent lung damage depends on the extent of infection when the patient presents for treatment. AIDS patients who develop disseminated MAC are usually in the end-stage of their disease.

Introduction

Mycobacterium avium complex (MAC) consists of environmental bacteria of the *M. avium* and *M. intracellulare* species. They are acid-fast, rod-shaped bacteria with dimensions ranging from 0.2 to 0.6 by 1.0–10.0 μm. Older literature frequently includes *M. scrofulaceum* in the group and calls it MAIS. These organisms are widely distributed in the environment and especially abundant in soil, marshland, streams, rivers, estuaries, and marine environments, though not groundwater (Martin, Parker, and Falkinham 1987). MAC generally causes three different types of diseases in humans: (1) pulmonary disease, (2) cervical lymphadenitis, and (3) disseminated MAC, both AIDS and non-AIDS related. Data suggest that the incidence of the first two types of MAC disease continues to be increasing (du Moulin et al. 1985; Howell, Heaton, Neutze 1997; Kubo et al. 1998; O'Brien, Geiter, and Snider 1987; Olivier 1998), and disseminated MAC disease is one of the most common opportunistic infections to strike AIDS patients.

Epidemiology

The epidemiology of the outcomes related to infection with nontuberculous mycobacteria varies greatly and is evolving. Disseminated MAC, which had been a common, AIDS-defining disease is decreasing in incidence due to antiretroviral drug therapies now available for AIDS patients. On the other hand, pulmonary MAC disease appears to be increasing in incidence in elderly women. The reason for this shift is unknown.

Pulmonary MAC

Pulmonary MAC disease (P-MAC) produces cough, fever, weight loss, and night sweats—not unlike tuberculosis. There may also be a relationship between subclinical MAC infection and the presence of nodules and bronchiectasis in the lungs of people who are otherwise healthy (Kubo et al. 1998; Olivier 1998). Also, people with other respiratory ailments may be infected with MAC without having the symptomatic criteria for diagnosis (Raju and Schluger 2000). P-MAC has traditionally been an opportunistic disease, affecting people with a predisposing lung condition, such as emphysema or healed tuberculosis. Because public health authorities generally do not require reporting on the isolation of MAC from clinical specimens, it is difficult to ascertain the prevalence of infection-related morbidity and rate of mortality in this country. Two sets of data, one from a Centers for Disease Control and Prevention (CDC) survey (O'Brien, Geiter, and Snider 1987) covering 1981 to 1983 and one from an Oregon health maintenance organization (Reich and Johnson 1991) covering 1975 to 1986, indicated a period prevalence of 1.28 cases/100,000 and an annual incidence of 0.98 cases/100,000, respectively. The CDC report of population-based data from ongoing surveillance in Houston and Atlanta suggest an incidence rate of 1.0/100,000 per year in the United States (CDC 1998). An Australian report estimated the yearly population-based incidence to be 2.1/100,000 (O'Brien, Currie, and Krause et al. 2000).

Iseman (1996) reports that although early studies of P-MAC found it to be more common in men with a history of tobacco use or lung disease (also echoed by Grange 1997), women without any evidence of preexisting conditions have been recently reported in the literature as infected with P-MAC (see Table 13-1). Of the P-MAC patients without recognized predisposing conditions (including smoking) that the literature has identified in the last decade, the overwhelming majority are elderly females (see Table 13-1). Studies in mice show an association between lack of endogenous estrogen and the development of P-MAC, which reinforces this epidemiological finding (Tsuyuguchi et al. 2001).

Though it appears that P-MAC results in more clinically significant infection in whites, not enough data exist to support a hypothesis that whites are more susceptible or exposed than nonwhites. However, a 1965 study of skin reactivity reported that blacks had a higher prevalence of reactivity than whites (Edwards and Smith 1965), which may indicate a higher rate of immunity.

AIDS patients, who are vulnerable to disseminated infection, are rarely seen with localized P-MAC (Hocqueloux et al. 1998).

Lymphadenitis

This is an uncommon disease seen primarily in children of preschool age, who present with nontender, enlarged cervical lymph nodes without other symptoms (Evans et al. 1998; Horsburgh 1996). In most cases, a single cervical lymph node is involved, and excision of the node usually cures

Table 13-1: Gender differences in P-MAC diagnoses

Number of Cases Without Pre-Disposing Factors	% Female	% Male	Reference
10	100	0	Kennedy and Weber 1994
53	91	9	Kubo et al. 1998
21	81	19	Prince et al. 1989
17	82	18	Reich and Johnson 1991

the disease (Grange 1997; Perlman, D'Amico, and Salomon 2001), though chemotherapy may also be added to the treatment (Evans et al. 1998). Risk of recurrence is approximately 5%. From 1981 to 1983, about 300 cases were reported annually in the United States (O'Brien, Geiter, and Snider 1987). Horsburgh estimates that the true number is closer to 500 cases per year. The fact that more cases have been reported in recent years may be a result of improved detection and awareness, though external factors cannot be ruled out (Howell, Heaton, and Neutze 1997). Case series have also been reported in Canada, Australia, Sweden, New Zealand, and the United Kingdom (Evans et al. 1998; Howell, Heaton, and Neutze 1997 in Horsburgh 1996). Lymphadenitis can also occur as part of a disseminated infection (Grange 1997).

Disseminated MAC

Disseminated MAC is usually associated with AIDS patients who have CD4 counts less than $100/mm^3$ but it can also occur in immunocompromised patients who do not have AIDS. In the United States, about 20 to 40% of AIDS patients develop MAC over the course of their illnesses (Horsburgh 1991; Nightingale et al. 1992), though that percentage may be lessening with the advent of effective antiretroviral drug protocols. Common symptoms of disseminated MAC include fever, anemia, night sweats, weight loss, abdominal pain, and diarrhea (Burman and Cohn 1996). Disseminated MAC is frequently centered in the gastrointestinal system.

In developing countries, and especially in Africa, AIDS patients are much less likely to develop disseminated MAC as their AIDS infection progresses. The reasons for this dichotomy are unclear, but published studies do not support the hypothesis that a lack of MAC organisms in the African environment results in little disease in late-stage African AIDS patients (Eaton et al. 1995; Morrissey et al. 1992). MAC was isolated in both soil and water samples from Uganda, and these isolates were similar to those found in US and European AIDS patients. Though MAC has been isolated in the African environment, it was not isolated in hospital water in Zaire or Kenya, which may provide clues to differences in exposure between developed and developing countries (von Reyn et al. 1993b). Some reports have speculated that prior low-level infection (with *M. tuberculosis* or any environmental mycobacteria strains) has resulted in broad mycobacterial immunity for populations in developing countries (von Reyn et al. 1996; Iivanainen et al. 1999). In support of the possible role of immunity, research indicates that occupational exposure to soil and water in the United States and Finland reduced people's risk of disseminated MAC (von Reyn et al. 1996).

Medical Treatment

The medical management of MAC-related disease (both pulmonary and disseminated) is extensive and complex. MAC is relatively resistant to antimicrobial agents, and the types of chemotherapies that have succeeded in treating tuberculosis are ineffective in treating MAC. Iseman (1996) summarized the descriptive literature regarding the efficacy of various chemotherapies to treat P-MAC. He noted that initial response ranged from 55 to 100% and relapse ranged from 10 to 38%. The drug protocols included two to six drugs, and though the American Thoracic Society recommended a 3- to 6-month treatment protocol, Iseman suggested that 18 months be the minimum time and that more than 24 months is appropriate for some patients. Resectional surgery is sometimes used to control P-MAC in intransigent cases. The results of surgical intervention vary, and the patient is still subject to intensive, lengthy follow-up drug therapy.

In AIDS patients, antimicrobial therapy is used both as prophylaxis and treatment of disseminated MAC. The US Public Health Service published guidelines for disseminated MAC prophylaxis for children and adults with HIV. A survey showed that health care providers are adhering to the guidelines for older children and adults but, because of possible misinformation, they do not follow them for children <1–2 years (Chong and Husson 1998). Part of the reason why drug therapies are challenged could be that many AIDS patients are infected with multiple strains of mycobacteria, which differ in their antibiotic susceptibility (Fry, Meissner, and Falkinham 1986). The emergence of two macrolide compounds, clarithromycin and azithromycin, has been instrumental in the treatment and prevention of MAC in AIDS and non-AIDS patients (Cohn 1997; Falkinham 1996; Griffith et al. 2001; Hewitt et al. 1999); however, an evolving resistance to these drugs, especially the more effective clarithromycin, is problematic (Benson 1997; Bermudez et al. 1998; Cohn 1997; Grosset and Ji 1997). These treatment regimens are also associated with significant adverse effects (Griffith 1999). Combinations of rifampicin, ethambutol, and isoniazid have been used with some success for non-AIDS related P-MAC, though relapse continues to be a problem (British Thoracic Society 2001). Fortunately, the incidence of MAC in AIDS patients has dropped dramatically due to the advent of combination antiretroviral therapy, which reduces viral load and keeps CD4 levels higher (Griffith 1999; Havlir et al. 2000; Palella et al. 1998; Williams, Currier, and Swindells 1999). Questions now emerge over how to adjust prophylaxis and treatment therapy in AIDS patients in light of the successes of antiretroviral therapy (Benson 1997; Detels et al. 2001).

Susceptible Subpopulations

MAC is an opportunistic infection that is considered an AIDS-defining disease. Before the use of antiretroviral therapies, AIDS patients commonly developed disseminated MAC at the end stage of disease. New therapies that keep CD4 counts high have reduced the susceptibility of AIDS patients to MAC and other opportunistic infections.

Because mycobacteria are commonly found throughout the environment, the fact that disease is rare indicates that normal host defense mechanisms must be strong. Though nontuberculous mycobacteria are similar to *M. tuberculosis,* they are significantly less virulent (Holland 2001). The study of susceptibility to these organisms in the immunocompetent population should be informative and offer more clues about immunity and risk factors.

In the immunocompetent population, typical susceptibilities to P-MAC have included male gender, predisposing lung condition, or smoking (Iseman 1996); however, elderly women with no history of lung disease have been reportedly susceptible (Iseman 1996; Kubo et al. 1998). Research in mice has suggested that a decrease in estrogen production affects resistance to MAC infection and the development of P-MAC disease (Tsuyuguchi et al. 2001). In addition to hormone-related susceptibility, researchers have looked at genetic factors in MAC infection and disease development. Frucht and colleagues (1999) studied a family with disseminated MAC susceptibility and found they suffered from a genetic defect affecting interleukin-12 (IL-12) regulation and interferon γ production. Other studies have shown that the frequency of certain human leukocyte antigen (HLA) alleles is associated with P-MAC disease progression in immunocompetent women (Kubo et al. 2000). Previous studies have linked genetic abnormalities affecting the interferon γ and IL-12 receptor pathway with mycobacterial infection (Altare et al. 1998; Newport et al. 1996).

Fifty-nine patients with P-MAC were assessed for their HLA phenotypes (Takahashi et al. 2000). The researchers found significant differences between the MAC patients and controls: HLA-A33 (28.8% in patients versus 12.5% in controls); HLA-DR6 (50.8% in patients versus 20.3% in controls); and haplotype A33-B44-DR6 (23.7% in patients versus 4.2% in controls). Interestingly, 29 of the 59 patients had tuberculosis history, but their phenotypes were no different from the phenotypes of the tuberculosis negative group, suggesting that MAC susceptibility has more to do with genetic composition than with tuberculosis history.

Transmission

Transmission probably occurs primarily through fine particle aerosolization and secondarily through ingestion. The likely routes of entry for mycobacteria are the respiratory and gastrointestinal tracts (Burman and Cohn 1996). Disseminated MAC may begin in the lungs and then spread to the gastrointestinal tract or colonize the gastrointestinal tract after ingestion of contaminated food or water. *M. avium* is acid tolerant and, therefore, able to withstand exposure in the stomach; preexposure to water increases its acid resistance (Bodmer, Miltner, and Bermudez 2000).

Unlike *M. tuberculosis*, person-to-person transmission is unlikely for MAC; this has been supported by studies of patients with pulmonary and disseminated disease (Carbonne et al. 1998; Horsburgh et al. 1994; Iseman et al. 1985). Though exact sources of exposure are unknown, they are most likely environmental in nature (Iseman 1996). In one study, samples taken from soil, dust, water, and aerosols found that only MAIS collected from aerosols shared characteristics unique to clinical forms of MAIS (Fry, Meissner, and Falkinham 1986).

Tap water was identified as the direct source of MAC infection in 15 simian immunodeficiency virus-inoculated macaques in a biolevel-3 facility (Mansfield and Lackner 1997). Isolates taken from the infected animals were similar to isolates from the facility's water distribution system. The lack of other contamination in the research facility and the similarity of the isolates provides strong evidence that drinking water can serve as a source of MAC infection.

MAC has been isolated from beef, pork, chicken, turkey, oysters, milk (pasteurized and unpasteurized), and frogs (Horsburgh 1996). Determining which environmental reservoir is the source of infection is difficult. Researchers collected tap water, food, and soil samples from the home environment of 290 persons infected with both HIV and mycobacteria (Yajko et al. 1995). They compared environmental strains with strains cultured from the patients. MAC was found in only 4 of 528 water samples and 1 of 397 food samples; in contrast, they recovered MAC from 55% of soil samples taken from potted plants in patients' homes. Analysis showed that some of the soil isolates were similar to isolates recovered from patients. These authors suggest that soil, not food or water, is a significant source of MAC exposure in this population. A recent study looked at the prevalence of nontuberculous mycobacteria in food samples from markets and farm stands (Argueta et al. 2000). Twenty-five of 121 samples tested positive using polymerase chain reaction (PCR), and the predominant species was *avium*. When these isolates were compared with 103 clinical isolates, one food isolate was identical and two were very close to clinical isolates (Yoder et al. 1999). The authors concluded that food (produce) may be a source of transmission. MAC colonization has also been associated with the consumption of raw and partially cooked fish and shellfish (Horsburgh et al. 1994).

MAC has also been isolated from swimming pools and whirlpools (Emde, Chomyc, and Finch 1992; Havelaar et al. 1985), and infection has been associated with the use of both (Embil et al. 1997; von Reyn et al. 1996). Recently, a case series of 10 immunocompetent patients with diffuse infiltrative lung disease was reported (Khoor et al. 2001). All of the patients were diagnosed with nontuberculous mycobacterial infection (90% of them with MAC) and all the patients had used a hot tub. Carson et al. (1988) detected nontuberculous mycobacteria in water samples at 83% of 85 hemodialysis centers in the United States. Infection in these patients had been associated with the water used in the hemodialyzers.

Researchers have debated whether MAC infection in AIDS patients is a reactivation of dormant infection acquired in childhood or the result of recent environmental exposure (Grange 1997; Peters et al. 1995; Singh and Yu 1994; von Reyn et al. 1996). Recent evidence lends more weight to the new exposure theory, though they may not be mutually exclusive. Most researchers support a new environmental exposure because

- the infection is equally distributed among patients in all HIV transmission groups;
- the incidence of MAC is higher in younger AIDS patients;
- MAC skin-test reactivity is too infrequent in healthy patients to account for the high incidence in AIDS patients;
- antecedent respiratory or gastrointestinal colonization has been documented in some patients (Hellyer et al. 1993; von Reyn et al. 1993a; Wayne, Young, and Bertram 1986); and
- although Wayne, Young, and Bertram (1986) reported no antibodies associated with mycobacteria, a recent study suggested that occupational exposure to soil and water are protective of MAC infection (von Reyn et al. 1996).

Skin testing has shown that exposure occurs in a variety of geographic areas and increases with age until about 20 years. Positive skin tests in community populations range from 7 to 60% (Horsburgh 1996). A study of 257,476 US military recruits in the 1950s showed an increase in reactivity from 20% of Northern recruits to 60% in Southeastern troops from coastal areas. The high prevalence of positive skin tests could result from cross-sensitization by other environmental mycobacteria or, more likely, a high ratio of subclinical infection to clinical disease.

Dose Response

MAC infects the body by attacking the intestinal or respiratory mucosa, resisting destruction by macrophages, and colonizing. It then spreads through the submucosal tissue. In disseminated MAC, the bacteria enter the bloodstream through lymphatic drainage (Horsburgh 1999). Recent data indicate that variations in morphology affect different strains' virulence and susceptibility to antibiotics (Reddy 1998).

Most of the exposure studies on MAC have been performed in beige mice. A lower infective dose could be expected in immunocompromised patients and those on antibiotic therapy than in healthy adults (Rusin et al. 1997) (see Table 13-2).

Occurrence in Water

MAC strains have frequently been isolated from environmental samples and water supply systems. Culturing can isolate samples but it is a difficult process; mycobacteria that grow rapidly in the first 7 days or so are discarded and the slower-growing mycobacteria (e.g., MAC) are then identified. Non-mycobacteria that grow over the colonies have to be killed by disinfection. Researchers are using selective media to try and directly isolate MAC organisms without these additional steps (du Moulin et al. 1988; von Reyn et al. 1993b). Individual strains are targeted by DNA probe analysis. Table 13-3 summarizes the literature on MAC isolated from various water sources.

Table 13-2: Infective dose response in beige mice

Dose	Route	Effects	Reference
1×10^4	5× oral doses on alternate days	26.9% bacteremia; 11.5% mortality within 4 weeks; 70% bacteremia; 100% disseminated disease at 8 weeks	Bermudez et al. 1992
1×10^8	5× oral doses	45% bacteremia; 26% mortality	Bermudez et al. 1992
3×10^7	1× oral	No systemic infection or mortality; spleen and lung isolation 6–8 weeks later	Gangadharam et al. 1989
7.5×10^6	1× intranasal	Lung infection in 1 day; spread of infection within 8 weeks; no mortality	Gangadharam et al. 1989

Geographic Distribution

MAC organisms have been isolated from water in many countries, both developed and developing, including Brazil (Telles et al. 1999), Uganda (Eaton et al. 1995; Morrissey et al. 1992), Japan (Ichiyama, Shimokata, and Tsukamura 1988), Canada (Embil et al. 1997), Finland (Katila et al. 1995), Spain (Sabater and Zaragoza 1993), Germany (Peters et al. 1995), and others. A recent study looking at the isolation of MAC from water in the United States, Finland, Zaire, and Kenya showed MAC in 22 of 91 samples (24%) distributed among all the geographic areas (von Reyn et al. 1993b). Of the overall samples taken from environmental sources and water supplies, higher rates of isolation came from developed versus developing countries ($p = 0.015$). MAC was more frequently found in water system samples in the United States and Finland, including hospital supplies (none was isolated from hospitals in Africa). The fact that most AIDS patients will be hospitalized in developed countries at some point during their infection and patients in developing countries may not be hospitalized might also help explain the difference in disease rate between the two populations.

Water Risk Factors

Mycobacteria are found in all types of water, both environmental and potable. Mycobacteria can multiply in oligotrophic water (Falcao, Valentini, and Leite 1993), and biofilms have been implicated as an important source of mycobacteria in oligotrophic water sources (Falkinham et al. 2001; Norton et al. 1999; Peters et al. 1995; Schulze-Robbecke 1993). Though organisms have been found in drinking water wells, it is unlikely that groundwater is a significant source (Martin, Parker, and Falkinham 1987; Falkingham, Norton, and LeChevallier 2001). Studies have shown that MAC flourishes in hot water systems (i.e., in hospitals and other institutions), which creates the opportunity for inhalation exposure through aerosolization from showers. MAC can grow at water temperatures from 52° to 57° C and can survive in temperatures up to 70° C (Iseman 1996; Schulze-Robbecke and Buchholtz 1992). Most hot water systems are set at 49–60° C, and institutional systems are frequently in the low part of the range because of concerns about scalding and energy use. Increased zinc levels have been a suggested risk factor in the growth of MAC (see Table 13-4); this could contribute to the high association between MAC and nosocomial infections since some hospitals use galvanized pipes made with zinc alloys (von Reyn et al. 1993b).

A 1999 study looked at water in Los Angeles County, California, as a source of MAC and non-MAC nontuberculous mycobacteria (Aronson et al. 1999).

Table 13-3: Isolation of MAC from water sources: Percent of positive samples/total number of samples and concentrations

Reference	Tap, %	Hot, %	Cold, %	Shower	Distribution, %	Raw, %
Aronson et al. 1999	91/101 (90)* 43/101 (43) MAC only: homes, businesses, hospitals	No difference between hot and cold (data not shown)				12/13 (92)* 5/13 (38) MAC only: reservoirs
Carson et al. 1988	14.5–195 CFU/mL					
du Moulin and Stottmeier 1986					0.5 CFU/mL	
du Moulin et al. 1988		1.0–500.0 CFU/mL (25)	1.0–2.0 CFU/mL (2)	>500.0 CFU/mL		
Eaton et al. 1995						3/7 (43) 3.3 CFU/mL
Falkinham, Norton, and LeChevallier 2001					(15) of 583 water samples 10–700,000 CFU/mL (29) biofilm	
Falkinham, Parker and Gruft 1980						(37) fresh (20) marine
Katila et al. 1995						(40)†

*Mycobacteria spp.
†MAIS organisms (*Mycobacterium avium, intracellulare, scrofulaceum*).

Table continued on next page.

Table 13-3: Isolation of MAC from water sources: Percent of positive samples/total number of samples and concentrations (continued)

Reference	Tap, %	Hot, %	Cold, %	Shower	Distribution, %	Raw, %
Kirschner, Parker, and Falkinham 1992						0.1–48.0 CFU/mL†
Montecalvo et al. 1998			(32)			
Norton et al. 1999					(6) plant effluent* (11) dead-end pipes (73) biofilms: <0.01–4300 CFU/cm^2	(57)*
Pelletier, du Moulin, and Stottmeier 1988						2.0–5.0 ground 0.3 surface
Peters et al. 1995		1/60 (2)	1/58 (2)			
von Reyn et al. 1993b		4/13 (31) 0–50%	3/17 (18) 0–60%		9/44 (20) 0–45%	13/47 (28) 12–38%
von Reyn et al. 1994		0.4–4.2 CFU/mL		1.2–5.2 CFU/mL	0.2–4.2 CFU/mL	
Yajko et al. 1995	4/528 (0.76)					

*Mycobacteria spp.
†MAIS organisms (*Mycobacterium avium, intracellulare, scrofulaceum*).

They found nontuberculous mycobacteria in 12 of 13 reservoirs, 45 of 55 home taps, 31 of 31 commercial building water sources, and 15 of 15 hospital sources. Most of these isolates were identified as either *M. avium* or *M. intracellulare.* They then compared the samples with clinical isolates from 17 hospital patients. Potable water containing *M. avium* in samples from three homes, two commercial buildings, one reservoir, and eight hospitals was related at some level to the clinical samples. The authors stressed that their sample included what was probably a fraction of the actual number of isolates in the environment. The isolation methods for environmental versus clinical samples are harsher and result in at least 50% of the isolates from the samples. Their limited sampling scheme may have also affected the number of isolates identified from the water. Indicator bacteria and chlorine levels did not affect the numbers of mycobacteria found. These data indicate that drinking water may be a source of exposure; the high levels of *M. avium,* particularly (93%) in the hospital samples, and their relevance to the patient samples suggest nosocomial spread.

In the most extensive study of MAC spp. in biofilms, 29% of 55 biofilm samples from eight systems tested positive for either *M. avium* or *M. intracellulare*; however, the average density for *M. avium* was <0.5 CFU cm^{-2} (Falkinham et al. 2001). The surface material of the pipe did not seem to affect the recovery, nor did the type of concentration of residual disinfection. In this study, *M. intracellulare* was rarely recovered from water samples in distribution systems but was abundant in the distribution biofilms (average 600 CFU cm^{-2}). Clearly, MAC organisms are easily found in biofilm regardless of the water source.

A recent study has shown that *M. avium* can enter and replicate in the waterborne amoeba *Acanthamoeba castellanii* (Cirillo et al. 1997), which can provide an environmental host to enhance its survivability in water and biofilm. Though *M. avium* replicated in amoebae at temperatures of 24° C, their highest growth rate occurred at 37° C. This may help explain why MAC is frequently associated with hot-water supplies. In addition, the study demonstrated that amoeba-grown *M. avium* organisms were more virulent in both in vitro and in vivo models.

Water Treatment

MAC organisms are more resistant to disinfection by chlorine than indicator bacteria due to their morphologic characteristics (Carson et al. 1978; Havelaar et al. 1985; Pelletier, du Moulin, and Stottmeier 1988). In Boston, Pelletier and colleagues (1988) evaluated the ability of free chlorine to inactivate several mycobacterial species, including MAC. They found that concentrations of 1.0 mg/L killed all the strains within 8 hours of exposure; however, 0.15 mg/L of chlorine had basically no biocidal effect. Within the city of Boston's water supply, the free chlorine residual ranged from less than 0.1 to 0.1 mg/L—clearly ineffective. Researchers in Japan found 3 minutes of exposure to 70° C water or ultraviolet light was effective, but mycobacterium strains exposed to chlorine at 4 mg/L for more than 60 minutes survived (Miyamoto, Yamaguchi, and Sasatsu 2000).

Recent data from 26 different water systems showed that mycobacteria were detected in 9% of samples with free chlorine and 18% of those using chloramines; however, the chloraminated systems had higher levels in the source water (Norton et al. 1999). In addition, 73% of the 55 biofilm samples from these systems tested positive for mycobacteria spp. This study also measured levels of assimilable organic carbon and biodegradable dissolved organic carbon, which did not appear to affect the levels of mycobacteria spp. Taylor et al. (2000) found *M. avium* to be resistant to chlorine, monochlorine, chlorine dioxide, and ozone. Concentration/time values were 580—1,200 times higher than that of *Escherichia coli*. In addition, they found that strains grown in water rather than in media were 10 times more resistant, which has definite application in drinking water systems.

Membrane filtration technology, such as micro-, ultra-, and nanofiltration, as well as reverse osmosis, has the capacity to remove water contaminants down to the ion level. Microfiltration, with a nominal pore size of 0.2 μm, can remove bacterial pathogens better than conventional water treatment, and ultrafiltration, with a nominal pore size of 0.01 μm, can

Table 13-4: Risk factors associated with MAC from water sources

Water Risk Factor	Reference
Presence of *A. castellanii*	Cirillo et al. 1997
Hot water more than cold water	du Moulin et al. 1988
Recirculation in hot water systems	du Moulin et al. 1988; von Reyn et al. 1994
Freshwater more than marine water	Falkinham, Parker, and Gruft 1980
Low pH, high organic content	Iivanainen et al. 1999
Zinc >0.75 mg/L warmer temperature, low dissolved oxygen, high humic acid, high fulvic acid	Kirschner, Parker, and Falkinham 1992
Water disinfected with chloramines rather than free chlorine	Norton et al. 1999
High turbidity in raw source water	Falkinham, Norton, and LeChevallier 2001
Rivers and streams higher than standing water	von Reyn et al. 1993b

achieve up to a 6-log pathogen removal (Najm and Trussell 1999; Taylor and Wiesner 1999). This technology will continue to gain in popularity as costs become more competitive and regulatory approval is assured.

Though it appears that traditional treatment such as sand filtration and coagulation–sedimentation reduce the numbers of mycobacterium organisms, the colonization of organisms in water distribution pipes and outlets allows for the bacteria's multiplication in the potable water system (Falkinham, Norton, and LeChevallier 2001; Norton et al. 1999; Pelletier, du Moulin, and Stottmeier 1988). Since organisms that get through the treatment process are unlikely to be affected by residual chlorine levels, this supply system colonization is a real concern. In addition, evidence suggests that MAC organisms are especially prevalent in recirculating hot-water systems, such as found in large institutions like hospitals (du Moulin et al. 1988; von Reyn et al. 1994).

Conclusions

MAC organisms cause three types of disease: P-MAC, lymphadenitis, and disseminated MAC. P-MAC seems to be increasing in incidence, though the reasons are not clear. Conversely, the disseminated disease in AIDS patients is decreasing because of new antiviral chemotherapies. Because both P-MAC and disseminated MAC are difficult to treat and are becoming resistant to antimicrobial drugs, symptomatic infection can be highly problematic and result in lengthy and complicated chemotherapy. Exposure to potable water through ingestion or inhalation has been implicated in the development of MAC-related illness. However, MAC is ubiquitous in the environment, and it is difficult to determine what percentage of MAC cases result from waterborne exposure.

MAC organisms are resistant to chlorine and biofilms within water distribution systems amplify their growth; therefore, treatment intervention at the plant level may not be productive. Research is ongoing in the occurrence and control of MAC as it relates to water systems. Additional research on why people without predisposing factors are affected by P-MAC and what percentage of infection is caused by exposure to water would help characterize the relationship between water and MAC-related disease.

Bibliography

Altare, F., A. Durandy, D. Lammas, J.F. Emile, S. Lamhamedi, F. Le Deist, P. Drysdale, E. Jouanguy, R. Doffinger, F. Bernaudin, O. Jeppsson, J.A. Gollob, E. Meinl, A.W. Segal, A. Fischer, D. Kumararatne, and J.L. Casanova. 1998. Impairment of mycobacterial immunity in human interleukin-12 receptor deficiency. *Science,* 280:1432–1435.

Argueta, C., S. Yoder, A.E. Holtzman, T.W. Aronson, N. Glover, O.G. Berlin, G.N. Stelma Jr., S. Froman, and P. Tomasek. 2000. Isolation and identification of nontuberculous mycobacteria from foods as possible exposure sources. *J. Food Prot.*, 63:930–933.

Aronson, T., A. Holtzman, N. Glover, M. Boian, S. Froman, O.G. Berlin, H. Hill, and G. Stelma. 1999. Comparison of large restriction fragments of *Mycobacterum avium* isolates recovered from AIDS and non-AIDS patients with those of isolates from potable water. *J. Clin. Microbiol.*, 37:1008–1012.

Benson, C.A. 1997. Disseminated *Mycobacterium avium* complex infection: implications of recent clinical trials on prophylaxis and treatment. *AIDS Clin. Rev.*, 271–287.

Bermudez, L.E., M. Petrofsky, P. Kolonoski, and L.S. Young. 1992. An animal model of *Mycobacterium avium* complex disseminated infection after colonization of the intestinal tract. *J. Infect. Dis.*, 165:75–79.

Bermudez, L.E., M. Petrofsky, P. Kolonoski, and L.S. Young. 1998. Emergence of *Mycobacterium avium* populations resistant to macrolides during experimental chemotherapy. *Antimicrob. Agents Chemother.*, 42:180–183.

Bodmer, T., E. Miltner, and L.E. Bermudez. 2000. *Mycobacterium avium* resists exposure to the acidic conditions of the stomach. *FEMS Microbiol. Lett.*, 182:45–49.

British Thoracic Society. 2001. First randomised trial of treatments for pulmonary disease caused by *M. avium intracellulare, M. malmoense,* and *M. xenopi* in HIV-negative patients: rifampicin, ethambutol and isoniazid versus rifampicin and ethambutol. *Thorax,* 56:167–172.

Burman, W.J., and D.l. Cohn. 1996. Clinical Disease in Human Immunodeficiency Virus-Infected Persons. In Mycobacterium avium-*Complex Infection: Progress in Research and Treatment.* Edited by J.A. Korvick and C.A. Benson. New York: Marcel Dekker.

Carbonne, A., N. Lemaitre, M. Bochet, C. Truffot-Pernot, C. Katlama, J. Grosset, F. Bricaire, and V. Jarlier. 1998. *Mycobacterium avium* complex common-source or cross-infection in AIDS patients attending the same day-care facility. *Infect. Control Hosp. Epidemiol.*, 19:784–786.

Carson, L.A., L.A. Bland, L.B. Cusick, M.S. Favero, G.A. Bolan, A.L. Reingold, and R.C. Good. 1988. Prevalence of nontuberculous mycobacteria in water supplies of hemodialysis centers. *Appl. Environ. Microbiol.*, 54:3122–3125.

Carson, L.A., N.J. Petersen, M.S. Favero, and S.M. Aguero. 1978. Growth characteristics of atypical mycobacteria in water and their comparative resistance to disinfectants. *Appl. Environ. Microbiol.*, 36:839–846.

Centers for Disease Control and Prevention (CDC). 1998. *Mycobacterium avium* Complex: Technical Information [Online]. Available: <www.cdc.gov/ncidod/dbmd/diseaseinfo/mycobacteriumavium_t.htm>. [cited November 11, 1999]

Chong, C.Y., and R.N. Husson. 1998. Lack of acceptance of guidelines for prevention of disseminated *Mycobacterium avium* complex infection in infants and children infected with human immunodeficiency virus. *Pediatr. Infect. Dis. J.*, 17:1131–1135.

Cirillo, J.D., S. Falkow, L.S. Tompkins, and L.E. Bermudez. 1997. Interaction of *Mycobacterium avium* with environmental amoebae enhances virulence. *Infect. Immun.*, 65:3759–3767.

Cohn, D.l. 1997. Prevention strategies for *Mycobacterium avium-intracellulare* complex (MAC) infection. A review of recent studies in patients with AIDS. *Drugs*, 54(Suppl)2:8–15.

Detels, R., P. Tarwater, J.P. Phair, J. Margolick, S.A. Riddler, and A. Muñoz. 2001. Effectiveness of potent antiretroviral therapies on the incidence of opportunistic infections before and after AIDS diagnosis. *AIDS*, 15:347–355.

du Moulin, G.C., I.H. Sherman, D.C. Hoaglin, and K.D. Stottmeier. 1985. *Mycobacterium avium* complex, an emerging pathogen in Massachusetts. *J. Clin. Microbiol.*, 22:9–12.

du Moulin, G.C., and K.D. Stottmeier. 1986. Waterborne mycobacteria: an increasing threat to health. *ASM News*, 52:525–529.

du Moulin, G.C., K.D. Stottmeier, P.A. Pelletier, A.Y. Tsang, and J. Hedley-Whyte. 1988. Concentration of *Mycobacterium avium* by hospital hot water systems. *JAMA*, 260:1599–1601.

Eaton, T., J.O. Falkinham, T.O. Aisu, and T.M. Daniel. 1995. Isolation and characteristics of *Mycobacterium avium* complex from water and soil samples in Uganda. *Tuber. Lung Dis.*, 76:570–574.

Edwards, L.B., and D.T. Smith. 1965. Community-wide tuberculin testing study in Pamlico County, North Carolina. *Am. Rev. Respir. Dis.*, 92: 43–54.

Embil, J., P. Warren, M. Yakrus, R. Stark, S. Corne, D. Forrest, and E. Hershfield. 1997. Pulmonary illness associated with exposure to *Mycobacterium-avium* complex in hot tub water. Hypersensitivity pneumonitis or infection? *Chest*, 111:813–816.

Emde, K.M., S.A. Chomyc, and G.R. Finch. 1992. Initial investigation on the occurrence of mycobacterium species in swimming pools. *J. Environ. Health*, 54: 34–37.

Evans, M.J., N.M. Smith, C.M. Thornton, G.G. Youngson, and E.S. Gray. 1998. Atypical mycobacterial lymphadenitis in childhood—a clinicopathological study of 17 cases. *J. Clin. Pathol.*, 51:925–927.

Falcao, D.P., S.R. Valentini, and C.Q.F. Leite. 1993. Pathogenic or potentially pathogenic bacteria as contaminants of fresh water from different sources in Araraquara, Brazil. *Water Res.*, 27: 1737–1741.

Falkinham, J.O., III. 1996. Molecular Epidemiology Techniques for the Study of *Mycobacterium avium*-Complex Infection. In Mycobacterium avium-*Complex Infection: Progress in Research and Treatment*. Edited by J.A.Korvick and C.A.Benson. New York: Marcel Dekker.

Falkinham, J.O., III, C.D. Norton, and M.W. LeChevallier. 2001. Factors Influencing numbers of *Mycobacterium avium*, *Mycobacterium intracellulare*, and other mycobacteria in drinking water distribution systems. *Appl. Environ. Microbiol.*, 67:1225–1231.

Falkinham, J.O., III, B.C. Parker, and H. Gruft. 1980. Epidemiology of infection by nontuberculous mycobacteria. I. Geographic distribution in the eastern United States. *Am. Rev. Respir. Dis.*, 121:931–937.

Frucht, D.M., D.I. Sandberg, M.R. Brown, S.M. Gerstberger, and S.M. Holland. 1999. IL-12-Independent costimulation pathways for interferon-gamma production in familial disseminated *Mycobacterium avium* complex infection. *Clin. Immunol.*, 91:234–241.

Fry, K.L., P.S. Meissner, and J.O. Falkinham III. 1986. Epidemiology of infection by nontuberculous mycobacteria. VI. Identification and use of epidemiologic markers for studies of *Mycobacterium avium, M. intracellulare,* and *M. scrofulaceum. Am. Rev. Respir. Dis.,* 134:39–43.

Gangadharam, P.R., V.K. Perumal, K. Parikh, N.R. Podapati, R. Taylor, D.C. Farhi, and M.D. Iseman. 1989. Susceptibility of beige mice to *Mycobacterium avium* complex infections by different routes of challenge. *Am. Rev. Respir. Dis.,* 139:1098–1104.

Grange, J.M. 1997. Environmental Mycobacteria: Opportunist Disease. In *Medical Microbiology: A Guide to Microbial Infections: Pathogenesis, Immunity, Laboratory Diagnosis and Control.* Edited by D. Greenwood, R. Slack, and J. Peutherer. New York: Churchill Livingstone, Inc.

Griffith, D.E. 1999. Risk-benefit assessment of therapies for *Mycobacterium avium* complex infections. *Drug Saf.,* 21:137–152.

Griffith, D.E., B.A. Brown, W.M. Girard, B.E. Griffith, L.A. Couch, and R.J. Wallace Jr. 2001. Azithromycin-containing regimens for treatment of *Mycobacterium avium* complex lung disease. *Clin. Infect. Dis.,* 32:1547–1553.

Grosset, J., and B. Ji. 1997. Prevention of the selection of clarithromycin-resistant *Mycobacterium avium-intracellulare* complex. *Drugs,* 54(Suppl)2:23–27.

Havelaar, A.H., L.G. Berwald, D.G. Groothuis, and J.G. Baas. 1985. Mycobacteria in semi-public swimming pools and whirlpools. *Zentralbl. Bakteriol. Mikrobiol. Hyg. [B],* 180:505–514.

Havlir, D.V., R.D. Schrier, F.J. Torriani, K. Chervenak, J.Y. Hwang, and W.H. Boom. 2000. Effect of potent antiretroviral therapy on immune responses to *Mycobacterium avium* in human immunodeficiency virus-infected subjects. *J. Infect. Dis.,* 182:1658–1663.

Hellyer, T.J., I.N. Brown, M.B. Taylor, B.W. Allen, and C.S. Easmon. 1993. Gastrointestinal involvement in *Mycobacterium avium-intracellulare* infection of patients with HIV. *J. Infect.,* 26:55–66.

Hewitt, R.G., G.D. Papandonatos, M.J. Shelton, C.B. Hsiao, B.J. Harmon, S.R. Kaczmarek, and D. Amsterdam. 1999. Prevention of disseminated *Mycobacterium avium* complex infection with reduced dose clarithromycin in patients with advanced HIV disease. *AIDS,* 13:1367–1372.

Hocqueloux, L., P. Lesprit, J.L. Herrmann, B.A. de La, A.M. Zagdanski, J.M. Decazes, and J. Modai. 1998. Pulmonary *Mycobacterium avium* complex disease without dissemination in HIV-infected patients. *Chest*, 113:542–548.

Holland, S.M. 2001. Nontuberculous mycobacteria. *Am. J. Med. Sci.*, 321:49–55.

Horsburgh, C.R., Jr. 1991. *Mycobacterium avium* complex infection in the acquired immunodeficiency syndrome. *N. Engl. J. Med.*, 324:1332–1338.

Horsburgh, C.R., Jr. 1996. Epidemiology of *Mycobacterium avium* Complex. In Mycobacterium avium-*Complex Infection: Progress in Research and Treatment*. Edited by J.A. Korvick and C.A. Benson. New York: Marcel Dekker.

Horsburgh, C.R., Jr. 1999. The pathophysiology of disseminated *Mycobacterium avium* complex disease in AIDS. *J. Infect. Dis.*, 179(Suppl)3:S461–S465.

Horsburgh, C.R., Jr., D.P. Chin, D.M. Yajko, P.C. Hopewell, P.S. Nassos, E.P. Elkin, W.K. Hadley, E.N. Stone, E.M. Simon, and P. Gonzalez. 1994. Environmental risk factors for acquisition of *Mycobacterium avium* complex in persons with human immunodeficiency virus infection. *J. Infect. Dis.*, 170:362–367.

Howell, N., P.A. Heaton, and J. Neutze. 1997. The epidemiology of nontuberculous mycobacterial lymphadenitis affecting New Zealand children 1986–95. *N. Z. Med. J.*, 110:171–173.

Ichiyama, S., K. Shimokata, and M. Tsukamura. 1988. The isolation of *Mycobacterium avium* complex from soil, water, and dusts. *Microbiol. Immunol.*, 32:733–739.

Iivanainen, E., P.J. Martikainen, P. Vaananen, and M.L. Katila. 1999. Environmental factors affecting the occurrence of mycobacteria in brook sediments. *J. Appl. Microbiol.*, 86:673–681.

Iseman, M.D. 1996. Pulmonary Disease Due to *Mycobacterium avium* Complex. In Mycobacterium avium-*Complex Infection: Progress in Research and Treatment*. Edited by J.A. Korvick and C.A. Benson. New York: Marcel Dekker.

Iseman, M.D., R.F. Corpe, R.J. O'Brien, D.Y. Rosenzwieg, and E. Wolinsky. 1985. Disease due to *Mycobacterium avium-intracellulare*. *Chest*, 87:139S–149S.

Katila, M.L., E. Iivanainen, P. Torkko, J. Kauppinen, P. Martikainen, and P. Vaananen. 1995. Isolation of potentially pathogenic mycobacteria in the Finnish environment. *Scand. J. Infect. Dis. Suppl.*, 98:9–11.

Kennedy, T.P., and D.J. Weber. 1994. Nontuberculous mycobacteria. An underappreciated cause of geriatric lung disease. *Am. J. Respir. Crit. Care Med.*, 149:1654–1658.

Khoor, A., K.O. Leslie, H.D. Tazelaar, R.A. Helmers, and T.V. Colby. 2001. Diffuse pulmonary disease caused by nontuberculous mycobacteria in immunocompetent people (hot tub lung). *Am. J. Clin. Pathol.*, 115:755–762.

Kirschner, R.A., Jr., B.C. Parker, and J.O. Falkinham III. 1992. Epidemiology of infection by nontuberculous mycobacteria. *Mycobacterium avium, Mycobacterium intracellulare,* and *Mycobacterium scrofulaceum* in acid, brown-water swamps of the southeastern United States and their association with environmental variables. *Am. Rev. Respir. Dis.*, 145:271–275.

Kubo, K., Y. Yamazaki, T. Hachiya, M. Hayasaka, T. Honda, M. Hasegawa, and S. Sone. 1998. *Mycobacterium avium-intracellulare* pulmonary infection in patients without known predisposing lung disease. *Lung,* 176:381–391.

Kubo, K., Y. Yamazaki, M. Hanaoka, H. Nomura, K. Fujimoto, T. Honda, M. Ota, and Y. Kamijou. 2000. Analysis of HLA antigens in *Mycobacterium avium-intracellulare* pulmonary infection. *Am. J. Respir. Crit. Care Med.*, 161:1368–1371.

Mansfield, K.G., and A.A. Lackner. 1997. Simian immunodeficiency virus-inoculated macaques acquire *Mycobacterium avium* from potable water during AIDS. *J. Infect. Dis.*, 175:184–187.

Martin, E.C., B.C. Parker, and J.O. Falkinham III. 1987. Epidemiology of infection by nontuberculous mycobacteria. VII. Absence of mycobacteria in southeastern groundwaters. *Am. Rev. Respir. Dis.*, 136:344–348.

Miyamoto, M., Y. Yamaguchi, and M. Sasatsu. 2000. Disinfectant effects of hot water, ultraviolet light, silver ions and chlorine on strains of *Legionella* and nontuberculous mycobacteria. *Microbios.*, 101:7–13.

Montecalvo, M.A., G. Forester, A.Y. Tsang, G.C. du Moulin, and G.P. Wormser. 1998. Colonisation of potable water with *Mycobacterium avium* complex in homes of HIV-infected patients. *Lancet,* 343:1639–1639.

Morrissey, A.B., T.O. Aisu, J.O. Falkinham III, P.P. Eriki, J.J. Ellner, and T.M. Daniel. 1992. Absence of *Mycobacterium avium* complex disease in patients with AIDS in Uganda. *J. Acquir. Immune Defic. Syndr.*, 5:477–478.

Najm, I., and R.R. Trussell. 1999. New and Emerging Drinking Water Treatment Technologies. In *Identifying Future Drinking Water Contaminants.* Washington, DC: National Academy Press.

Newport, M.J., C.M. Huxley, S. Huston, C.M. Hawrylowicz, B.A. Oostra, R. Williamson, and M. Levin. 1996. A mutation in the interferon-gamma-receptor gene and susceptibility to mycobacterial infection. *N. Engl. J. Med.*, 335:1941–1949.

Nightingale, S.D., L.T. Byrd, P.M. Southern, J.D. Jockusch, S.X. Cal, and B.A. Wynne. 1992. Incidence of *Mycobacterium avium-intracellulare* complex bacteremia in human immunodeficiency virus-positive patients. *J. Infect. Dis.*, 165:1082–1085.

Norton, C., M. LeChevallier, J.O. Falkingham III, M. Williams, and R. Taylor. 1999. *Mycobacterium avium* in drinking water. Presented at *AWWA International Symposium on Waterborne Pathogens,* August 29–September 1, 1999, Milwaukee, WI.

O'Brien, D.P., B.J. Currie, and V.L. Krause. 2000. Nontuberculous mycobacterial disease in northern Australia: a case series and review of the literature. *Clin. Infect. Dis.*, 31:958–967.

O'Brien, R.J., L.J. Geiter, and D.E. Snider Jr. 1987. The epidemiology of nontuberculous mycobacterial diseases in the United States. Results from a national survey. *Am. Rev. Respir. Dis.*, 135:1007–1014.

Olivier, K.N. 1998. Nontuberculous mycobacterial pulmonary disease. *Curr. Opin. Pulm. Med.*, 4:148–153.

Palella, F.J., Jr., K.M. Delaney, A.C. Moorman, M.O. Loveless, J. Fuhrer, G.A. Satten, D.J. Aschman, and S.D. Holmberg. 1998. Declining morbidity and mortality among patients with advanced human immunodeficiency virus infection. HIV Outpatient Study Investigators. *N. Engl. J. Med.*, 338:853–860.

Pelletier, P.A., G.C. du Moulin, and K.D. Stottmeier. 1988. Mycobacteria in public water supplies: comparative resistance to chlorine. *Microbiol. Sci.*, 5:147–148.

Perlman, D.C., R. D'Amico, and N. Salomon. 2001. Mycobacterial infections of the head and neck. *Curr. Infect. Dis. Rep.*, 3:233–241.

Peters, M., C. Muller, S. Rusch-Gerdes, C. Seidel, U. Gobel, H.D. Pohle, and B. Ruf. 1995. Isolation of atypical mycobacteria from tap water in hospitals and homes: is this a possible source of disseminated MAC infection in AIDS patients? *J. Infect.*, 31:39–44.

Prince, D.S., D.D. Peterson, R.M. Steiner, J.E. Gottlieb, R. Scott, H.L. Israel, W.G. Figueroa, and J.E. Fish. 1989. Infection with *Mycobacterium avium* complex in patients without predisposing conditions. *N. Engl. J. Med.*, 321:863–868.

Raju, B., and N.W. Schluger. 2000. Significance of respiratory isolates of *Mycobacterium avium* complex in HIV-positive and HIV-negative patients. *Int. J. Infect. Dis.*, 4:134–139.

Reddy, V.M. 1998. Mechanism of *Mycobacterium avium* complex pathogenesis. *Front Biosci.*, 3:d525–d531.

Reich, J.M., and R.E. Johnson. 1991. *Mycobacterium avium* complex pulmonary disease. Incidence, presentation, and response to therapy in a community setting. *Am. Rev. Respir. Dis.*, 143:1381–1385.

Rusin, P.A., J.B. Rose, C.N. Haas, and C.P. Gerba. 1997. Risk assessment of opportunistic bacterial pathogens in drinking water. *Rev. Environ. Contam. Toxicol.*, 152:57–83:57–83.

Sabater, J.F., and J.M. Zaragoza. 1993. A simple identification system for slowly growing mycobacteria. II. Identification of 25 strains isolated from surface water in Valencia (Spain). *Acta. Microbiol. Hung.*, 40:343–349.

Schulze-Robbecke, R. 1993. Mycobacteria in the environment. *Immun. Infekt.*, 21:126–131.

Schulze-Robbecke, R., and K. Buchholtz. 1992. Heat susceptibility of aquatic mycobacteria. *Appl. Environ. Microbiol.*, 58:1869–1873.

Singh, N., and V.L. Yu. 1994. Potable water and *Mycobacterium avium* complex in HIV patients: is prevention possible? *Lancet*, 343:1110–1111.

Takahashi, M., A. Ishizaka, H. Nakamura, K. Kobayashi, M. Nakamura, M. Namiki, T. Sekita, and S. Okajima. 2000. Specific HLA in pulmonary MAC infection in a Japanese population. *Am. J. Respir. Crit. Care Med.*, 162:316–318.

Taylor, J.S., and M. Wiesner. 1999. Membranes. In *Water Quality and Treatment: A Handbook of Community Water Supplies.* 5th ed. Edited by American Water Works Association. New York: McGraw-Hill.

Taylor, R.H., J.O. Falkinham III, C.D. Norton, and M.W. LeChevallier. 2000. Chlorine, chloramine, chlorine dioxide, and ozone susceptibility of *Mycobacterium avium*. *Appl. Environ. Microbiol.*, 66:1702–1705.

Telles, M.A., M.D. Yates, M. Curcio, S.Y. Ueki, M. Palaci, D.J. Hadad, F.A. Drobniewski, and A.L. Pozniak. 1999. Molecular epidemiology of *Mycobacterium avium* complex isolated from patients with and without AIDS in Brazil and England. *Epidemiol. Infect.*, 122:435–440.

Tsuyuguchi, K., K. Suzuki, H. Matsumoto, E. Tanaka, R. Amitani, and F. Kuze. 2001. Effect of oestrogen on *Mycobacterium avium* complex pulmonary infection in mice. *Clin. Exp. Immunol.*, 123:428–434.

von Reyn, C.F., R.D. Arbeit, A.N. Tosteson, M.A. Ristola, T.W. Barber, R. Waddell, C.H. Sox, R.J. Brindle, C.F. Gilks, A. Ranki, C. Bartholomew, J. Edwards, J.O. Falkinham, III, and G.T. O'Connor. 1996. The international epidemiology of disseminated *Mycobacterium avium* complex infection in AIDS. International MAC Study Group. *AIDS*, 10:1025–1032.

von Reyn, C.F., T.W. Barber, R.D. Arbeit, C.H. Sox, G.T. O'Connor, R.J. Brindle, C.F. Gilks, K. Hakkarainen, A. Ranki, and C. Bartholomew. 1993a. Evidence of previous infection with *Mycobacterium avium-Mycobacterium intracellulare* complex among healthy subjects: an international study of dominant mycobacterial skin test reactions. *J. Infect. Dis.*, 168:1553–1558.

von Reyn, C.F., J.N. Maslow, T.W. Barber, J.O. Falkinham III, and R.D. Arbeit. 1994. Persistent colonisation of potable water as a source of *Mycobacterium avium* infection in AIDS. *Lancet*, 343:1137–1141.

von Reyn, C.F., R.D. Waddell, T. Eaton, R.D. Arbeit, J.N. Maslow, T.W. Barber, R.J. Brindle, C.F. Gilks, J. Lumio, and J. Lahdevirta. 1993b. Isolation of *Mycobacterium avium* complex from water in the United States, Finland, Zaire, and Kenya. *J. Clin. Microbiol.*, 31:3227–3230.

Wayne, L.G., L.S. Young, and M. Bertram. 1986. Absence of mycobacterial antibody in patients with acquired immune deficiency syndrome. *Eur. J. Clin. Microbiol.*, 5:363–365.

Williams, P.L., J.S. Currier, and S. Swindells. 1999. Joint effects of HIV-1 RNA levels and CD4 lymphocyte cells on the risk of specific opportunistic infections. *AIDS*, 13:1035–1044.

Yajko, D.M., D.P. Chin, P.C. Gonzalez, P.S. Nassos, P.C. Hopewell, A.L. Reingold, C.R. Horsburgh, M.A. Yakrus, S.M. Ostroff, and W.K. Hadley. 1995. *Mycobacterium avium* complex in water, food, and soil samples collected from the environment of HIV-infected individuals. *J. Acquir. Immune Defic. Syndr. Hum. RetroVirol.*, 9:176–182.

Yoder, S., C. Argueta, A. Holtzman, T. Aronson, O.G. Berlin, P. Tomasek, N. Glover, S. Froman, and G. Stelma Jr. 1999. PCR comparison of *Mycobacterium avium* isolates obtained from patients and foods. *Appl. Environ. Microbiol.*, 65:2650–2653.

CHAPTER 14

Characterizing Microbial Risk

by Rebecca T. Parkin

Introduction

Risk characterization is the final step of the microbial risk assessment paradigm. At this phase of the process, relevant and scientifically reliable results of the host–pathogen and exposure profiles are synthesized into a coherent, complete, and useful summary for risk management decision-making (USEPA 2000a). In addition, the science policy decisions are made and the uncertainties associated with those decisions are described (USEPA 2000a). Risk synthesis relies on modeling techniques; exposure measurement; and epidemiological, clinical, toxicological, and population knowledge available at the time of the assessment (USEPA 2000a; Soller, Eisenberg, and Oliveri 1999; Teunis and Havelaar 1999). The success of risk characterization depends on the previous steps of the risk assessment paradigm. These steps include the clarity of the problem statement, the data available to address the risk problems of concern, the validity of the conceptual model and statistical methods used to conduct and test the model, the analysis of uncertainties, and the nature and extent of assumptions made to complete the risk estimates (Rhomberg 1995; Soller, Eisenberg, and Oliveri 1999; Teunis and Havelaar 1999). A multidisciplinary team of experts is often needed if this phase of risk assessment is to be conducted successfully. In this chapter, we describe the goals of risk characterization, examine the steps within this phase of risk assessment, and discuss the unique issues that must be addressed in microbial pathogen risk management problems.

Goals of Risk Characterization

Components

In the risk characterization step, the potential risks of adverse health outcomes and the uncertainties about those risk estimates are presented (Rhomberg 1995). The estimates, uncertainties, and major contributing factors to the estimates are discussed in the contexts of problem setting, current scientific knowledge, uncertainties, and assumptions (Figure 14-1) (Soller, Eisenberg, and Oliveri 1999). Risk management options are examined, the sources of uncertainty and variability in the model are evaluated, and the host–pathogen findings of the previous steps in the risk assessment are integrated through systematic consideration and interpretation (ILSI 2000). Ideally, the risk characterization step links the conceptual model used in the analyses with the original risk problem. The data and discussions presented in this phase must respond specifically to the questions raised in the problem-formulation step. For example, if high-risk subgroups were identified in the problem statement, the probability of risks for these groups needs to be addressed separately and explicitly. Details may include who would be protected from the risks and at what levels of exposure and to what extent that protection could be realized. Typically, the risk probabilities are presented for each of a series of risk management scenarios. The expected public health impacts of the adverse health outcomes are discussed, supported by appropriate measurements (Teunis and Havelaar 1999).

Stakeholders

One purpose of risk characterization is to provide the breadth and depth of information stakeholders need to participate effectively in risk management decision processes. The US Environmental Protection Agency's (USEPA's) risk characterization policy states that the final product must be clear, consistent, transparent, and reasonable (USEPA 1995). The policy calls for the explicit identification and potential impacts of all major sources of uncertainty, an exploration of a range of exposure scenarios, the use of multiple risk descriptors, estimation of the population health risks for subgroups, and identification of the most sensitive variables in the model, with their impacts on the final estimates. In addition, the findings must be scientifically sound, thoroughly critiqued by the assessor, and relevant to the problems identified at the beginning of the risk assessment process and to the range of stakeholders involved in the risk management decision making. USEPA also calls for risk assessors to address 13 key elements and write the characterization with enough detail so that other experts can reconstruct the steps taken to reach the reported result (USEPA 2000a).

- Link the conceptual model with the problem statement
- Respond to questions raised in the problem-formulation phase
- Integrate the results from the previous steps of the process
- Present risks of adverse health outcomes, as specified in the problem statement
  - —General population
  - —Subpopulations
- Describe the level of confidence in the risk estimates
  - —Evaluate sources of uncertainty
  - —Evaluate sources of variability
- Discuss the risk estimates in the contexts of the
  - —Problem setting
  - —Current scientific knowledge
  - —Assumptions made
  - —Uncertainties
- Discuss the public health impacts of the adverse health outcomes
- Examine risk management options
  - —Present risk estimates for each of a series of risk management options

Figure 14-1 Goals of microbial pathogen risk characterization

Methods

To achieve these goals, the risk assessor should use a clear framework to organize and present the risk probability estimates by risk scenario and/or subpopulation of concern; describe the impacts of the data quality, modeling processes, uncertainties, assumptions, and defaults used on the final risk estimates; and project the potential impacts of the methodological limitations on the risk decisions to be made (USEPA 2000a). The primary methods used by risk assessors to conduct risk characterizations include inference tools and assumptions (Rhomberg 1995; Haas, Rose, and Gerba 1999), epidemiologically based mathematical parameterization and modeling techniques (Soller, Eisenberg, and Oliveri 1999; Teunis and Havelaar 1999), and evaluations of the sources of uncertainty and variability in the final risk estimate (Haas, Rose, and Gerba 1999). When extensive data are available, quantitative risk estimates can be generated. More often, data are missing so that semi-quantitative or qualitative risk characterizations are completed.

Various approaches can be used to characterize the risks to susceptible subpopulations. On the simplest level, point estimates may be calculated. However, increasing awareness of the many sources of variability that affect susceptibility, exposures, and outcomes has led to common use of interval estimates (Haas, Rose, and Gerba 1999). The probability of response across the spectrum of outcome severity (e.g., the range of acute to chronic outcomes) can be estimated, and specific subgroups more likely than others to appear in specific areas of the spectrum can be identified. The quality of the data and extent of assumptions made for different subpopulations are important to point out, especially if data are of highly variable quality for different groups. Because of the number of potentially influential factors and the complexity of the relationships between factors and outcomes, a subsection specifically focused on the issues related to susceptibility is an appropriate way to ensure that subgroup concerns are not missed or inadequately considered.

Steps in Characterizing Risks

The International Life Sciences Institute (ILSI) paradigm provides a series of steps for completing a microbial pathogen risk characterization (ILSI 2000). The two major steps are risk estimation and risk description. The latter step includes five substeps: description of uncertainty and variability, evaluation of the variables, discussion of how well the risk assessment addresses the initial problem statement, evaluation of the risk management options, and analysis of the risk management options.

In risk estimation, the probability of risks at given levels of exposure are estimated from the health and exposure characterization data developed earlier in the risk assessment process. These estimates of the type, severity, latency, and magnitude of adverse health effects may be calculated on an individual, subgroup, or population scale (USEPA 2000b; Soller, Eisenberg, and Oliveri 1999; Teunis and Havelaar 1999). Quantitative estimates are usually preferred but may not be achievable with available data or modeling methods. The assumptions used to make the estimations need to be identified and their potential impacts discussed (ILSI 2000). If the assumptions could have differential impacts on susceptible subpopulations, their impacts on subgroup risk estimations need to be described separately (USEPA 2000b).

In risk description, the adverse health outcomes (nature, severity, and consequences) and the weight of the evidence for those outcomes are examined and documented. The focus of this substep is the evaluation of the impacts of variable inputs, including various population and exposure scenarios (ILSI 2000). The evaluation must be closely tied to the original risk problem statement to ensure that key issues are addressed. To understand the potential impacts of changing the value of any particular variable in the model, the assessor must explore the sources of uncertainty and variability in the data and evaluate the sources' impacts on the risk estimates. The estimates are then examined to determine the weight of the evidence (i.e., the confidence the assessor has in the veracity of the estimates). If the data are determined to be of high quality and sufficient to establish a causal chain between the exposure and response, the assessor can consider whether the chain is biologically plausible (ILSI 2000). For example, Soller and colleagues (1999) used outbreak data to assess how closely their risk estimates related to a real microbial pathogen risk event. Although some researchers (Hunter 2001) have questioned the value of using outbreak data to estimate the actual scope of microbial pathogen outbreaks, no better option has been proposed for risk characterization purposes. In many cases, this plausibility assessment may only be possible in qualitative terms. In addition, it is important to note that there is no generally accepted definition of "biological plausibility" (Weed and Hursting 1998). As a result, assessors should take care in describing the criteria they use to evaluate the data for causality.

The second substep of risk description has received considerable methodological attention in recent years. The effort to conduct a sensitivity analysis of the variables used in the statistical model requires multivariate methods, advanced computational systems, a clear problem statement, and thorough analytic reasoning abilities (Soller, Eisenberg, and Oliveri 1999; ILSI 2000). The assessor must consider the variables' roles in or proximal to the causal chain of events and determine the importance of the variables' impacts on the final risk estimate. It may be that some factors are not at all important to some subgroups but are key factors to others. In efforts to understand the impacts of variables, the Monte Carlo statistical method (Gilks, Richardson, and Spiegelhalter 1996) has been used to simulate the distribution of parameter values from several hundreds or thousands of populations that could have been studied. The distributions of many variables and the potential impacts of values at the extreme ends of the distributions must be considered. Once the contribution of each variable to the risk estimate has been evaluated, the assessor describes which variables explain most of the risk estimate and whether the contribution of specific variables changes according to subpopulation (Teunis and Havelaar 1999). Assessors also indicate which variables would benefit from more data and thereby could reduce uncertainties in future models.

In the third substep of risk description, the assessor is charged with discussing how well the risk estimation process has addressed the original risk problem formulation. Issues typically considered in this substep include whether the data available were well-matched to the risk problem, whether a sufficient conceptual model could be constructed from existing scientific knowledge, whether appropriate statistical methods were available, and whether the risk assessment was able to address all of the components of the original risk problem.

Next, the risk assessor considers the range of risk management options available to control the risks estimated. Typically, different exposure scenarios relevant to the risk problem are examined with and without simulated controls. A focus on the variables that contribute the most to the risk estimates can reveal the more effective risk management approaches. Sometimes, new risk management options emerge at this point in the analysis, based on the detailed understanding gained in the prior substep.

To complete the risk characterization, risk options' advantages and disadvantages are compared; the risk options may be prioritized based on criteria previously set by the stakeholders. The risk assessor may identify the option that appears best from a scientific viewpoint or recommend that decision-makers consider a limited range of the options examined.

Overall, risk characterization demands advanced scientific reasoning by the assessor. It is a challenging step that requires comprehensive knowledge of microbial pathogen risk scenarios, disease processes, and outbreak mechanisms (as in Soller, Eisenberg, and Oliveri 1999; Teunis and Havelaar 1999). A number of problems emerge during this phase that need to be addressed systematically and thoroughly to ensure that the final estimate is a credible product. For example, in the risk estimation process, the complexities of the human immune system and secondary spread of disease in human populations need to be fully understood. The dynamics of infectious disease and change of individuals' susceptibility status need to be incorporated in the risk estimation model. Time-dependent and exposure-dependent processes may require complex modeling methods and statistical computing approaches. The definition and measurement of variables or the relationships between variables may not seem difficult until this stage in the assessment process is reached. An earlier lack of clarity or limited reliability of measurements may raise numerous questions that must be answered before the risk characterization can be completed. It may not be possible to construct a comprehensive conceptual model or collect the large amount of data needed to fully operationalize a model quantitatively. Decisions needed at this point will relate to the model's scope, the modeling assumptions, and the scale or type of data needed to use the model. A complete understanding of the impacts of the sources of uncertainty and variability can be difficult. It is best facilitated by having a structured approach for considering issues thoroughly and in a logical sequence.

Unique Issues in Microbial Risk Characterization

When critiquing the first version of ILSI's microbial risk assessment paradigm, Soller and coworkers (1999) and Teunis and Havelaar (1999) identified problems that are common to risk assessment frameworks but also found several important issues that are unique to microbial risk problems. They recognized that the clarity of the problem statement in the problem-formulation phase is pivotal to the entire process. Without a clear statement, assessors can misinterpret managers' primary concerns. Soller and colleagues (1999) pointed out that clarification of the assessment's goal, breadth, and scope are important foundations for the conceptual model (the schematic description of the exposure and risk process) and for the organization of the data available to generate risk estimates using the model.

Teunis and Havelaar (1999) noted the importance of having clearly stated exposures and outcomes, including definitions of the outcomes in terms of severity and duration. They stated that it is particularly important in modeling microbial pathogen outcomes to understand the clinical course of the adverse health outcomes, to conceptualize the disease process with appropriate population movement between susceptibility states, and to develop a statistical paradigm that is stochastic and based on conditional probabilities. Although there are limited data on the treatability of pathogens in drinking water and on subpopulation consumption patterns, it may be dangerous to use models that are too simplistic. Teunis and Havelaar (1999) pointed out that detailed data on the population's age distribution and immune status and better methods for typing pathogens and estimating consumption in subpopulations are needed.

Several factors make microbial pathogen risks different from either chemical or radiological hazards in drinking water, and this makes microbial pathogen risk characterization unique. One set of factors relates to the characteristics of the actual microorganisms. For example, microbial pathogens replicate and some die before others. There is a wide variation in the virulence of organisms, so data from even a similar organism may not be useful for risk assessment purposes. Some pathogens remain viable in the environment over long periods of time, possibly because they are resistant to a wide range of ambient temperatures and destruction by other pathogens or predators. Others are readily treated or damaged in the environment, significantly reducing the total number of infectious microorganisms in a water sample, despite a high concentration of organisms in a source (RPR 2000).

Another factor that makes microbial pathogen risk characterization unique is the secondary spread of disease among humans. Microbial pathogens and their infectious capacities may change during their passage through one human to another. People may not need to be in contact with the original source of the pathogens to become ill but may only need contact with another individual who is infected. The dynamic nature of both pathogen and population characteristics affect secondary spread, making it particularly challenging to model statistically (Soller, Eisenberg, and Oliveri 1999). Stochastic methods are necessary but require more data, computing power, and technical ability (Teunis and Havelaar 1999).

In addition, the movement of people in and out of different states of susceptibility may be very important to estimating vulnerable subpopulations' risks and illnesses both at a point in time as well as later in life when long-term sequelae may emerge from early-life exposures. Few risk assessors to date have attempted to model such important population dynamics (Soller, Eisenberg, and Oliveri 1999).

The variable clinical course of many pathogen-related disease states (relapses of acute illnesses, adverse sequelae, potentially related chronic health outcomes, etc.) are typically poorly understood but may result in elevated mean symptom duration times and variability. These skewed statistics may not be relevant to most of the population but may be critical to selected subgroups (Teunis and Havelaar 1999). Appropriate disease models are crucial to organizing events into a plausible chain using conditional probabilities (Soller, Eisenberg, and Oliveri 1999; Teunis and Havelaar 1999).

Summary

Characterization of waterborne microbial risks requires different strategies and methods than are used for chemical and radiological hazards. Our limited scientific knowledge of dose–response relationships and the unusual statistical nature of the outcome data make it especially difficult to model many microbial disease processes at this time. Advancement of available statistical models will require much more data and knowledge than is foreseeable in the near future. However, efforts now to begin working out the issues that are unique to microbial risk assessment will be invaluable in uncovering additional issues and identifying promising methods to test.

Bibliography

Gilks, W.R., S. Richardson, D.J. Spiegelhalter, eds. 1996. *Markov Chain Monte Carlo in Practice.* London: Chapman and Hall.

Haas, C.N., J.B. Rose, and C.P. Gerba. 1999. *Quantitative Microbial Risk Assessment.* New York: John Wiley & Sons.

Hunter, P. 2001. Problems in assessing risk from microbial pathogens in drinking water. Presentation at the *Microbial/Disinfection By-Products Health Effects Symposium.* Lisle, IL: American Water Works Association.

International Life Sciences Institute (ILSI). 2000. *Revised Framework for Microbial Risk Assessment.* Washington, DC: ILSI.

Rhomberg, L.R. 1995. *A Survey of Methods for Chemical Health Risk Assessment Among Federal Regulatory Agencies.* Washington, DC: National Commission on Risk Assessment and Risk Management.

Risk Policy Report (RPR). 2000. EPA Experts Crafting Microbial Risk Assessment Framework.

Soller, J.A., J.N. Eisenberg, and A.W. Oliveri. 1999. *Evaluation of Pathogen Risk Assessment Framework.* Oakland, CA: Eisenberg, Oliveri & Associates, Inc.

Teunis, P.F.M., and A.H. Havelaar. 1999. Cryptosporidium *in drinking water: Evaluation of the ILSI/RSI quantitative risk assessment framework.* Report No. 284 550 006. Bilthoven, The Netherlands: National Institute of Public Health and the Environment.

US Environmental Protection Agency (USEPA). 1995. *Policy for Risk Characterization.* 1–14. Washington, DC: USEPA. Appendix A in EPA. 2000. *Risk Characterization Handbook.* EPA 100-B-00-002. Washington, DC: USEPA.

US Environmental Protection Agency (USEPA). 2000a. *Risk Characterization Handbook.* EPA 100-B-00-002. Washington, DC: USEPA.

US Environmental Protection Agency (USEPA). 2000b. *Report to Congress: EPA Studies on Sensitive Subpopulations and Drinking Water Contaminants.* EPA 815-R-00-015. Washington, DC: USEPA.

Weed, D.L., and S.D. Hursting. 1998. Biologic plausibility in causal inference: current method and practice. *Am. J Epidemiol.*, 147:415–25.

CHAPTER 15

Risk Communication

by Rebecca T. Parkin and Martha A. Embrey

Introduction

Risk communication is a widely used but not always consistently understood term. Although there is greater awareness that the issues involved in good risk communication are more complicated than once thought, much confusion remains about what risk communication is, what it takes to communicate effectively, and why more rigorous attention to communication is important. Risk assessors, managers, and stakeholders increasingly realize that communication issues can significantly affect the outcome and success of a risk management program (NERAM 2000).

Risk communication has been conceptualized on several different levels—interpersonal, organizational, societal, and cultural. Each approach provides a different but complementary perspective on communication. Although important in the field of risk communication, the organizational (e.g., Chess 2001), societal (e.g., Kasperson et al. 1988), and cultural (e.g., Douglas and Wildavsky 1982) levels will not be emphasized in this chapter.

Risk Perception

On the individual level, risk is viewed and defined differently by different people; an individual's concept of risk is partially influenced by his or her cognitive and affective states, values, and beliefs. Also, concepts about "safety" and what risks are safer relative to other risks are important to an individual's risk decision-making framework (Slovic 2000). Perceptions of risk are known to be influenced by many factors, such as the familiarity of the risk, fears associated with the risk, past events that are recalled and linked to the risk, underestimation of potential harm from the risk to one's self versus others, trust in the source of the risk information, demographic and other personal characteristics, knowledge related to the risk, etc. Considerable research has demonstrated that expert and nonexpert views of risk may differ in important ways and that those differences may have significant repercussions on risk communication strategies (e.g., Kahneman, Slovic, and Tversky 1982; Slovic 1987; Weinstein 1987; Stallen and Tomas 1988, Morgan et al. 2002).

A mental model is a well-established concept in psychology and has been the focus of extensive research (Atman et al. 1994; Bostrom, Fischhoff, and Morgan 1992; Johnson-Laird 1983; Morgan et al. 2002). A person's "mental model" can be thought of as a complex web of deeply held beliefs that operates below the conscious level to affect how an individual defines a problem, reacts to issues, and makes decisions about messages and options concerning topics that come to their attention through communications. Mental models tend to prevent people from seeing alternative perspectives and define boundaries of thought and action, thereby limiting people to familiar patterns of reasoning and action. It is well established that people's mental models vary in important but often unpredictable ways and that risk decisions are strongly affected by these mental models (Fischhoff and Downs 1997; Morgan et al. 2002). Experts and stakeholders alike have challenges associated with the way in which they think about how to communicate about risk. However, over the last 30 years, we have learned that speculation about stakeholders' interests and priorities is naïve at best and often risky when designing communication strategies. Stakeholders typically address decisions from a different conceptual framework (mental model) than experts and they use different terms.

Challenges

People given the same environmental risk information will come to different conclusions based on their beliefs, perceptions, and values. Risk communication, then, is challenging in part because of those different interpretations. Risk mitigation usually depends on voluntary compliance, and the ability to communicate to all stakeholders is important to obtain compliance. However, US society values independent rights—personal control is important. For example, 1,225 people were surveyed regarding their perceptions of who had the ability to mitigate health risks. The survey showed that the participants believed that certain risks were under individual control (such as AIDS transmission, heart disease, and automobile accidents), but that experts had more ability to control risks such as chemical waste and air and water pollution (O'Connor, Bord, and Fisher 1998).

Traditional, one-way information and education strategies do not work well when stakeholders must make important tradeoffs, risks and their acceptability are uncertain or controversial, the credibility of organizational leaders is low or under pressure, communications are technical and hard to understand, messages are readily misunderstood, forced solutions are not necessary, and program results must be sustained over time. It is clear that effective communication strategies, their key elements, and key messages must work within the context of stakeholders' mental models, particularly in lower trust–higher concern environments.

Changing Societal Contexts

During the 1990s, there were many important societal changes. Not only did detection and information technologies improve, but the population's drinking water practice and disease outbreak patterns shifted. New public laws, mandates, and policies resulted from increased demands for public participation in risk management processes and greater concerns for at-risk subpopulations. These changes created a more complex and dynamic setting in which water-related risk issues had to be assessed and addressed.

Management Processes

For example, in the United States and Canada, new risk management paradigms were developed to reflect the increased recognition that public participation and risk communication play important roles throughout the risk management process (CSA 1997; P/C 1997). Public concerns and increasing activism regarding drinking water and susceptible subpopulation issues also resulted in stakeholders expecting more and earlier participation in public decision-making processes (NRC 2001). Although efforts had been made in both the United States and Europe to include public representatives in risk management processes, the public is increasingly calling for openness and transparency. Through their personal experiences with stakeholders and implemented programs, risk managers learned that the open, ongoing exchange of ideas and information is fundamental and essential to the risk management process (NERAM 2000). Stakeholder participation involves and informs the public and is founded on the concept that people are capable of making sound decisions. The overall trend in risk management is to develop more inclusive processes that are based on the effective application of risk communication knowledge and methods (NRC 2001).

Epidemiologic Impacts

Several well-publicized disease outbreaks involving many thousands of people increased the awareness and urgency of improving risk communication strategies for waterborne hazards. The cryptosporidiosis outbreak in

Milwaukee in 1993 changed the public's view of water contamination and outbreaks (Griffin, Dunwoody, and Zabala 1998). After identifying municipal drinking water as the source of an enteric disease outbreak source, Milwaukee issued a boil-water advisory, then lifted it a week later, when the water did not test positive for the oocysts. A survey was conducted 5 months later: at that time, 29% of county residents believed that their tap water was still not safe, and 27% of the residents reported that they had less confidence in the city government because of the *Cryptosporidium* problem (Griffin et al. 1998). Currently, there are debates about the effectiveness of boil-water notices. Some people ignore the notices and do not change their behavior (Angulo et al. 1997), others change their behavior but do not change it back once the risk has been controlled. The level of people's trust in the message source and clarity of the information provided to them are probably important explanatory factors for these responses.

The HIV/AIDS epidemic has increased the number of immunocompromised persons in the population. As the Milwaukee outbreak demonstrated, immunocompromised persons were at greater risk than the general population for severe adverse health outcomes from exposure to waterborne pathogens. This outbreak led many officials and the public to become more concerned about the vulnerability of sensitive subpopulations to waterborne microbial pathogens.

Consumption Patterns

In addition, over the last decade or two, people have changed their normal customs concerning drinking water. There is a continuing increase in the number of people who use some alternative to unfiltered tap water, often citing health, safety, or taste concerns. Forty percent of 200 Toronto residents polled in a survey used an alternative at least occasionally (Auslander and Langlois 1993), and 12% reported that they boiled their water regularly because of health concerns and not in response to any particular message. These figures compare with 7% of Toronto residents using alternatives in 1983 and 19% in 1988. Fifty-one percent (n = 1,447) of a random US sample reported drinking bottled water, at least occasionally (IBWA 2000). In 1999, almost 47% of greater Washington, D.C., area residents reported that they do not drink water directly from the faucet (Waters 1999).

Technological Developments

Risk factors for waterborne infections are deduced primarily from outbreak surveillance data; however, in the United States, only a fraction of the estimated water-related outbreaks are reported through passive surveillance.

In the past several years, advances in molecular detection techniques have increased our knowledge about foodborne and waterborne causes of disease, allowing the association of certain pathogens with biological and exposure-related susceptibilities in their hosts. Now, molecular techniques, such as polymerase chain reaction and pulsed field gel electrophoresis, can link sample isolates from the stools of affected people to a water or food vehicle to an animal fecal source of contamination. Using these sophisticated techniques, we have much more power to trace contamination to its source.

As Internet access has increased, the complexity of communication pathways has become more important in health risk-related situations. Federal agencies primarily use the Internet and Web sites to disseminate water-related and disease information, while nongovernmental organizations and individuals use them to promote their views and information. This use of technology outside of traditional communication formats has distressed many risk managers because they feel they are not able to effectively identify stakeholders or anticipate their issues and needs in a timely manner.

Legal Mandates

The US Congress has incorporated some risk communication requirements into specific legislation. For drinking water, the 1996 amendments to the Safe Drinking Water Act (SDWA) require community water systems to provide annual informational materials on water quality to their customers. Known as Consumer Confidence Reports (CCRs), these reports comprise the primary public notification provisions in the SDWA. They are designed to give consumers information on the source of their drinking water, the process by which it is treated and delivered to them, and the quality of the water. In addition, the 1996 SDWA Amendments require the US Environmental Protection Agency to identify groups that may be at greater risk than the general public from contaminated drinking water. The water suppliers are required to communicate risk to those people, and the CCRs must mention cryptosporidiosis specifically. The one-way nature of this type of communication may, however, be problematic; one media report on the CCRs noted they "are supposed to be easy to read, but they aren't always... [Consumers should] be prepared to spend some time to understand [them]..." (Wein 2001). A review of 430 CCRs from across the United States provides evidence that some of the problems with CCRs' readability stem from the uninviting framing of the information provided, the complexity and lack of clarity of the contaminant table, inaccessibility of key information (e.g., warnings required for vulnerable subpopulations), inappropriate or ineffective design and literacy approaches, and the lack of non-English versions (CSADW 2000).

Challenges of Risk Communication

Two challenges of environmental risk communication are to identify specific groups that may be at greater risk and to understand the information needs of all stakeholders. Delivering the details in a way that people understand is important but not absolute. The temptation to disseminate information without interaction and assume that people are satisfied still lingers; however, the risk assessment community recognizes that the more interaction there is among the interested parties, the better the chance for success in the entire risk process.

Definitions

Another challenge is to ensure that "risk communication" is understood adequately to address water-related risks in current societal contexts. In the 1980s, most environmental health risk communicators thought of communication as a "tell" process—one that did not require interaction with stakeholders (Fischhoff 1995). During this period, risk communication primarily entailed disseminating information based on the belief that the public would recognize the value of the information and act on it to reduce risks. However, people did not react as the experts had expected (Leiss 1996). The impacts of this early, limited conceptualization of communication are still being felt, even though many efforts have been made to expand the definition and practice of risk communication.

Risk communication has often been defined with a focus on the individual scale (Morgan et al. 2002) but it can also be conceived on other scales as well. For example, the National Research Council (NRC) convened environmental health risk communication experts who developed the following definition, which was used as the basis for this chapter (NRC 1989):

> Risk communication is an interactive process of exchange of information and opinion among individuals, groups, and institutions. It involves multiple messages about the nature of risk and other messages, not strictly about risk, that express concerns, opinions, or reactions to risk messages or to legal and institutional arrangements for risk management. (p. 21)

This definition not only indicates that communication requires interaction between at least two entities but also that the interacting entities may be on different scales—individual, group, and institutional. It notes that risk communication is about more than facts; it also involves opinions and judgments that have important impacts on how risks are perceived and managed (Slovic 1987). Ideally, risk communication brings science and policy together in a deliberative, interactive process so that risks can be managed more effectively (NRC 1996).

Purposes

Just as definitions have expanded, the purposes identified for risk communication have changed as well. Based on a review of the environmental health literature, Covello and colleagues (1987) identified four types of risk communication: (1) information and education, (2) behavior change and protective action, (3) disaster warning and emergency notification, and (4) joint problem-solving and conflict resolution. The earliest forms of public health risk communication, such as boil-water notices, were primarily designed to motivate individual protective actions. Later, environmental health risk communicators focused on increasing the public's awareness and knowledge about technical aspects of complex risk issues to generate support for large-scale risk management strategies. However, processes in which experts maintained exclusive control became insufficient and unacceptable in the United States as stakeholders increasingly demanded participation in decision-making processes (Fischhoff 1995).

The goals of risk communication include

- Increasing awareness and understanding about technical and other issues
  - —Primarily among decision makers and people who will be affected by risk decisions and
  - —Secondarily among other interested parties.
- Providing meaningful, relevant, accurate, and clear information to
  - —Reduce misconceptions,
  - —Expand understanding about the full range of sound options for action,
  - —Provide ways to identify and monitor risk situations and design appropriate responses to the risks found,
  - —Enable stakeholders to make decisions effectively on personal and societal levels, and
  - —Help people take protective actions under emergency conditions.
- Meeting stakeholders' needs for information sharing and inclusion.
- Fostering meaningful interactions for deliberative and analytic risk management processes and thereby
  - —Build trust and confidence among stakeholders and decision makers,

—Facilitate greater participation in group processes, and
—Develop more consensus and support.

- Motivating collaborative actions to reduce and manage risks.

At a workshop of water-related health risk communication experts, the participants agreed, however, that programs are typically designed to achieve more than one goal and accomplish multiple purposes (Parkin, Embrey, and Hunter 2002).

Risk Communication in the Risk Management Process

Risk communication is now recognized as a critical component of every step of the risk management paradigm (NERAM 2000). Initially, risk communication was tacked on to the risk assessment process as an afterthought along with risk management. Over time, risk managers began to need help in communicating environmental risks to the public. There was recognition that public input into the goals and mechanics of risk assessment helped create trust in the process overall and in risk managers in particular. "Public confidence that risk managers are addressing real concerns, as opposed to going through a process perfunctorily, is critical to the future of risk assessment as an activity capable of improving the quality of life" (NRC 1994).

Current risk communication research efforts aim to improve risk analysis by providing a basis for understanding, anticipating public responses to hazards, and improving the communication of risk information among lay people, technical experts, and decision makers. Risk managers who uphold and regulate health and safety must understand how people think about and respond to risk. Experience shows that merely disseminating information without reliance on communication principles can lead to ineffective health messages and public health actions (Angulo et al. 1997; Harding and Anadu 2000; Owen et al. 1999).

The Presidential/Congressional Commission on Risk Assessment (1997) recognized the increased need for risk communication as a necessary component of risk analysis. They emphasized that communication, such as continuing relationships with citizen advisory panels, must begin before important decisions are made. The practice of risk communication is moving from distributing risk information to citizens to building relationships between plant managers and residents, agencies, the public, etc. However, in general, many citizens still believe that they are at greater risk from environmental hazards than they were in years past.

The Canadian Standards Association published *Q850 Risk Management: Guideline for Decision-Makers* in 1997, after a 6-year, multi-contributor effort. This model is similar to that of the Presidential/Congressional Commission's model but is more elaborate. Integrating risk perception and risk communication throughout the decision process fosters well-informed decisions and actions on the part of the stakeholders. These Canadian Q850 guidelines emphasize the inclusion of communication at the beginning of and during the risk assessment/management process (see Figure 15-1)—not just at the end—and using communication to build trust and alignment with stakeholders and their priorities. The Q850 model is one of the more sophisticated risk management and communication models available.

Types of Risk Communication

When the focus of risk communication is to warn the public about a particular environmental health risk, the task is to notify as well as motivate people to act to mitigate the risk. However, people tend to underestimate these sorts of risks and fail to take any protective actions (Wiedemann and Schütz 1999). Also, targeting the messages to the right audience is challenging: one must identify the populations at risk and design appropriate out-

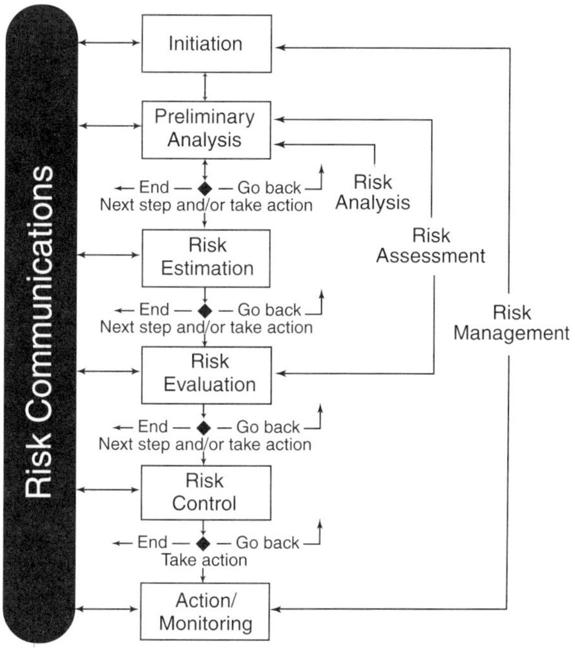

Canadian Standards Association, 1997. Used with permission.

Figure 15-1 Q850 Risk management model

reach campaigns (usually brochures or media outlets). However, these methods may reach a very small percentage of the intended audience and actually inform only a small proportion of the people reached. Even if the right audience obtains the risk information, they may see themselves as personally not at risk or they may not understand what action to take. A California study (in Harding and Anadu 2000) of 900 consumers found that 80% of the respondents did not take any action in response to public notification regarding drinking water safety. The author of the study speculated that this resulted from the lack of preventive measures detailed in the notification. In addition, resources are often not available to those responsible for outreach to construct a carefully researched risk communication strategy.

One-Way Communication

The original one-way form of risk communication is sometimes referred to as the "tell" (or monologue) method, because they do not include an interactive process. These are very limited forms of communication; in fact, some experts do not consider one-way methods as communication at all. In one-way risk communication, experts define the issues and goals and decide what needs to be said and how to say it. They also take and keep the lead on information dissemination and decision making for stakeholders. The expert typically portrays him/herself as the exclusive holder of key information that others need. The source of the one-way messages rarely shows any need to know who the information receivers are or what their reactions are to the information. One-way methods include press releases, press conferences, radio or television interviews, public notices (e.g., boil-water notices, fish advisories, CCRs, etc.), mobile megaphones, broadcast faxes, brochures and posters, public hearings where comments are received but not answered during the session, and scripted public meetings. "Tell" methods are not suitable for all, or in fact many, risk situations (see Table 15-1).

Two-Way Communication

Two-way or "dialogue" forms of communication are those in which experts involve and learn from stakeholders and try not to maintain exclusive control over the process. These methods are often, but not always, characterized by entities who have direct contact with each other and know or at least see each other while their interaction is taking place. Methods have included interactive town meetings, open houses before public meetings, consensus conferences, citizen advisory panels, community or neighborhood-level meetings or networks, meetings of stakeholder representatives, door-to-door contacts, telephone trees and hot lines, call-in television or radio shows, and e-mail for the public and experts to ask questions and receive information.

Many-Way Communication

Many-way (or "polylogue") types of risk communication have become more common as Internet communication has become more accessible. Particularly in the United States and Europe, the numbers of stakeholders actively engaged in environmental health risk issues and the volume of stakeholders' interactions have rapidly increased. More people have access to and potential roles in the decision-making processes, but their active involvement is related to their sense of how relevant the issue is to their own interests. As accessibility to the process increases, it becomes more difficult but still very important for experts to identify stakeholders' attributes, perceptions, and priorities that may be brought into the decision-making process. This change in the risk communication environment has led many experts to feel a severe lack of control and has raised their concerns about how to effectively manage risks and risk management processes. Very few risk communicators have learned how to manage this new, many-way form of communication or how to integrate its dynamics with more traditional two-way methods.

Table 15-1: Suitability of one-way and two-way methods for communicating risks (Butte, Thorne, and Parkin 2000)

Attribute of the Risk Situation	One-way (Tell) Methods	Two-way (Dialogue) Method
Nature of the risk	Familiar, acceptable to stakeholders	Uncertain, controversial, and less acceptable
Risk setting	Lower-level concerns, higher trust among parties	Higher-level concerns, lower trust among parties
Risk information	Important, available, uncomplicated	Unavailable, incomplete, technical, can be misunderstood
Risk information processing	Easy for stakeholders	More difficult
Risk management methods	Acceptable to stakeholders	Questioned by stakeholders, forced decisions will be unacceptable
Risk management organization	Well-respected, credible	Low credibility, under pressure
Tradeoffs	No important tradeoffs to be made by stakeholders	Important, potentially unacceptable tradeoffs
Decision timeframe	Short	Stakeholders can influence the timing to extended time
Risk reduction implementation	Short-term	Long-term
Continuous improvement	Not urgent	Important

In addition, very little research has been done to understand the impacts of Internet communications on stakeholders' risk perceptions and decision-making processes. Initial design and research work has been conducted to merge the mental model with technical Web site assessment methods (Parkin et al. 2001). We are not aware of any studies to date that have examined Internet impacts on water-related risk perceptions or decision-making processes.

Lessons Learned

From the past decades of risk communication practice, one thing has become abundantly clear. Good risk communication cannot always improve situations or remove all conflicts about risks, but bad communication almost always makes a risk management situation worse (NRC 1989). Regardless of the form of communication chosen, two tasks are critical to fostering success. One is to establish and sustain trust and openness between the parties communicating. The other is to ensure that the stakeholders' issues and priorities are known, respected, and addressed (NRC 1989). Other lessons that apply to all forms of risk communication include

- Risk communication must be audience-focused—based on sound information about the audience's values, preferences, and needs.
- The credibility and expertise of the information source should be established early.
- Risk messages must be clear, useful, and relevant to the stakeholders. Messages should not only describe the risk situation accurately but also give recipients specific steps that they can take to reduce their risk.
- Communication must be timely for the stakeholders.
- Knowledge gaps and scientific uncertainties must be acknowledged.
- Communication should be conducted in multiple ways that are familiar to the stakeholders and that meet the stakeholders' needs.
- Sufficient resources are necessary to sustain communication until the risk problem is effectively resolved.
- Getting risk communication right the first time is less expensive and difficult than fixing the damage from poor processes later.

Several other important points are known about two-way and many-way communication. For example, feedback about the communication process should be obtained on an ongoing basis and responded to in a timely manner. Constructive and respectful relationships between stakeholders need to be developed and sustained throughout the risk management process. It is especially important to provide information to stakeholders that fits with their mental model of the risk, is pivotal to their decision-making framework, and is meaningful to them. A "data dump" by experts can do more harm than good by increasing conceptual confusion and decreasing trust. Understanding what people *need* to facilitate their decision making is critical (Morgan et al. 2002).

In addition, the risk communication statement of purpose and goals must be clarified early in the risk management process. Evaluation of communication efforts should be planned early with stakeholder input and implemented both during and after a communication process.

Methods Used to Address Water-Related Risk Issues

Important lessons about water-related risk communication have been learned in recent years; however, new solutions to past problems need to be designed, implemented, and evaluated. While emphasis has been placed on improving tools and techniques to control risks, the potential impacts on poorly scoping the risk problem should not be ignored.

Problem Definition

Several reports have discussed the advantages and disadvantages of communications between risk assessors and managers, particularly in defining the scope of the risk issue and interpreting the results of an assessment (e.g., NRC 1983). These reports have also focused on the distinctions between science and policy decisions and who should have the lead in each area. Traditionally, attention to communication between assessors and stakeholders during the assessment phase of a risk management issue has been very limited (NRC 2001). In some cases, the public was not involved in the early phases of defining the scope of a risk problem or determining what information sources may be available to conduct a preliminary risk analysis.

In risk situations that are ongoing, public meetings may be held or citizen advisory panels convened to obtain insights about stakeholders' concerns. The Internet may be used to exchange information and views between advisory group meetings. During an emergency, a standing citizen panel that is familiar and well respected by the community it serves can play an important role.

Compared to utility or government officials, the panel members may be more likely to be seen by citizens as valid representatives of their interests and as credible sources of input and information.

A few water utilities have begun using Web sites to post their monitoring data as a way to provide real-time information to their consumers and build greater awareness of day-to-day contaminant levels and regulatory compliance. If an exceedence does occur, it is anticipated that a better-informed community will have a broader context in which to place the data and understand the related level of risk.

Risk Management

From the collective body of research that is now available, it has become clear that risk management must deal with more than risk information; it must also address risk perceptions, cognitive processes, and social interactions (Bennett and Calman 1999; Fischhoff and Downs 1997; Slovic 2000). Ignoring any of these dimensions will reduce the success of a risk management strategy.

Formal risk communication is used in a variety of ways to improve water-related risk management. Frequently used one-way methods include fish consumption advisories, boil-water notices, and other provisions of the SDWA amendments (SDWA 1996), which require public notification when treatment violations occur. Violations that can cause acute illness require public notification through television and radio broadcasts within 72 hours, followed by a newspaper notice within 14 days and a mail or hand-delivered notice within 45 days. Additional consumer right-to-know efforts include the distribution of annual CCRs. According to the regulation, CCRs must include the following information:

- Source of the drinking water
- Whether the water meets federal standards
- Potential health effects if the standards are violated
- Potential sources of any contamination found
- Where consumers can go for more information on water quality
- Educational information for susceptible people on avoiding *Cryptosporidium*

In the case of a boil-water advisory or a fish-consumption warning, there may be little opportunity for officials to interact with the intended audience. For example, if a boil-water advisory is issued, the timing of the message can be essential to the public's perception of trust in the source. It may not be clear which entity has the lead authority to declare a boil-water alert, which may result in confusion and delayed release. State or local health departments, environmental protection departments, or both may have the authority, depending on the state. In addition, the criteria for issuing and rescinding the orders are not consistent (Gostin et al. 2000). In Milwaukee, for example, health officials delayed the release of information, including the issuance of a boil-water order, even though the outbreak appeared to be waterborne. This delay resulted in the public's outrage and long-lasting distrust of government officials (Sly 2000; Griffin, Dunwoody, and Zabala 1998). A false alarm is easier to correct and results in the public's perception of the source as being appropriately concerned for their welfare and taking the initiative to protect it. Though quick action to alert communities may enhance trust in the source, it still may not affect people's willingness to adopt risk-reduction behavior (O'Donnell, Platt, and Atson 2000); 25% of people failed to boil their water after they learned about an alert because they did not believe it to be true (Angulo et al. 1997).

Most monitoring for specific pathogens in water is done after there is evidence suggesting an outbreak. The United Kingdom has instituted standard monitoring for cryptosporidial oocysts in drinking water supplies, which has raised the question of what to do when oocysts are detected but no cases of illness are indicated. The relationship between pathogens found in the course of regular water monitoring and actual health risks is still very unclear, but health officials developed guidelines on what sort of actions to take when oocysts are found in the water supply. Depending on the follow-up information available, under these guidelines, health officials have a spectrum of options available for public notification that range from taking no action to issuing a boil-water advisory (Hunter 2000). As risk communication methodology advances and the limitations of one-way communication methods are demonstrated, risk communicators will increasingly seek out ways to design and implement two-way, interactive processes.

Lessons Learned

Major lessons learned when communicating about water-related risks are that there is no single audience, no single message, and no single method that works well in all situations or even in situations that appear to be the same (Parkin, Embrey, and Hunter 2002). It cannot be assumed that everyone knows even basic concepts, such as how to boil water. Instead, communicators must determine what people actually need to know to make informed choices and take effective action. Although stakeholders are increasingly using Internet and Web-based services, risk communicators observe that individual and interactive methods are often more effective for achieving risk communication purposes. Despite this knowledge and changing societal contexts, water-related risk communication programs remain heavily dependent on written materials such as bus placards, post-card campaigns, pamphlets, and brochures (Parkin, Embrey, and Hunter 2002).

Effective risk communication requires more than the ability to speak and write well; not everyone is readily trained, skilled at, or comfortable with providing risk communication services. Increasingly, it requires expertise in tasks such as planning strategically and comprehensively, analyzing the nature of risk problems, identifying the range of stakeholders involved, effectively assessing their issues and priorities, designing messages that will address their concerns, recognizing and developing communication methods that are appropriate for the specific target audiences and risk problem, facilitating stakeholder dialogues, and fostering decision-making processes that will meet stakeholders' needs. In other words, risk communication often requires a team approach, one that involved people who are experts with a range of knowledge, abilities, and skills. The more complex the risk problem, the behaviors needed to reach the risk management goals, and the communication environment are, the greater the need will be for a wide range of abilities and skills that can be brought to bear on risk communication issues.

Developing Effective Water-Related Risk Communication Strategies

Waterborne microbial risks involve many concepts and contexts that differ from chemical or radiological risks. As a result, it would be inappropriate to apply communication strategies for nonmicrobial risks to microbial pathogens. Effective strategies must be based on well-designed and well-executed information and analytic processes.

Unique Aspects of Water-Related Microbial Pathogen Risk Communication

As with the traditional paradigm, microbial risk assessment includes an evaluation of the hazard (in this case, pathogen), the host, and the means of exposure. Because the relationship between the pathogen and the host is so dynamic, these areas are frequently discussed in terms of host–pathogen interaction or health effects. This dynamism also brings up unique issues that affect the risk assessment itself and, therefore, the communication of that risk.

Recognizing and diagnosing waterborne disease from microbial agents is challenging. It can most easily be done when the outbreak is large enough that health care providers recognize it. In normal situations, cases are sporadic, making etiology and an accurate clinical diagnosis more difficult to determine. In the United States, physicians do not routinely test stools for pathogens, particularly given the disincentives for additional testing provided by many managed care organizations. Furthermore, many cases of waterborne disease are subclinical. Yet, subclinically infected individuals can transmit the disease to others. When an outbreak is identified, the role of waterborne exposure is difficult to estimate for most pathogens. Those that are transmitted via the fecal–oral route can contaminate food as well as water or be spread by person-to-person contact.

With microbial pathogens, one is typically not only at risk from exposure to the primary source (e.g., drinking water) but to others who have been infected from contact with that source. Secondary spread or transmission is usually highest with a pathogen that has a low infective dose and lengthy fecal shedding. Therefore, though the source of the infection is no longer contaminated or people have been alerted to not drink their water without precautions, the transmission of infection can continue throughout the community. As a result, persons who once thought they were not at risk may become at risk. The challenges for the risk communicator are to continuously identify changes in the at-risk population and inform and influence those subgroups with appropriate and effective communication messages and methods.

Other strategic considerations include the introduction of risk/risk tradeoff debates—this is particularly true in the area of microbial disinfection of drinking water and disinfection by-products, which have been shown to have their own spate of health effects. The complexity of these issues presents significant challenges for risk communicators.

Developing Appropriate Strategies

Although risk communication occurs throughout the risk management process, it may take different forms at different steps in that process. To determine what forms of communication will be needed, risk managers must be able to recognize what type of risk situation they will be addressing, who is affected by the risk decisions, and what their information and other needs are. A manager can begin to develop an understanding of risk communication priorities by assessing the nature and context of the risk situation, the appropriate type of decision-making process for the situation, and what forms of communication will best support that process. Clearly, in emergency conditions, an organization must have communication strategies that were developed in advance and have resources readily available to implement those strategies. However, when decision settings involve longer time frames, a more deliberate and explicit diagnosis of the risk situation is possible. In either case, a series of steps to obtain needed information and use that information to address different types of risk problems can be used as a procedural framework.

An NRC committee has identified five broad classifications of risk decisions: (1) unique and wide-impact, (2) routine and narrow-impact, (3) repeated and wide-impact, (4) generic hazard and dose–response characterizations, and (5) decisions about policies for risk analysis (NRC 1996). The NRC used these five categories as guides, not rigid markers, to indicate what types of processes may be best suited for which types of risk decisions.

For example, unique and wide-impact decisions were characterized as those that affect people over broad geographic and social scales, involve many parties and perspectives, and require a lengthy and open decision process (e.g., setting the national standard for arsenic in drinking water). These decisions were more likely to be linked to broad inclusion of affected and interested parties in an explicit process.

Routine and narrow-impact decisions were described as every-day (thousands per year) emission releases or minimal-scale design permit decisions and are addressed more often through simple, generic decision processes. However, the NRC cautioned that the more inclusive and explicit processes are more likely to result in negative reactions from stakeholders than are the more implicit, limited processes typically used to make routine decisions. When in doubt, organizations were advised to be as inclusive as possible in the risk analysis process.

The committee described an eight-step method for diagnosing risk situations and matching appropriate decision methods to those situations. The steps are

1. Diagnose the kind of risk and the state of knowledge about the risk.
2. Describe the organization's legal mandate related to the risk.
3. Describe the purpose of the risk decision.
4. Describe the affected parties and barriers to their participation and anticipate public reactions.
5. Estimate resource needs and the expected timetable.
6. Plan organizational needs to implement the appropriate decision process.
7. Develop a preliminary process design.

8. Summarize and discuss the diagnosis within the organization, clarifying expectations, potential problems, and organizational commitments.

Risk managers and assessors should be involved in each step of the diagnostic process to ensure that potential problems and conflicts are recognized and addressed early. Organizations must balance the level and intensity of effort to which they commit with the resources and time available for each decision.

Specific questions to ask were listed within each step. For example, under the first step, questions lead the diagnostician to identify whether there are highly susceptible subpopulations that may be affected, whether risk is inequitably distributed, whether affected parties are likely to participate, and whether limited levels of knowledge about the risk may contribute to fears and potential secondary and tertiary effects. Though the answers to these questions can only be preliminary at this stage, they should provide a stronger basis for the decision-making process.

The result of the eight-step process is a draft procedure for each step in the risk management process. The proposal should include: a clear sequence of steps, who is involved, when and where the steps take place, any relevant rules for the steps, and the expected interim and final products. However, no procedure or risk communication plan can be final until affected parties have provided their input in developing the process. Organizations must remain open, flexible, and responsive to stakeholders' contributions. Even when it is difficult to achieve, the success of the decision and communication processes depends on the affected parties' satisfaction with and participation in the processes.

Development of the risk messages and communication methods will be facilitated by the information that is gathered to develop the decision process. A key issue in risk communication is identifying audiences who are or may be affected by the risk or who have interests in the risk. Preliminary information about stakeholders is collected in the first diagnostic step. Once the risk situation and the issues and needs of the stakeholders have been characterized, specific risk communication goals and strategies can be prepared. As for any strategy, clarity of the program's purpose, participants, timing, content, and anticipated results must be explicitly stated. The communication plan should be discussed within the organization to ensure that sufficient resources and coordination in the decision process are achievable.

Finally, it is important to allow for resources for ongoing feedback and evaluation of the communication process. When stakeholders have a routine way in which to provide their insights about the effectiveness of communications as they occur, the quality of the communication messages and methods can be improved in a responsive and timely manner. Changes that reflect the stakeholders' input are visible demonstrations of the organizations' respect for affected parties' concerns and are valid ways of building trust and confidence, which are critical to sustaining effective, long-term decision-making processes; constructive relationships; and risk management programs.

Summary

Risk communication is not a simple task that everyone can perform. Effective strategies and programs depend on sound information about the nature of the risk, its societal context, and the people who are affected by it. Scientific and technical knowledge, expertise, and judgment, often in the face of considerable uncertainty, are needed to develop strategies. Because there are many types of risk communication and many different potential methods and messages, diagnosis of the risk situation is critical to developing and implementing a meaningful risk communication plan. In most circumstances, risk communication requires considerable interpersonal skills, the willingness to develop trusting relationships with stakeholders, and the ability to leverage capabilities among a variety of interested and committed parties. Additionally, a team approach that involves people with complementary knowledge, abilities, and skills is required to effectively manage the increasing complexity of water-related risk communication. Appropriate investment of resources and time are important to ensuring successful and cost-effective immediate, long-term, and future risk communication strategies and programs.

Bibliography

Angulo, F.J., S. Tippen, D.J. Sharp, B.J. Payne, C. Collier, J.E. Hill, T.J. Barrett, R.M. Clark, E.E. Geldreich, H.D. Donnell Jr, and D.L. Swerdlow. 1997. A community waterborne outbreak of salmonellosis and the effectiveness of a boil-water order. *Am. J. Pubic Health*, 87(4):580–584.

Atman, C.J., A. Bostrom, B. Fischhoff, and M.G. Morgan. 1994. Designing risk communications. *Risk Anal.*, 14(5):779–788.

Auslander, B.A., and P.H. Langlois. 1993. Toronto tap water: perception of its quality and use of alternatives. *Can. J. Public Health*, 84(2):99–102.

Bennett, P., and K. Calman, eds. 1999. *Risk Communication and Public Health*. Oxford: Oxford University Press.

Bostrom, A., B. Fischhoff, and M.G. Morgan. 1992. Characterizing mental models of hazardous processes. *J. Social Issues*, 48(4):85–100.

Butte, G., S. Thorne, and R. Parkin. 2000. *Anthrax Vaccine Health Risk/ Benefit Communication: Orientation to Strategic Risk Management.* Washington, DC: United States Veterans Administration.

Campaign for Safe and Affordable Drinking Water (CSADW). 2000. *Measuring Up: Grading the First Round of Drinking Water Right to Know Reports.* Washington, DC: Campaign for Safe and Affordable Drinking Water.

Canadian Standards Association (CSA). 1997. *Risk Management: Guideline for Decision-Makers, A National Standard for Canada*, CAN/CSA-Q850-97, Etobicoke, Ontario: Canadian Standards Association.

Chess, C. 2001. Organizational theory and the stages of risk communication. *Risk Anal.*, 21(1):179–188.

Covello, V.T., P. Slovic, and D. von Winterfeldt. 1987. Risk communication: a review of the literature. *Risk Abstracts,* 3(4):171–182.

Douglas, M., and A. Wildavsky. 1982. *Risk and Culture: An Essay on the Selections of Technological and Environmental Dangers.* Berkeley, CA: University of California Press.

Fischhoff, B. 1995. Risk perception and communication unplugged: twenty years of progress. *Risk Anal.*, 15(2):137–145.

Fischhoff, B., and J.S. Downs. 1997. Communicating foodborne disease risk. *Emerg. Infect. Dis.*, 3(4):489–495.

Gostin, L.O., Z. Lazzarini, V.S. Neslund, and M.T. Osterholm. 2000. Water quality laws and waterborne diseases: *Cryptosporidium* and other emerging pathogens. *Am. J. Public Health*, 90:847–853.

Griffin, R.J., S. Dunwoody, and F. Zabala. 1998. Public reliance on risk communication channels in the wake of a *Cryptosporidium* outbreak. *Risk Anal.*, 18(4):367–375.

Harding, A.K., and E.C. Anadu. 2000. Consumer response to public notification. *Jour. AWWA*, 92(8):32–41.

Hunter, P.R. 2000. Advice on the response from public and environmental health to the detection of cryptosporidial oocysts in treated drinking water. *Commun. Dis. Public Health*, 3:24–27.

International Bottled Water Association. 2000. *Beverage Consumption in 2000*. New York: Yankelovich Partners.

Johnson-Laird, P.N. 1983. *Mental Models*. Cambridge, MA: Harvard University Press.

Kahneman, D., P. Slovic, A. Tversky, eds. 1982. *Judgment Under Uncertainty: Heuristics and Biases*. New York: Cambridge University Press.

Kasperson, R.E., O. Renn, P. Slovic, H.S. Brown, J. Emel, R. Goble, J.X. Kasperson, and S. Ratick. 1988. The social amplification of risk: a conceptual framework. *Risk Anal.*, 8:177–187.

Leiss, W. 1996. Three phases in the evolution of risk communication practice. *Annals AAPSS*, 545:85–94.

Morgan, M.G., B. Fischhoff, A. Bostrom, and C.J. Atman. 2002. *Risk Communication: A Mental Models Approach*. Cambridge, UK: Cambridge University Press.

National Research Council (NRC). 1983. *Risk Assessment in the Federal Government: Managing the Process*. Washington, DC: National Academy Press.

National Research Council (NRC). 1989. *Improving Risk Communication*. Washington, DC: National Academy Press.

National Research Council (NRC). 1994. *Science and Judgment in Risk Assessment*. Washington, DC: National Academy Press.

National Research Council (NRC). 1996. *Understanding Risk: Informing Decisions in a Democratic Society*. Washington, DC: National Academy Press.

National Research Council (NRC). 2001. *Classifying Drinking Water Contaminants for Regulatory Consideration*. Washington, DC: National Academy Press.

Network for Environmental Risk Assessment and Management (NERAM). 2000. *Environmental Health Risk Management: A Primer for Canadians.* Waterloo, Ontario: Institute for Risk Research, University of Waterloo.

O'Connor, R.E., R.J. Bord, and A. Fisher. 1998. Rating threat mitigators: faith in experts, governments, and individuals themselves to create a safer world. *Risk Anal.*, 18(5): 547–556.

O'Donnell, M., C. Platt, and R. Aston. 2000. Effect of a boil-water notice on behaviour in the management of a water contamination incident. *Commun. Dis. Pub. Health*, 3(1):56–59.

Owen, A.J., J.S. Colbourne, C.R.I. Clayton, and C. Fife-Schaw. 1999. Risk communication of hazardous processes associated with drinking water quality—a mental models approach to customer perception, Part 1—a methodology. *Water Sci. Technol.*, 39(10–11):183–188.

Parkin, R., M. Embrey, G. Butte, S. Thorne, and B. Fischhoff. 2001. *Anthrax Vaccine Risk/Benefit Communication Initiative: Phase One: Developing the Expert Model.* Final Contract Report. Contract No. V101 (93) P-1684. Washington, DC: United States Departments of Veterans Administration and of Defense.

Parkin, R.T., M.A. Embrey, and P. Hunter. 2002. Water-Related Risk Communication: Lessons Learned. Submitted for peer reviewed publication.

Presidential/Congressional Commission on Risk Assessment and Management (P/C). 1997. *Framework for Environmental Health Risk Management: Final Report.* Vol. 1. Washington, DC: Presidential/ Congressional Commission on Risk Assessment and Risk Management.

Safe Drinking Water Act Amendments of 1996, 42 *US Code* 300 (1974).

Slovic, P. 1987. Perception of risk. *Science*, 236:280–285.

Slovic, P., ed. 2000. *The Perception of Risk.* London: Earthscan Publication, Ltd.

Sly, T. 2000. The perception and communication of risk: a guide for the local health agency. *Can. J. Public Health*, 91(2):153–156.

Stallen, P.J.M., and A. Tomas. 1988. Public concerns about industrial hazards. *Risk Anal.*, 8:235–245.

Waters, W.F. 1999. *Water Consumption in the Washington, DC Area: Perceptions and Practice.* Report to the EPA Office of Water, December 1999.

Wein, H. 2001. What Are We Drinking? *The Washington Post* (March 13), p. HE10.

Weinstein, N.D. 1987. *Taking Care: Understanding and Encouraging Self-Protective Behaviour.* New York: Cambridge University Press.

Wiedemann, P.M., and H. Schütz. 1999. Risk communication for environmental health hazards. *Zentralbl Hyg Umweltmed.*, 202:345–359.

APPENDIX A

Biological Terrorism and Potable Water

With the renewed focus on anthrax and other biological agents as tools of terrorism, the water industry has received questions regarding the vulnerability of municipal water supplies. Several resources that detail the threat to drinking water sources from bioterrorism are listed at the end of this appendix.

The Centers for Disease Control and Prevention (CDC) published a series of recommendations on how to prepare and respond to biological and chemical terrorist attacks (CDC 2000). The CDC classifies biological agents by their risk to national security based on the agent's (1) ability to be disseminated or transmitted from person to person, (2) ability to cause high mortality, (3) ability to cause public panic, and (4) need for special action in response to a release. The highest priority agents include:

- Variola major (smallpox)
- *Bacillus anthracis* (anthrax)
- *Yersinia pestis* (plague)
- *Clostridium botulinum* toxin (botulism)
- *Francisella tularensis* (tularaemia)
- Filoviruses (Ebola hemorrhagic fever, Marburg hemorrhagic fever)
- Arenaviruses (Lassa fever, Argentine hemorrhagic fever)

Their next level of priority includes five waterborne pathogens:

- *Salmonella* spp.
- *Shigella dysenteriae*
- *Escherichia coli* O157:H7

- *Vibrio cholerae*
- *Cryptosporidium parvum*

Though most emphasis has been on aerosolized dissemination of biological agents, contamination through food or water is probably an easier way to distribute a pathogen (Khan et al. 2001). Of course, the impact from this type of exposure would be different. Though generally centralized distribution of municipal drinking water would make widespread dissemination easy, the drinking water treatment process, including filtration and disinfection, would reduce or eliminate the effects of some, if not all, pathogens. The contamination of groundwater supplies is more of a concern because groundwater typically undergoes less treatment than surface water. In addition, adulteration of the water supply after treatment—such as in the distribution system—would obviously bypass these treatment safeguards. Khan and colleagues (2001) cite five factors that would help protect municipal water from bioterrorism efforts: (1) dilution; (2) specific inactivation from disinfectants, such as chlorine; (3) nonspecific inactivation from sunlight, hydrolysis, etc.; (4) filtration; and (5) the relatively small amount of water that people drink straight from the tap. They report that most of the usual waterborne pathogens, with the exception of *Cryptosporidium*, would be neutralized by one of these five factors.

The potential *Cryptosporidium* has to cause significant morbidity and mortality through a drinking water supply was illustrated through the largest waterborne disease outbreak in US history, which occurred in Milwaukee in 1993. Large numbers of *C. parvum* oocysts in Milwaukee's surface water source overwhelmed the city's conventional treatment barriers, causing cryptosporidiosis in more than 400,000 and death to at least 54 (Kramer et al. 1996). *Cryptosporidium* oocysts are extremely resistant to chlorine disinfection, and depending on the strain, it may take only a few oocysts to cause infection—traits that make this protozoan especially dangerous as a waterborne pathogen.

Pathogenic elements that would affect an agent's efficacy as a bioterrorist weapon include a small infectious dose, easy direct transmission from person to person (called secondary spread), and the capacity to cause public panic (e.g., the panic that would ensue if medical treatment was not available and the mortality rate was high). The military would also be concerned about the pathogen's ability to incapacitate its human host for an extended period. As mentioned, *Cryptosporidium* is easily transmitted through water and from person to person. Generally, it causes mild to moderate gastrointestinal symptoms; however, in people with immunodeficiencies, especially AIDS, cryptosporidiosis can be fatal.

Tables A-1 and A-2 list some of the pathogens of concern, particularly for drinking water utilities, and their characterization in terms of their susceptibility in water and human susceptibility factors. Additional information can be found in the resources listed at the end of this appendix.

Table A-1: CDC's target pathogens and their susceptibility in water

Biological Agent	Stable in Water	Chlorine Tolerance
Variola major (smallpox)	Unknown	Unknown
Bacillus anthracis (anthrax)	2 years (spores)	Spores resistant
Yersinia pestis (plague)	16 days	Unknown
Clostridium botulinum toxin (botulism)	Stable	Inactivated at 6 ppm for 20 minutes
Francisella tularensis (tularaemia)	Up to 90 days	Inactivated at 1 ppm for 5 minutes
Filoviruses (Ebola hemorrhagic fever, Marburg hemorrhagic fever)	Unknown	Unknown
Arenaviruses (Lassa fever, Argentine hemorrhagic fever)	Unknown	Unknown
Salmonella spp.	8 days in freshwater	Inactivated
Shigella dysenteriae	2 to 3 days	Inactivated at 0.05 ppm for 10 minutes
Escherichia coli O157:H7	1 or more months—especially in cold water	Easily inactivated
Vibrio cholerae	Survives well	Easily inactivated
Cryptosporidium parvum	Stable for days or months	Resistant

Adapted from Burrows and Renner (1999) and Khan et al. (2001).

Table A-2: CDC's target pathogens and human disease characteristics

Biological Agent	Secondary Spread*	Chemotherapy Available
Variola major (smallpox)	Potentially high	No
Bacillus anthracis (anthrax)	No	Effective if caught early
Yersinia pestis (plague)	High	Effective if caught early
Clostridium botulinum toxin (botulism)	None	Effective if caught immediately
Francisella tularensis (tularaemia)	No	Effective if caught early
Filoviruses (Ebola hemorrhagic fever, Marburg hemorrhagic fever)	Moderate	No
Arenaviruses (Lassa fever, Argentine hemorrhagic fever)	Moderate	Moderately effective for Lassa fever
Salmonella spp.	Moderate	Effective
Shigella dysenteriae.	High	Effective
Escherichia coli O157:H7	High	No
Vibrio cholerae	Negligible	Moderately effective
Cryptosporidium parvum	High	No

Adapted from Burrows and Renner (1999), USAMRIID (2001), and Khan et al. (2001).

*Based on infectious dose and/or aerosol transmission potential.

Resources

American Public Health Association (APHA). 2000. *Control of Communicable Diseases Manual.* 17th ed. Edited by J. Chin. Washington, DC: APHA.

Burrows W.D, and S.E. Renner. 1999. Biological warfare agents as threats to potable water. *Environ Health Perspect.*, 107(12):975–984.

Centers for Disease Control and Prevention (CDC). 2000. Biological and chemical terrorism: strategic plan for preparedness and response. *MMWR Recommendations and Reports*; 49(RR04):1–14.

Franz D.R., P.B. Jahrling, A.M. Friedlander, D.J. McClain, D.L. Hoover, W.R. Byrne, J.A. Pavlin, G.W. Christopher, T.J. Cieslak, A.M. Friedlander, and E.M. Eitzen, Jr. 1997. Clinical recognition and management of patients exposed to biological warfare agents. *JAMA*, 278(5):399–411.

Khan, A.S., D.L. Swerdlow, and D.D. Juranek. 2001. Precautions against biological and chemical terrorism directed at food and water supplies. *Pub. Health Rep.*, 116:3–14.

Kramer, M.H., B.L. Herwaldt, G.F. Craun, R.L. Calderon, and D.D. Juranek. 1996. Surveillance for waterborne-disease outbreaks—United States, 1993–1994. *MMWR*, 45(1):1–33.

US Army Center for Health Promotion & Preventive Medicine. 2000. The medical NBC battlebook. USACHPPM Tech Guide 244. Aberdeen Proving Ground, MD: USACHPPM.

US Army Medical Research Institute of Infectious Diseases. 2001. USAMRIID's Medical Management of Biological Casualties Handbook. 4th ed. Fort Detrick, MD: USAMRIID [Online]. Available: <http://www.usamriid.army.mil/education/bluebook.html>. [cited January 10, 2002]

Glossary

Aerosolize: To become part of water vapor.

Asymptomatic: Without exhibiting symptoms.

Bacteria: Any of a group of single-celled organisms that are often grouped into colonies or motile by means of flagella that have a range of biochemical and sometimes pathogenic properties.

Biofilm: A buildup of organic material—frequently occurring in water distribution pipes.

Boil-water advisory: A statement to the public advising people to boil tap water before drinking or using (e.g., brushing teeth).

CFU: Colony-forming unit (metric for bacterial cultures).

Coagulation: The process of adding chemicals to water to destabilize charges on naturally occurring particles to facilitate their aggregation and removal by flocculation or filtration.

Coliforms: All aerobic and facultative anaerobic, gram-negative, nonsporeforming, rod-shaped bacteria that ferment lactose with gas formation within 48 hours at 95° F. They are used as an indicator of microbial contamination in water.

Colony: Visible growths of microorganisms in a nutrient medium.

Community/noncommunity water system: A public water system can be classified as community or noncommunity. A community system serves at least 15 connections or 25 residents year-round. A noncommunity system serves the public but does not serve the same people year-round (e.g., summer camp, trailer park).

Conjunctivitis: Inflammation of the conjunctiva (the mucous membrane that lines the inner surface of the eyelid and exposed surface of the eyeball).

Consumer Confidence Report (CCR): A document required by law to be sent from water purveyors to customers with information regarding their community's drinking water quality.

Conventional treatment: A series of processes, including chemical coagulation, flocculation, sedimentation, and filtration, used to treat water.

Cross connection: A connection between a drinking water system and an unapproved water supply or other source of contamination.

Cyst: The infectious stage of *Giardia* and some other protozoan and bacterial parasites that have protective walls, which facilitate their survival in the environment.

Disinfection: Destruction of organisms, usually by the use of bactericidal chemicals.

Disinfection by-product: A compound formed by the reaction of a disinfectant, such as chlorine, with organic material in the water supply. Trihalomethanes are disinfection by-products.

Distribution system: Water pipes, storage reservoirs, tanks, and other means used to deliver drinking water to consumers or store it before delivery.

DNA: Deoxyribonucleic acid. Any of various nucleic acids that are the molecular basis of heredity. They are located in cell nuclei and are constructed of a double helix held together with hydrogen bonds.

Endemic: Prevalent in a particular location or people.

Endocarditis: Inflammation of the endocardium (the membrane that lines the interior of the heart).

Enteric: Of the intestine.

Enterotoxin: A toxin produced by bacteria that is specific to intestinal cells and causes gastroenteritis.

Enzyme-linked immunosorbent assay (ELISA): A test that uses antibodies or antigens coupled to an enzyme as a means of detecting an antigenic match. If the target antibody or antigen is present, the test is positive.

Epidemiology: The study of the incidence and distribution of a disease in a population.

Fecal coliform bacteria: Bacteria found in the intestinal tracts of humans and animals. Their presence indicates fecal contamination and, therefore, potential presence of pathogens.

Filtration: The process of removing suspended particles from water by passing it through one or more permeable membranes or media of small diameter (e.g., sand, anthracite, or diatomaceous earth).

Finished water: Water that has passed completely through a treatment plant.

Flocculation: The water treatment process after coagulation that uses stirring to cause suspended particles to form larger, aggregated masses (floc). The floc is removed from the water by a separation process (e.g., sedimentation, flotation, or filtration).

Fluorescent antibody test: An assay that identifies an organism using a fluorescently labeled antibody specific to the organism.

Fomite: An inanimate object that serves to transfer infectious organisms from one individual to another.

Gastroenteritis: Inflammation of the lining of the stomach and the intestines.

Genera (pl.) or **genus** (sing.): Taxonomic categories ranking below a family and above a species.

Genome: One haploid set of chromosomes.

Groundwater: Fresh water beneath the Earth's surface usually found in aquifers that supply wells and streams. Groundwater is a major source of drinking water in the United States. Groundwater can also be under the direct influence of surface water, which makes it susceptible to the same contaminants.

Hepatotoxin: A substance that is toxic to the liver.

Homology: Similarity of nucleotide or amino-acid sequence in nucleic acids indicating a correspondence in structure and evolution.

Immunocompromised or **immunosuppressed:** A reduced immune response of an organism to pathogens.

Maximum contaminant level (MCL): The maximum permissible level of a contaminant in finished water. The maximum contaminant level goal (MCLG) is the maximum level of a contaminant at which no adverse effect on health is known, plus a margin of safety. MCLGs are unenforceable.

Meningitis: Inflammation of the meninges of the brain or spinal cord—usually caused by a bacterial or viral infection.

Morbidity: The quality of affect by disease.

Mortality: The proportion of deaths in a population.

Neonate: Newborn.

Nephelometric turbidity unit (ntu): The unit identifying turbidity (the presence of particulate matter) in water. Turbidity is used as an indicator of water quality.

Neurotoxin: A substance that is toxic to nerve cells.

Noncommunity/community water system: A public water system can be classified as community or noncommunity. A community system serves at least 15 connections or 25 residents year-round. A noncommunity system serves the public but does not serve the same people year-round (e.g., summer camp, trailer park).

Nosocomial: Disease acquired in a hospital.

Oncogenic: Causes tumors.

Oocyst: The infectious stage of *Cryptosporidium* and some other coccidian parasites with a protective wall, which facilitates survival in the environment.

Pathogen: An agent that causes disease—especially a microorganism.

Pathogenesis: The origin and development of disease.

Pathogenic: Capable of causing disease.

pfu: Plaque-forming unit. The metric used when culturing viruses.

Picocurie (pCi): Measure of radioactivity equivalent to 0.037 disintegrations per second. The US Environmental Protection Agency has determined a level of 4 pCi/L due to radon in indoor air requires action to remediate.

Polymerase chain reaction (PCR): The testing of short stretches of genetic material to compare with the known genetic sequences in the microbe. Only certain laboratories can perform this test, including the Centers for Disease Control and Prevention. It is considered the "gold standard" for detection but cannot indicate the viability of the organism.

Protozoa: Any of the subkingdom or phylum (protozoa) of chiefly motile and single-celled organisms, which include some pathogenic parasites of humans and animals (e.g., *Cryptosporidia* and *Giardia*).

Public water system: A system, classified as either a community water system or a noncommunity water system, that provides piped water to the public for human consumption and is regulated under the Safe Drinking Water Act.

Raw water: Surface water or groundwater that has not been treated.

Recreational water: Water used for recreation—could be a natural body of water, such as a lake or river, or swimming pool or hot tub.

Residual chlorine: Amount of available chlorine remaining after a given contact time under specified conditions. The residual serves to disinfect finished water that becomes contaminated in the distribution system.

Reverse osmosis: The application of pressure to liquid across a semipermeable membrane to produce demineralized water.

Safe Drinking Water Act (SDWA): Legislation passed by Congress in 1974 to ensure safe drinking water for consumers. The SDWA amendments of 1996 required, among other things, a greater emphasis on public outreach and information, as with the creation of the Consumer Confidence Report.

Sedimentation: The act of allowing material to settle to the bottom of liquid. This is a typical step in conventional drinking water treatment.

Self-limiting: Limiting itself, as in a disease that clears without treatment.

Sepsis: The presence of bacteria (bacteremia), other infectious organisms, or their toxins in the blood (septicemia) or in other tissues of the body.

Seroepidemiology: The study of disease outbreak and spread using blood tests to look for antibody response.

Serotyping: A test to determine a group of related microorganisms distinguished by their antigenic composition.

sp. and **spp:** Abbreviations for species (singular and plural, respectively).

Species: A taxonomic classification, ranking after genus and consisting of organisms capable of interbreeding.

Sporogony: The formation of spores as part of reproduction that is a characteristic of some organisms (e.g., microsporidia).

Surface water: All water naturally open to the atmosphere (rivers, lakes, reservoirs, etc.).

Taxonomy: The science and process of classifying organisms into established categories.

Titer: The concentration of a substance in solution determined by titration (the process of adding a substance to a standard reagent of known concentration in carefully measured amounts until a reaction is complete).

Turbidity: The parameter of having suspended matter (e.g., clay, silt, or plankton) in the water that results in a loss of clarity. This parameter is measured in nephelometric turbidity units and is an indicator for contamination.

Virulence: The capacity of a pathogen to overcome the body's defenses and produce disease.

Virus: A microorganism smaller than a bacteria, with a single nucleic acid surrounded by a protein coat, which cannot grow or reproduce apart from a living cell. A virus invades living cells and uses their chemical machinery to keep itself alive and to replicate.

Water quality indicator: A microbial, chemical, or physical parameter that indicates the potential risk for infectious diseases associated with use of water for drinking or recreation. The best indicator is one whose density or concentration correlates best with health hazards associated with a given type of hazard or pollution. Fecal coliforms are traditionally the most common water quality indicators.

Watershed: An area from which water drains to a single point—in a natural basin, the area contributing water to a given place or a given point on a stream.

Zoonosis or **zoonotic:** A disease that can be transmitted from animals to humans (e.g., rabies).

Index

Note: *f.* indicates figure; *t.* indicates table.

A

B

C

D

E

I

J

K

L

N

R

S

T

U

V

W

Z